The Complete Guide to
CONTRACTING YOUR HOME

4TH EDITION

by KENT LESTER and DAVE McGUERTY

BETTERWAY HOME

an imprint of F+W Media, Inc.

CINCINNATI, OHIO

www.popularwoodworking.com

Read This Important Safety Notice

To prevent accidents, keep safety in mind while you work. Use the safety guards installed on power equipment; they are for your protection.

When working on power equipment, keep fingers away from saw blades, wear safety goggles to prevent injuries from flying wood chips and sawdust, wear hearing protection and consider installing a dust vacuum to reduce the amount of airborne sawdust in your woodshop.

Don't wear loose clothing, such as neckties or shirts with loose sleeves, or jewelry, such as rings, necklaces or bracelets, when working on power equipment. Tie back long hair to prevent it from getting caught in your equipment.

People who are sensitive to certain chemicals should check the chemical content of any product before using it.

Due to the variability of local conditions, construction materials, skill levels, etc., neither the author nor Popular Woodworking Books assumes any responsibility for any accidents, injuries, damages or other losses incurred resulting from the material presented in this book.

The authors and editors who compiled this book have tried to make the contents as accurate and correct as possible. Plans, illustrations, photographs and text have been carefully checked. All instructions, plans and projects should be carefully read, studied and understood before beginning construction.

Prices listed for supplies and equipment were current at the time of publication and are subject to change.

Metric Conversion Chart

to convert	to	multiply by
Inches	Centimeters	2.54
Centimeters	Inches	0.4
Feet	Centimeters	30.5
Centimeters	Feet	0.03
Yards	Meters	0.9
Meters	Yards	1.1

Distributed in Canada by Fraser Direct
100 Armstrong Avenue
Georgetown, Ontario L7G 5S4
Canada

Distributed in the U.K. and Europe by David & Charles
Brunel House
Newton Abbot
Devon TQ12 4PU
England
Tel: (+44) 1626 323200
Fax: (+44) 1626 323319
E-mail: postmaster@davidandcharles.co.uk

Distributed in Australia by Capricorn Link
P.O. Box 704
Windsor, NSW 2756
Australia

Visit our Web site at www.popularwoodworking.com.

Other fine Popular Woodworking Books are available from your local bookstore or direct from the publisher.

17   16   15   6   5

Library of Congress Cataloging-in-Publication Data

Lester, Kent, 1953-
  The complete guide to contracting your home / by Kent Lester and Dave McGuerty. – 4th ed.
    p. cm.
  Rev. ed. of: Complete guide to contracting your home / Dave McGuerty & Kent Lester.
  Includes bibliographical references and index.
  ISBN 978-1-55870-871-6 (pbk. : alk. paper)
  1. House construction--Amateurs' manuals. 2. Building--Superintendence--Amateurs' manuals. 3. Contractors--Selection and appointment--Amateurs' manuals. I. McGuerty, Dave, 1954- II. McGuerty, Dave, 1954- Complete guide to contracting your home. III. Title.
  TH4815.M27 2009
  692'.8--dc22
                          2009029856
Acquisitions editor: David Thiel, david.thiel@fwmedia.com
Senior editor: Jim Stack, jim.stack@fwmedia.com
Cover Designer: Brian Roeth
Interior Designer: Doug Mayfield
Production coordinator: Mark Griffin
Illustrator: Kent Lester

Kent Lester

About the Authors

Kent Lester provides computer consultation and programming services to business and construction firms. He also authored *The Compete Guide to Being Your Own Remodeling Contractor.*

Dave McGuerty is a management/data processing consultant to international accounting firms.

Forward

The Complete Guide to Contracting Your Home has been around for over 20 years and is now in its fourth release. We strive to keep the material up to date, but new construction technologies, methods, and changes in the marketplace are happening at an exponential rate.

In an effort to keep you informed, we have created a new web site called www.selfmadehomes.com, where you can find updates on construction technology, supplemental forms and checklists, and a forum where you can interact with other aspiring "self made" builders and homeowners.

Check back often and give us yourw suggestions for making this the best home contracting site on the Internet.

We hope you will enjoy the site and this book!

Sincerely,
Kent Lester

Disclaimer

The forms and contracts are provided as is, for example purposes only. Construction laws and regulations vary considerably from state to state, so make sure to have any legal documents you plan to use for construction purposes reviewed by a lawyer familiar with regulations in your area.

The authors take no claim or responsibility for the use of these documents.

Image Contributors

The authors would like to thank the following for their kind assistance:

ACP backsplash and ceiling
Alan Mascord
Alson bath fixtures
American Standard
Aristrokraft Cabinetry
Armstrong
Atrium Windows
Boise
Boral Bricks
Bosch
Building America
Carrier
Certainteed
Clopay Garage Doors
ClosetMaid
David Wiggins
Delta Dry
Delta faucets
Dupont
Eagle roof tile
Ecostar faux slate
EnergyStar.gov
Framerite Steel Framing
GE Appliances
Grace Waterproofing
Greenguard
Hurd
Hy-Lite
Icynene
Impasse Termite
Kitchenaid
Kitchens
Kohler
Koma Trim
Lennox
Leviton
Louisiana Pacific
Mann
Mannington

Marvin
MaxiTile
Merillat
Milwaukee
Moderna
Nutone-Broan
Oatey
Omega Cabinetry
Owens Corning Roof and ICF
Panasonic Vent fans
PATH
Peachtree
Portland Cement Association
Progress Lighting
Ridgid
Rinnai
Rubbermaid Closets
Schulte closets
Sears
Senco tools
Sentricon
Sherwin-Williams
Silestone
Simpson Strongtie
Southern Pine Council
Southern Pine Institute
SRS Energy
Superior Walls
Tamko
Tile Roofing Institute
TimberTech decking
TREX
Universal Forest Products
Viega
Vortec Drain wrap
Winguard
Wolmanized lumber
Yorktowne Cabinets

Table of Contents

What this book will do for you

Be Your Own Contractor

Many would-be homebuilders are discouraged from attempting to build their own homes because of the fear and "mystique" of construction practices. This book will expose the professional policies and procedures that builders use to manage their own projects so that you can become the contractor of your own "self made home."

This Book's Step-By-Step Method

Whether you are a homebuilder or an individual who wants to build your own home, this manual will guide you toward your goals. Its workbook format provides a simple and complete method to manage your construction project from beginning to end. The steps in this book tell you what is happening at each stage of construction, even if you are not the person completing each step. The work you choose to perform is up to you.

What Is Contracting?

Contracting is the process of coordinating the efforts of skilled tradesmen (carpenters, masons, electricians, plumbers, etc.) and materials in an efficient manner. This book will show you how to work with subcontractors and material suppliers to become a successful project manager, whether you plan to build one house or a hundred.

Contracting Objectives
- Save Time
- Save Money
- Reduce Frustration

Save Time

This book keeps you organized with a detailed master plan that divides work into discrete, manageable tasks. Efficiency is gained by completing tasks at exactly the right moment.

Save Money

This book can easily help save 25% to 42% over a comparable home purchased through a real estate agent. Here's how:

Real estate commission	up to 7%
Builder markup	up to 20%
Savings on material purchases	up to 4%
Cost saving construction	up to 2%
Doing work yourself	up to 9%
Total	up to 42%

Reduce Frustration

This book will:
- EQUIP you with the knowledge to effectively manage a residential construction project.
- HELP you avoid major problems and pitfalls.
- PROVIDE you with the benefit of years of experience.
- ADVISE you when to avoid tackling tasks which are best left to specialists.
- SHOW you how to specify and inspect the work performed by others.

After all, learning from bad experiences doesn't mean they have to be your own!

INTRODUCTION

The cornerstone of the American Dream is to own your own home, one that reflects your personality and provides the atmosphere and conveniences that describe you and your family as individuals. In our highly technical world, we are spending more and more time at home. So why would you want to live in a nondescript, cookie-cutter home when you can live in one that reflects your unique personality?

Has the American Dream become harder to reach? The current economy has made it more and more difficult for homeowners to realize their dreams. In recent years, homeowners have opted to buy far more house than they can afford. The idea behind this concept was to build an investment for one's future, but can a home ever be considered an investment in the truest sense? You can only realize the return on your investment when you sell it, and then you must replace your "home" investment with yet another home. The "home as investment" strategy only reaps its reward when one decides to downsize, perhaps at retirement age. Often, a large home only adds a huge burden of debt.

The best way to increase your investment is not to build beyond your means, but to achieve cost savings. Remember the old adage, "a penny saved is a penny earned." With total control of a building project, you can increase your investment percentage by reducing labor and material costs. You can substitute your work for that of other contractors and subcontractors, thereby building "sweat equity" into the value of your home.

Contracting vs. Building

To achieve a self-made home, you do not have to be a do-it-yourself expert. But you will have to be organized, tenacious, and fearless. This book is not about picking up a hammer and building the house. Few people have the skills or the time to construct a home from the ground up. Instead, this book focuses on the job of the contractor. The contractor is the organizer, project manager, and negotiator. If you work for a corporation or own your own business, you probably possess many of these necessary skills.

The biggest reason for contracting your own home is to get exactly the house you desire. Professional home builders make their money by building homes as quickly and as efficiently as possible. Most homes are built speculatively. That means each one is designed to attract the largest number of potential buyers. They are vanilla and lack the personality of a custom home. If you want a truly unique abode, one that reflects your personality and that of your family, then you will want a custom home.

But custom homes are expensive. There is an incredible volume of material and labor that must be coordinated to construct a modern home. For each factory worker, builder, sub-contractor, lawyer, financier, architect, designer, building inspector, craftsman, real estate agent, appliance salesman, interior designer, plumber, electrician, and salesperson that touches your project, there will be a markup in the price. Reduce the number of middle-men, and you reduce costs.

Another reason for building your own home is to make use of the latest innovations and products in the industry. Builders tend to be very conservative because they don't want to alienate customers. But that means that most modern homes are 5 to 10 years behind the latest technology by the time they are built.

Flexibility is another advantage of being your own contractor. When you make last minute changes to your design, the builder will charge you a pretty penny to make them. When you make the changes yourself, the costs are usually far less, although as we will discuss later, it is strongly recommended that you plan ahead so the changes are made before the first nail is driven.

The greatest lament of homeowners comes from the compromises made to afford a new home. To keep costs within line, many niceties must be sacrificed. This can be doubly frustrating since you will likely have the opportunity to build the home of your dreams only once or twice in a lifetime. To live with those compromises for half your life can be a frustration indeed.

Quality construction is key. If you are the project manager, you will make certain that it meets your standards. When you buy a speculative home, many important construction details have already been hidden behind walls and trim. Shortcuts and shoddy construction can come back to haunt you in future years. If you are there, watching every aspect of the construction process and checking the quality of construction, you can save yourself time and expense later as your home ages.

You won't have to worry about wooden concrete forms left embedded in the garage, or out-of-plumb walls that will leave gaps behind your kitchen cabinets. You won't be afraid of improperly compacted earth causing cracks in your foundation. You won't have to worry about cold spots where insulation was improperly installed. You won't be frustrated by the lack of electrical outlets in key locations.

Finally, there is the sense of satisfaction and accomplishment that comes from watching your dream move from an idea, to a model, and then finally, to a place where you will spend much of your life.

Who Can Be A Homebuilder?

A homebuilder is anyone who desires to be one. Many states have no examinations or licensing requirements for builders. Check with local building authorities in your area to determine if you must meet any special requirements. The homebuilder, or contractor, is the person in charge of overseeing the entire construction project, coordinating the efforts of professional sub-contractors and labor, and managing materials. The term "homebuilder" is somewhat of a misnomer since a homebuilder normally does not even have to raise a hammer to build a home.

Should You Consider Contracting Your Own Home?
Make no mistake; contracting your own home will be one of the most demanding projects you will undertake. If you have a strong constitution, a knack for organization, and the ability to bargain, then contracting your home can be a worthy challenge.

Managing a construction project is like running a bureaucracy. With each layer of this complex system, there is paperwork to be completed, costs to be added, time to be wasted, delays to be scheduled, limitations to be overcome, and frustrations to be multiplied. This will take considerable time and energy, but the rewards are vast. This book will teach you the skills necessary to get the job done, quickly and efficiently.

Reasons for Not Contracting
If you decide not to contract your own home, here are a few excuses you can use:
- I have plenty of money. I don't need to save money by contracting myself.
- I'm not particular about getting exactly what I want in a home, representing the single largest financial commitment of my life.
- I don't know how.
- It seems too complicated — I don't know where to begin.
- I can't even hammer a nail straight.
- It's not worth the trouble.
- I don't have the time.
- It's not my thing.

The last excuse is the only valid one. Building isn't for everyone. Many will use one of the above excuses or custom make their own. To make up an excuse is to rationalize why one pays "retail" for a home. Sadly enough, the money saved by contracting your own home can make all the difference in affording one or not.

The second most valid excuse is: "I Don't Have the Time."

Most of us who can afford to build a home have full-time jobs. So how can one build a home in the evenings and on weekends? This is resolved in the How to Be a Builder section.

Contracting Not Your Thing?
Contracting your own home is a daunting experience. If, after reading this book you decide to turn this responsibility over to a professional, take heart. By investing your time and energy in learning the concepts of *The Complete Guide to Contracting Your Home,* you will become an experienced client. You will be able to understand the concepts of home construction, speak the lingo, spot the quality issues, and bargain with the best of the trade.

Your knowledge will be respected, however grudgingly, by whomever you hire to complete your custom home.

Builders, by far, prefer to work with intelligent and experienced customers, who understand the challenges faced in building a great home. Whether you contract yourself, or hire someone, the knowledge presented in this book will prepare you for the adventure of a lifetime.

Let's get started!

HOW TO USE THIS BOOK

How This Book Differs from Other Construction Manuals
There are plenty of books on construction techniques: how to frame a home, lay bricks, mix concrete, cut beveled miter joints, etc. Yet, most publications fail to cover the practical day-to-day details of managing a construction project. They fail to provide the necessary forms and step-by-step instructions helpful in guiding you through the minefield of sub-contractors and scheduling.

The Complete Guide to Contracting Your Home is different. It provides you with the information and skills to manage and execute a successful project whether you do the work yourself or hire professional help.

Using the Book's Five Sections
The Complete Guide to Contracting Your Home breaks the construction process down into five distinct sections. They are:

1. Preface
The section you are reading now. This prepares you to get the most from the other sections by establishing a consistent project management workflow.

2. Design and Planning
This is the most critical stage of any project. The decisions you make here will affect everything that happens later: how long it takes to complete your project, the money you save, the level of workmanship, and the goals you want to reach. Concepts such as lot and plan selection, financial and legal issues, dealing with members of the construction industry, and other topics are important to gain a proper perspective on the project. Some decisions will be difficult or impossible to correct later:
- Where will your dream house be located, in town or the countryside?
- How is it situated in terms of the sun?
- What is the size?
- What style of design?
- Cost vs. function
- What about room to grow?

Changes during planning are child's play to correct with an eraser. Mistakes made during actual construction require a sledgehammer, can cost thousands, and take years off your life. Make your mistakes early and often, and learn from them. By the time you dig the first shovel of dirt, you should have all contingencies covered. Take your time during this phase and you'll save time later.

3. Project Management

This section will teach you how to become an effective project manager. You will learn the critical skills necessary to:

- Bargain with sub-contractors
- Draft airtight contracts that protect you from liability and the unknown
- Work part time to finish your project
- Avoid the critical pitfalls that can interrupt your journey
- Handle the emotional strains that inevitably crop up
- Keep meticulous records so you know precisely how much you have spent

When you finish this section, you will be a builder. Don't let anyone tell you otherwise. It's also the section that will determine the relationships you develop with subcontractors and other tradesmen.

4. Construction

This covers the actual construction process. Chapters follow the rough sequential order of the project. The Steps and Master Plan will help to smooth over tasks that fall out of sequence.

Please Note: This section is not intended as a hands-on guide for the actual construction of your house. There are plenty of excellent books on the market that cover the craft of homebuilding. Check the appendix for our recommendations. Instead, The Complete Guide to Contracting Your Home *focuses on the skills necessary to stay on schedule and budget, and to determine quality workmanship. You will learn the general process that takes place, but more importantly, how to tell the difference between shoddy construction techniques and craftsmanship that you can be proud to call your own.*

5. Appendix

Here, you will find valuable reference materials:

- Listings of books on architecture and construction.
- An address listing of major construction suppliers and manufacturers.
- A MASTERPLAN of the every construction step to guide you through the entire project.
- Every form and contract you need to deal with sub-contractors and suppliers.
- A materials list for estimating.
- A glossary of common construction terms so you can think and talk like a builder.

Note: Since address listings change so frequently, you can visit our website at www.selfmadehomes.com for updated listings.

Using the Workbook Sub-Sections

Most of the Project Management and Construction chapters include workbook sub-sections. These include:

1) Steps
2) Inspection Checklists
3) Specifications

1. Steps

The steps section takes you through the chapter (and your construction project) in sequential order. Sometimes, certain steps are dependent on the completion of other steps in other chapters before they themselves can be completed. In project management, this is called the critical path. You will find references to the other chapters when this occurs.

The MasterPlan in the Appendix provides a master worksheet of all the steps in all chapters. You can use the MasterPlan as the key reference blueprint for your entire project. When viewing a step in the MasterPlan, you can easily refer back to the detailed step description in its chapter by using the Step coding: each step is prefaced by a two letter code that corresponds to the chapter heading where the step was originally written. For instance, Step FR1 in the MasterPlan refers back to the first step in the Framing chapter. This encoding makes it easy for you to drill down to the precise level of detail needed to track and complete your project.

2. Specifications

As we will discuss in later chapters, you may have very different ideas about workmanship from your sub-contractors. The best way to avoid confusion and frustration is to spell out the specific standards you expect for each job. These specifications should be added to your original sub-contractor agreement so there is no possibility of confusion. The Specifications listed in each chapter give you suggestions for describing these details. You should consider adding some or all of these specifications to your contracts. The specifications will also teach you what to expect from each tradesman.

CONCRETE — SPECIFICATIONS (SAMPLE)

General

- All concrete, form and finish work is to conform to the local building code.
- Concrete will not be poured if precipitation is likely or unless otherwise instructed.
- All form, finishing and concrete work MUST be within ¼" of level.
- All payments to be made five working days after satisfactory completion of each major structure as seen fit by builder.

PRE-CONSTRUCTION STEPS (SAMPLE)

PC1 BEGIN construction project scrapbook. This is where you will begin gathering all sorts of information and ideas you may need later. Start with a large binder or a file cabinet. Break into sections such as:

- Kitchen ideas
- Bath ideas
- Floor plans
- Landscaping ideas
- Sub-contractor names and addresses
- Material catalogs and prices

Your "scrapbook" will probably evolve into a small library, a wealth of information used to familiarize yourself with the task ahead, and serve as a constant and valuable source of reference for ideas to use in the future.

PC2 DETERMINE size of home. Calculate the size of home you can afford to build. Refer to the HOUSE PLANS section in this book for details on this step.

PC3 EVALUATE alternate house plans. Consider traffic patterns, your lifestyle and personal tastes. Refer to the HOUSE PLANS section, presented earlier, for details on evaluating house plans.

PC4 SELECT house plan or have house plan drawn. If you use an architect, he will probably charge you a certain fee per square foot. Make sure you really like it before you go out and buy a dozen copies of the blueprints!

PC5 MAKE design changes. Unless you have an architect custom design your home, there will always be a few things you might want to change — kitchen and kitchen cabinets, master bath, a partition, a door, etc. This is the time to make changes. Refer to the material on making changes in the HOW TO BE A BUILDER and HOUSE PLANS sections.

Concrete Supplier

- Concrete is to be air-entrained ASTM Type I (General Purpose), 3,000 psi after 28 days.
- Concrete is to be delivered to site and poured into forms in accordance with generally accepted standards.
- Washed gravel and concrete silica sand to be used.
- Each concrete pour to be done without interruption. No more than one half hour between loads of concrete to prevent poor bondage and seams.
- Concrete to be poured near to final location to avoid excessive working. Concrete will not be thinned at the site for easier working.

Formwork

- Bid is to perform all formwork per attached drawings including footings for:
  - Exterior walls

- Monolithic slab
- Poured walls
- Bulkheads
- Garage
- AC compressor slabs
- Patio
- Basement pier footings

- All footing forms to be of 2" or thicker wood or steel.
- All forms to be properly oiled or otherwise lubricated before being placed into service.
- All forms, washed gravel, reinforcing bars to be supplied by form sub-contractor unless otherwise specified in writing. All of these materials to be included in bid price.
- All forms will be sufficiently strong so as to resist bowing under weight of poured concrete.
- Form keys to be used at base of footings and for brick ledge.
- Expansion joints and expansion joint placement to be included in bid. Expansion joints: Driveway every 15'. Sidewalks every 5'.
- Pier footings and perimeter footings to be poured to exactly the same level.
- All concrete to be cured at the proper rate and kept moist for at least three days.
- Foundation and garage floor to be troweled smooth with no high or low points.
- Finished basement floor will slope toward drains.
- Garage floor, driveway and patios will slope away from dwelling for proper drainage.
- Concrete to be poured and finished in sections on hot days in order to avoid premature setting.

Concrete Block
- All courses to be running bond within ¼" level.
- Top course to be within ¼" level.
- Standard 12" block to be used on backfilled basement area.
- 4" cap block to finish all walls.
- Horizontal reinforcing to be used on every three courses.
- Trowel joints to be flush with block surface.

3. Inspection Checklists
The inspection checklists make it easy to double-check sub-contractor workmanship. The checklist items will help you to identify best practices and insure a quality job.

EXCAVATION — INSPECTION (SAMPLE)

Clearing
- ☐ Underground utilities left undisturbed.
- ☐ All area to be cleared is thoroughly cleared and other areas left as they were originally.
- ☐ All felled trees removed or buried. All remaining trees are standing and have no scars from excavation. Firewood cut and stacked if requested.

Excavation
- ☐ All area to be excavated is excavated including necessary work space:
  - Foundation area (extra 3' at perimeter)
  - Porch and stoop areas
  - Fireplace slab
  - Crawl space cut and graded with proper slope to insure dry crawl space.
- ☐ Trash pit dug according to plan if specified.
- ☐ All excavations done to proper depth with bottoms relatively smooth. Should be within 1" of level.

Checklist Prior to Digging and Clearing
- ☐ All underground utilities marked.
- ☐ All trees and natural areas properly marked off.
- ☐ Planned clearing area will allow access to site by cement and other large supply trucks. Room to park several vehicles.
- ☐ All survey stakes in ground.

Checklist Prior to Backfill
- ☐ Any needed repairs to foundation wall complete.
- ☐ Form ties broken off and tie holes covered with tar (poured wall only).
- ☐ All foundation waterproofing completed and correct. This includes parging, sprayed tar, poly, etc.
- ☐ Joint between footing and foundation sealed

4. Master Plan

This is your guide. Everything you will encounter appears on the master plan.

MASTERPLAN

STEP FRAMING	DESCRIPTION	DATE STARTED	DATE COMPLETED	DATE INSPECTED	COMMENTS
FR1	CONDUCT standard bidding process (material)				
FR2	CONDUCT standard bidding process (labor)				
FR3	DISCUSS aspects of framing with crew				
FR4	ORDER special materials				
FR5	ORDER first load of framing lumber				
FR6	INSTALL sill felt				
FR7	ATTACH sill plate with lag bolts				
FR8	INSTALL support columns				
FR9	SUPERVISE framing process				
FR10	FRAME first floor joists and subfloor				
FR11	FRAME basement stairs				
FR12	POSITION all first floor large items				
FR13	FRAME exterior walls (first floor)				
FR14	PLUMB and line first floor				
FR15	FRAME second floor joists and subfloor				
FR16	POSITION all large second floor items				
FR17	FRAME exterior walls (second floor)				
FR18	PLUMB and line second floor				
FR19	INSTALL second floor ceiling joists				
FR20	FRAME roof				
FR21	INSTALL roof decking				
FR22	PAY framing first installment				
FR23	INSTALL lapped tar paper				
FR24	FRAME chimney chases				
FR25	INSTALL pre-fab fireplaces				
FR26	FRAME dormers and skylights				
FR27	FRAME tray ceilings, skylights and bays				
FR28	INSTALL sheathing on exterior walls				
FR29	INSPECT sheathing				
FR30	REMOVE temporary bracing				
FR31	INSTALL exterior windows and doors				
FR32	APPLY deadwood				
FR33	INSTALL roof ventilators				
FR34	FRAME decks				
FR35	INSPECT framing				
FR36	SCHEDULE loan draw inspection				
FR37	CORRECT any framing problems				
FR38	PAY framing labor (second payment)				
FR39	PAY framing retainage				

OTHER SUGGESTIONS

Before Your Project Begins:

• READ the glossary section to familiarize yourself with common construction terms. Check back to the glossary whenever you hear a new term or phrase.

• FOLLOW through the MasterPlan to get a feel for the ebb and flow of the project.

• REFER to the Sources section in the Appendix for additional names and addresses of construction oriented manufacturers, associations, and suppliers.

During Your Project:

• FOLLOW the Steps outlined in the Project Management section. Certain steps may not pertain to your project. For example, if you are building on a slab foundation, excavating a basement will not be applicable. Common sense and judgment is essential in this respect. The execution of many steps will naturally vary from project to project. Do not even think of starting your project until you are familiar with the steps covered in this book.

• REFER to the sample specifications for ideas on what to include in your specifications when dealing with tradesmen. These are primarily intended to remind you of items easily overlooked and to illustrate the level of detail you should use.

• REFER to the Inspection Checklists for ideas on how and what to inspect to help insure you are getting good workmanship as your project progresses. It may take ten years to learn how to lay bricks properly, but only ten minutes to check the work for quality craftsmanship.

• USE the forms provided in the Appendix when applicable:
  • House plan evaluation checklist
  • Lot evaluation checklist
  • Lien waivers
  • Home construction contract
  • Material and labor take-off sheets
  • Purchase orders
  • Sub-contractor specification form
  • Change order

Additional Helpful Material

In conjunction with this book, several other documents will provide you with a wealth of information. They are:

Your Local Building Code — These are the rules you must abide by while building in your area. Obtain copies from your county building inspector and from the International Code Council. (www.iccsafe.org)

Modern Carpentry: Essential Skills for the Building Trades — A general purpose text with excellent illustrations covering the actual techniques of construction. By Willis H. Wagner.

Means Illustrated Construction Dictionary – A comprehensive dictionary of more than 20,000 construction terms, accompanied by illustrations. By R.S Means Co.

Architectural Graphic Standards for Residential Construction: The Architect's and Builder's Guide to Design, Planning, and Construction Details – A visual guide devoted to construction standards of residential structures, with information on energy efficiency, HVAC, home wiring, accessibility, indoor air quality, and more. By The American Institute of Architects

 **Design & Planning**

THE LOT

The lot you purchase will determine almost everything else about your home, from its size and orientation, to its price and style. It's a location and setting you had better be comfortable with. There's an old saying about the three most important factors when buying a lot: location, location and location. While this stretches the truth, it does make an important point.

You can always expand or remodel an existing home. You can never fix a poor lot decision. (Unless you want to spend the money and effort to move a house.) So don't finalize your home plans until you have made your lot choice.

Considerations

If you have not purchased your lot yet, you may wish to ask yourself the following questions:

• Would you prefer to build in a planned development (subdivision), where home values will be more stable and controlled, or on a one-of-a-kind lot? If you are purchasing a lot in a subdivision, make sure that the building covenants work for you; not against you. Do the covenants allow you to build the home you want? When you want? Do the covenants allow you to sub-contract the home yourself? Do they protect against low priced homes which could affect your resale value? Don't build in an area where homes are more than 15 or 20% lower in value than your planned home.

• Would you prefer a large lot, which may require lots of time and money in maintenance, or a smaller one that may cost less and involve less maintenance?

• Is the lot and surrounding area suitable for the type home you are planning to build? Consider this from both a physical and an esthetic point of view.

• Will the lot support a basement, slab or crawl space if you plan to have one? A better location is almost always worth spending a little more money.

Essential Advice

Don't attempt to save money by buying a "less desirable" lot. Curb appeal can affect the value of your home by 25% or more. Saving $2,000 in lot costs could cost you $10,000 in home resale value. Don't step over a dollar to pick up a dime! Don't overbuild for the area; build in an area of comparable value. Make sure that the lot and the neighboring homes are worthy of the home you plan to build. In fact, try to find a lot in a neighborhood with homes of higher value than the one you plan to build. They will pull up the value of your own house.

Obtain a title search of the property before purchasing to ensure free and clear title to the land. Make sure public sewer access is available if the owner states it. Don't take his word for it.

It is wise to have a general idea of the style and size of home you plan to build while looking at lots. It would be a shame to find out that the home you want to build won't even fit on the lot after considering zoning and set-backs. Also, you may want to see if nearby homes are of comparable size.

Lot Evaluation

Review the following checklists of strengths and weaknesses. While many of them seem obvious, almost every lot will have some weaknesses. Your challenge is to find a lot with weaknesses that are not deal breakers. Think of each weakness as an opportunity to negotiate a lower offering price. Once you own the land, weaknesses are your problem.

When evaluating a lot, walk it thoroughly to check for soft spots, drain pipes, lay of the land, etc. Realize that all the trees on a lot may well disappear after the foundation area is cleared. Lots in older neighborhoods may have a collection of beautiful tall trees, but they can often pose problems for new construction. Trees on undeveloped lots usually grow tall and slender, as they compete for light with surrounding trees. Densely packed trees also tend to support and protect one another from wind and ice. Once you begin to clear out the dense growth, those majestic, tall trees may become accidents waiting to happen. Without the support of their neighbors, spindly trees are prone to collapse, and may choose your house for their final resting place. Consider carefully if you will be able to keep these heirloom trees.

Corner Lots – Good or Bad?

Although some homeowners enjoy corner lots, they are generally not preferred by speculative homebuilders because they lack backyard privacy and often have enlarged side setbacks. The "non-square" shape complicates the design and placement of the house.

Riparian Rights

The definition of riparian water rights is a system of allocating water use among those who possess land near to or adjacent to the water source. It is an ancient real estate concept that has its origins in English common law. Stated simply, riparian rights allow landowners "reasonable use" of adjacent lakes and rivers for purposes of fishing, boating, and other activities.

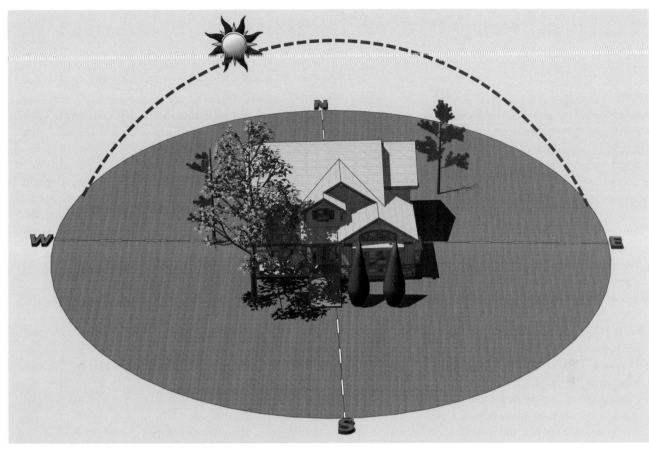

With the summer's high sun, shade trees and eaves cast shadows over south facing windows, lowering temperatures 10 degrees or more.

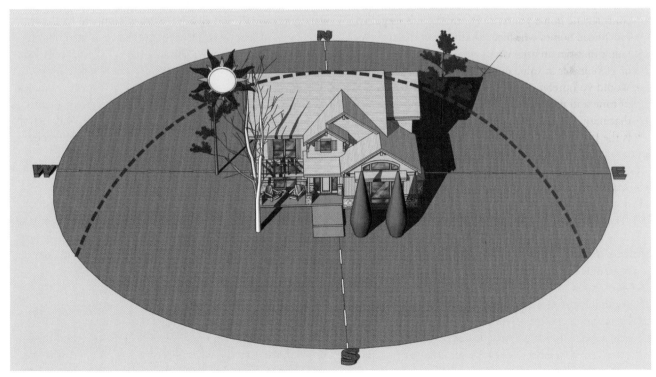

In winter, the sun is much lower in the sky, sending light into south facing windows.

Riparian rights also have an important effect on water that may flow across your land. You are generally not allowed to change the grade and slope of your land in a way that diverts the watershed to adjoining property. Look for evidence of excessive water drainage across your lot, like trenches or gullies. If your lot "appears" extremely fertile with dense undergrowth, it may be because it's the watershed for the entire neighborhood. If that's the case, you will be severely limited in how you divert the water away from the foundation of your home. If your drainage grading diverts the water to a neighbor's property, you may have a lawsuit pending.

Flood Plain

A related caution concerns the hundred-year flood plain. Many "out-lots" in older neighborhoods may seem too good to be true until you discover that the lot is prone to flooding. Check with your county engineer to obtain a map of the hundred year flood plain in your area. If your lot contains some of this flood plain, you could be severely limited to certain areas where structures can be built. On the flip side, if the lot is big enough, you may be able to obtain a bargain if you can find a house plan that makes good use of the buildable portions of the lot and protects you from any future flooding.

Percolation Test

If you plan to install a septic tank, you will need room for the drainage field. This area must be relatively free of rocks and boulders and must pass a percolation test. The percolation test checks the permeability of the soil. An engineer will visit the site and bury an open ended container in the ground. He then monitors the length of time it takes for a standard quantity of water to drain from the container. If the soil contains dense clay, the lot may not be appropriate for a septic installation.

Solar Orientation

If you are planning an energy efficient or solar design, the orientation of your home on the lot will be critically important. Take a compass with you when inspecting lots and find the orientation of the lot north to south. Check the lot in Google maps to get a better feel for its orientation, not only to the sun, but to neighboring homes. Why? Because solar orientation can have a great influence on energy bills. Southern facing walls and roofs will receive the most sunlight year round, but especially in the winter. Northern facing walls will receive the least sun and heat. Depending on where you live, seasonal variations will influence how you orient a home.

If you live in a northern climate, you'll want southern facing walls oriented to collect the most sunlight for natural warmth. If you live in the south or Midwest, you'll probably want to avoid a lot of windows on southern facing walls. However, the home's design and orientation of the lot will play a large part in this decision. If your favorite house plan has large windows on the front façade, it will be difficult to achieve the right solar balance if your lot requires that the front face south.

If you want the option to install solar collectors, you'll probably want the back or sides of the home to face south in order to hide the collectors from the street. Check out your neighbor's homes. Will their roof lines cast shadows across the area where you plan to install solar electric or solar water heaters? What about existing trees? The choice of house plan and lot are inextricably linked. Think of them as a single decision.

Topography

As with solar orientation, the topography of the lot will affect home design. Are you planning a full basement? If so, then low lying, flat lots are out of the question, especially if ground water lies near the surface. This is why there are almost no basement homes in areas like New Orleans with its high water table. Sloping lots work best for basements, but make sure to check for the presence of large rocks that may make excavation difficult.

Conversely, a ranch style home on a slab foundation needs a relatively level lot. Slab foundations are relatively rare in northern climates because of the danger of "frost heave." Freezing ground can lift and crack shallow foundations. Latitude and topography greatly influences the styles of homes across the country and these two factors should play a large part in your lot and design decisions.

Finding a Lot

Finding the right lot may involve researching the sources below and others as well:

- Local real estate agents and/or multiple listing services
- For sale ads in newspapers
- For sale signs on land
- Legal plats in county courthouses or other local government offices.
- Satellite view in Google maps. (This can be a great time saver.)

Purchasing the Lot

Review the considerations and the checklists below. Ask the seller to provide a legal description of the lot, prepared by a registered surveyor, showing all legal easements and baselines. Negotiate legal and other closing costs to be paid by or split with the seller. If you must pay all or part of the legal fees, insist on appointing an attorney of your choice. Cash talks — especially when purchasing from an individual rather than an organization. Look for owner financing.

Title Search

Important! Never buy a lot without a thorough and recent title search. Insist on and pay for a title search and title insurance if the seller refuses to pay. Title insurance is imperative to protect you from liens or other clouds on the title. If there is a dispute over land ownership or tax liens, the land could be tied up in litigation for years, costing you lots of time and money. When applying for a construction loan at a local lending institution, you must have "free and clear title" to the land. This insures that no dispute over title will interfere with the completion of financing.

STRENGTHS: CHECKLIST

- ☐ Lot slopes to allow for basement (if desired).
- ☐ Lot has rear and side privacy.
- ☐ On sewer line. Check with a plumber or city engineer to locate the elevation of the nearest drain line. Your lowest drain pipe should be slightly higher than the elevation of that sewer line, or you may have to install "flush-up" toilets in lower levels.
- ☐ On cul-de-sac or in other low traffic areas.
- ☐ Well-shaped square, rectangular or otherwise highly usable.
- ☐ Trees and woods. Hardwoods and some larger trees.
- ☐ Good firm soil base. Important for a solid foundation.
- ☐ No large subterranean rocks that will require blasting.
- ☐ Not in flood plain.
- ☐ Safe, quiet atmosphere.
- ☐ Attractive surroundings such as comparable or finer homes and attractive landscaping.
- ☐ Level or gradual slope up from road.
- ☐ Good drainage.
- ☐ Lot located in stable, respected neighborhood.
- ☐ Just outside of city limits. May pay less local taxes.
- ☐ Easy access to major roadways, highways and thoroughfares.
- ☐ Convenient to good schools, shopping centers, parks, swimming/tennis, firehouse and other desirable amenities.
- ☐ Situated in an area of active growth.
- ☐ Provides attractive view from dwelling site.
- ☐ Fully usable. No ditches, ruts or irregular surfaces.
- ☐ Seller will finance partially or in full.
- ☐ In area zoned exclusively for single family dwellings.
- ☐ Underground utilities. Phone, electrical, cable TV.
- ☐ Natural features in areas that will remain undisturbed.
- ☐ Swim, tennis, golf and other amenities.
- ☐ Private backyard area. If there are no trees to block the neighbor's house, check for room to plant trees or shrubs.

WEAKNESSES: CHECKLIST

- ☐ No sewer, must install septic tank. Must pass a percolation test.
- ☐ On or very near major thoroughfare. Yellow or white lines down the center of the road. Speed limit over 35 mph.
- ☐ Odd-shaped (too narrow or too shallow). Either will cause limited front or side yards.
- ☐ No trees or barren.
- ☐ Rocky. May have large rocks to excavate or dynamite.
- ☐ Sandy soil. May require special foundation work. May not support a large foundation. Excavation difficult.
- ☐ In flood plain. Normally involves expensive flood insurance and difficult resale.
- ☐ Near airport, railroad tracks, landfill, exposed electric power facilities, swamps, cliffs or other hazardous area. Near large power easements, commercial properties or radio towers.
- ☐ Slopes down from road and/or steep yard. Hard to mow. Difficult access by car when iced over. Water drains toward home. May require excessive or unusual excavation and/or fill dirt.
- ☐ Flat yard. Less than 2% slope may be difficult to drain.
- ☐ Water collects in spots. May require expensive drainage landscaping and/or fill dirt. May result in a wet/damp basement.
- ☐ Lot located in or near unstable/declining neighborhood.
- ☐ Just inside city limits. May have to pay more local tax.
- ☐ Relatively inaccessible by major roadways.
- ☐ Isolated from desirable facilities such as shopping centers, swimming/tennis, parks, etc.
- ☐ Lot provides unattractive views, either from prospective dwelling site or from road.
- ☐ Shallow lots leaving a small area for the backyard.
- ☐ Evidence of dumping or burying of garbage that may leave depressions.
- ☐ Creek, gully or deep valley running near center of lot.
- ☐ Area zoned near commercial, duplex, quadruplex or other high-density, multi-family dwelling.
- ☐ Area populated by less expensive homes.
- ☐ Utilities above ground or not yet established. Unsightly electrical wires.

DESIGNING YOUR HOME

Measure Twice, Cut Once

This time-honored builder's adage sums up the key focus and purpose of this book. It is never more important to understand this, than now: during the design phase of your project. In fact, the cost, timing, and success of your homebuilding adventure will be determined right here, right now. If you spend the majority of your time analyzing and designing your home from the ground up, you can help to ensure a hassle free and successful experience.

As we will discuss later on, you can use the latest technology to design and build your home in "virtual space" before you ever drive the first nail. Make the mistakes now, not later. By building a virtual version of your house on paper (and possibly on the computer) you can experience and solve all the potential mistakes before the cash register starts ringing. Once ground is broken, you become a slave to scheduling, construction loan interest, weather, and inspectors.

How Much Home Can You Afford?

Before you build or purchase your "dream home" you must determine what features you can realistically afford. Many future homeowners start out by designing a dream home with all the features they have ever wanted. The size and expense of the average American home has doubled in recent decades, leading to a glut of "McMansions." The recent recession and collapse of the housing market has brought the American Dream back to reality. People are starting to question the need for huge homes with huge energy bills and stifling property taxes. The recent trend toward "green" construction has revived the idea of quality vs. quantity in our living experience.

Alan Mascord, a home designer and leader in energy efficient living, puts it this way, "efficient living does not mean radically changing your lifestyle. Whether you prefer traditional, contemporary, minimalist, or more eclectic living styles-there is a way to live more efficiently. The key is to be educated about the purchases you make. It's not the bricks and boards which make a house a home. A house is more deeply colored by its content — the people, the activities, the sharing of a house — these are the elements that foster a feeling of flow."

Before you spend valuable dollars on blueprints and architects' fees, determine the exact amount of home you can afford to purchase. A standard rule of thumb is that your principal and interest (P&I) payments should not exceed 25%-35% of your Gross Annual Income. Forget about what banks and mortgage lenders will offer you. It's much wiser to start conservatively. Contracting your own home will allow you to save enough money to stay in the "comfort zone" of affordability, one that will help you avoid the stifling mortgage debts that have ruined the American dreams of so many homeowners in recent years.

Work Backwards

To find a home design you can afford, work backwards from your salary. Use the Mortgage Payment Table to help determine the total amount of mortgage your monthly salary will allow you to cover financially. These figures will be approximate, but they will help you to establish a size and cost limit.

Example:

John earns $57,000 per year and has just bought a subdivision lot for $22,000. How much can he spend on house construction and still meet his other expenses? The bank is willing to finance 80% of the value of the house, so John will be making a 20% down payment. The interest rate is 7% and the mortgage term is 30 years.

Based on a mortgage payment of about 25% of John's total income, divide John's yearly salary by 12 to obtain the monthly income and then divide it by 4 (25%). This will give you the total payment amount available to put toward monthly mortgage payments.

$57,000÷12 = $4,750/4 = $1,187.50

Look up the payment amount in Table 1 for a 30 year loan at 7% interest to determine the amount of the mortgage your monthly payments will cover. (The table gives figures in $100 per month increments, so round up your available salary. Answer: $15,013) Multiply this by your available monthly payment in hundred dollar increments to obtain the total mortgage amount that 25% of your salary would cover.

Monthly Gross Income in $100 increments = Total
   Mortgage Amount
(Rounded up) $1,200÷$100 = 12 × $15,031 = $180,372
   total mortgage

Since the bank will finance this much mortgage at 80%, the total price of the house will be the mortgage amount plus the 20% down payment John will contribute to the purchase. Work backwards to obtain the total house value.

$180,372÷80% = $225,465

Now take the total value of the house and subtract the cost of the land, construction loan cost and closing costs to determine the total amount of money you can afford to spend on the construction of your home.

Total value of home	$225,465
Land cost	- $22,000
Construction interest	- $6,500
Closing costs	- $8,800
Construction cost	**$ 188,165**

This amount should be the upper limit that John is willing to spend to construct the home of his dreams, in this example.

Cost per Square Foot

For decades, this was standard rule of thumb that builders and homeowners used to quickly estimate the cost of a home. It was mathematical shorthand that was always fraught with risk. Today, construction technologies and costs of appliances and materials vary so greatly that cost per square foot is an almost useless figure. Your contribution to sweat equity and other cost saving techniques will skew this figure even more.

So, how can you conveniently determine the range of house plans that are appropriate for your budget? The best way is through comparables. Check out homes selling in your area and their "real" sales price (not the list price). Calculate the square footage and determine if the appliances and finishes are comparable to the ones you plan for your own home. If you then factor in the savings you expect to achieve through sweat equity and other techniques, then you will have a very rough guideline.

Important: Only after you have completed your detailed cost take-off, will you know the true cost of the house you have chosen. Adding energy efficient and green features may alter the final price significantly, but may be worth it, because of savings down the road. See the Saving Money chapter for more information.

Be Conservative

There will always be unforeseen issues that will drive up the cost of construction beyond your original estimate. It is much better to choose a house that is well within your affordability zone, rather than pushing the envelope. You can always expand the house later, using your new-found contracting skills.

AMOUNT OF MORTGAGE PER $100 MONTHLY PAYMENT

	10 YR. LOAN	15 YR. LOAN	20 YR. LOAN	25 YR. LOAN	30 YR. LOAN	35 YR. LOAN
3.0%	10,356	14,481	18,031	21,088	23,719	25,984
3.5%	10,113	13,988	17,243	19,975	22,269	24,196
4.0%	9,877	13,519	16,502	18,945	20,946	22,585
4.5%	9,649	13,072	15,807	17,991	19,736	21,130
5.0%	9,428	12,646	15,153	17,106	18,628	19,814
5.5%	9,214	12,239	14,537	16,284	17,612	18,621
6.0%	9,007	11,850	13,958	15,521	16,679	17,538
6.5%	8,807	11,480	13,413	14,810	15,821	16,552
7.0%	8,613	11,126	12,898	14,149	15,031	15,653
7.5%	8,424	10,787	12,413	13,532	14,302	14,831
8.0%	8,242	10,464	11,955	12,956	13,628	14,079
8.5%	8,065	10,155	11,523	12,419	13,005	13,389
9.0%	7,894	9,859	11,114	11,916	12,428	12,755
9.5%	7,728	9,576	10,728	11,446	11,893	12,171
10.0%	7,567	9,306	10,362	11,005	11,395	11,632
10.5%	7,411	9,047	10,016	10,591	10,932	11,134
11.0%	7,260	8,798	9,688	10,203	10,501	10,673
11.5%	7,113	8,560	9,377	9,838	10,098	10,245
12.0%	6,970	8,332	9,082	9,495	9,722	9,847
12.5%	6,832	8,113	8,802	9,171	9,370	9,476
13.0%	6,697	7,904	8,536	8,867	9,040	9,131
13.5%	6,567	7,702	8,282	8,579	8,730	8,808
14.0%	6,441	7,509	8,042	8,307	8,440	8,506
14.5%	6,318	7,323	7,813	8,050	8,166	8,223
15.0%	6,198	7,145	7,594	7,807	7,909	7,957

Table 1: Affordable mortgage amounts

Home Design Plan Sources

Here are some sources of house plans:
- Publishers of Plan Books (stock plans)
- Residential Architects (custom and stock plans)
- Home Designers (custom and stock plans)
- Stock plans included with computer planning software

Home Plan Publishers

Several publishers specialize in printing plan books. These books often have a theme and may include plans from a large number of different designers. Some of these publications have been around for many years and feature plans which can be out of date when it comes to modern amenities. They provide a good start for prospective homebuilders however, because they offer many different solutions to the same design challenges. Plan books often focus on certain styles of architecture, such as traditional, contemporary, southwestern, or Western European style homes, etc. These books are readily available in bookstores and magazine stands. Complete blueprints can be ordered directly from the plan service. Since these plans are marketed in bulk, you benefit from economies of scale and resulting low plan costs. Plans may be purchased by the set.

Residential Architects

Architects are certified by the AIA (American Institute of Architects) and often provide the most comprehensive plan design services. Local architects provide design skill and knowledge of local architectural styles. They usually develop a portfolio of plans that will reflect a certain style, but their bread and butter usually comes from custom plan designs. If you have an unusual or challenging home design, architects have the skills to handle any structural or design issues. They have been trained in engineering and fully understand the principles of load bearing structures. Most architects have commercial experience as well, and can often suggest unique solutions to problems.

All that training comes at a price. Architects usually charge the highest fees for custom design work. On the other hand, you can often find bargains hidden in their stock portfolios. They must compete with the other plan and design services. Look for plans that match your goals closely, and keep customizations (and the resulting costs) to a minimum. If you chose an architect, make sure he or she has a general sense of your budget.

Home Designer

A less expensive alternative to an architect is a home designer. He or she is probably not a licensed architect, (otherwise, the certification would be on the shingle) but can be fully capable of preparing excellent home designs at a lower cost than architects. The skill level of home designers can vary tremendously, so it's best to look for ones who have been certified by the American Institute of Building Design. (AIBD) Choose designers who have large, popular portfolios. That popularity usually ensures quality and satisfaction. Ask local builders and new homeowners to recommend their favorite designers. Ask designers for references and whether they provide engineering and load bearing calculations for their plans.

Computer Plan Software

There are some retail computer packages that provide future homeowners with a new tool. These design programs are amazingly powerful and easy to use, allowing you to play "what if" to your heart's content. But be careful. Just because these programs allow for extensive creativity, does not guarantee that your design choices will be structurally sound or meet building codes. These software programs should be used as creative tools only, to explore design possibilities and to visualize them in three dimensions. If you develop a rough plan to your liking, you will then want to have an architect or home designer draft the final design. Commercial blueprints are the working drawings used by sub-contractors, inspectors, loan officers, truss manufacturers and other tradesmen. Retail software is incapable of rendering useful substitutes. Depend on the expertise of your architect or designer for these critical documents.

Combine Stock and Custom Designs

The hybrid approach is most common and the best bang for the buck. Find a plan that is close to the home of your dreams and have the original architect or draftsman merge your ideas into it for you. This minimizes the time they have to invest in planning and analysis and keeps costs down considerably. For an existing home plan, the architect or designer has already worked out the technical issues. Be aware that home designs are copyrighted material, representing hours of analysis and labor by the original creator. Avoid the temptation to create a "copycat" design. The original author of the home plan is far more familiar with the plan's details than a total stranger. The resulting "hybrid" plan from the creator will be a far superior to any attempt at copying by others.

Make Use of Other Engineering Services

If your house will make extensive use of floor or roof trusses (see the framing chapter) you can make use of the engineers employed by the companies that manufacture them. Each truss design must be custom engineered and certified. If you are unsure about the structural integrity of your design or length of a truss span, discuss the issue with the manufacturers. Their structural engineers can go a long way toward solving these issues.

WHAT TYPE OF HOME SHOULD YOU BUILD?

Fit the Home to Your Lifestyle

Before deciding on a plan, make a list of features that you find most compelling for your style of living. Use this criteria while evaluating plans.

- Do you like a formal setting or a more rustic environment?
- Is a formal dining room or living room important to you?
- Should the Master Bedroom be isolated from the living area? Does it provide privacy from other bedrooms or entertainment areas?
- What are your family considerations? Do you need a nursery near the master bedroom?
- What about a family room near the kitchen so you can keep an eye on the kids?
- Do you like an open living style or private niches?
- Is the kitchen near the garage or driveway for easy access when bringing in groceries?
- Do the stairs direct traffic through the entertainment areas, or do they allow private access?
- Can kitchen messes be seen from sitting or living areas?
- Do the windows cast a glare on televisions? Will the TV be mounted on the wall?
- Will home's solar orientation cause glare on faces at sunrise and sunset?
- Are closets positioned to act as sound buffers between bedrooms and entertainment centers?
- Are laundry areas near the source of dirty cloths?
- Do outside sitting areas have privacy from neighbors?

Choose a Design That Fits the Lot

Make sure the design is appropriate for the lot you have purchased. Full basement plans are difficult to achieve on flat lots or in areas with a high water table. Ranch plans may crowd the edges of smaller lots. The lay of sloping land will determine whether or not a plan needs to be reversed to allow access to basement doors and garages. By analyzing the topography of your lot first, you can often filter out a lot of inappropriate plans right off the bat.

Home Styles

The style of your home, more than anything, reflects your personality. American home styles have inherited influences from all over the world and there are hybrid designs too numerous to describe within the scope of this book. You should refer to some of the classic American styles when choosing a design that will be distinctive and reflect "you." Alan Mascord, designer and founder of Alan Mascord Design Associates in Portland, has this advice, "Try to pick a design that distinguishes your house from your neighbors, but also one that does not clash with your community or seem out of place for your region of the country. Many designs incorporate elements that are unique to the climate of a particular area and will often look out of place in other climate zones."

Steep roofs became popular in northern climates because of their snow shedding properties. Flatter roofs are more appropriate in windy areas prone to tornados and hurricanes. Some elements, such as stucco, work better in hot, dry areas but perform poorly in areas with excessive moisture and freezing temperatures. Porches and overhangs become essential sun blockers in warmer climates. Study the different designs in your area and consider the reasons why these designs became popular. Be distinctive with your own design, but don't choose a design that simply doesn't work in your area. Here is a very short list of classic American home styles:

Craftsman bungalow. Courtesy of Alan Mascord Design Associates.

Craftsman

Popular on the West Coast and Midwestern states, it features low roof lines with overhangs and distinctive, square columns. It originated from the English Arts and Crafts movement and signaled a departure from the ornate and pretentious styles of the Victorian era. It reflects a simple, rustic design with lots of natural wood materials. Also known as the Bungalow style, it was strongly influenced by the work of Frank Lloyd Wright.

French country. Courtesy of Alan Mascord Design Associates.

European Revival

This style has inherited an amalgam of different European influences, including English Tudor, French Countryside, European Cottage and many others. Soldiers brought these influences back to America after the two world wars. It features steep hip roofs and gables that often frame a grand entrance foyer. This style is popular in the present day throughout the country, especially in Southern suburbs.

Georgian with porch. Courtesy of David E. Wiggins, Architect.

Georgian

Also related to the Federal and Greek Revival styles common in the early 1800s. It represented an Early American appreciation of Greek and Roman styles and featured a symmetrical structure and window placement. It's most commonly represented by the classic styles of Williamsburg, Virginia. Common on the East Coast and in the Southern States, Georgian styles influenced the large Colonial Revival homes common to the plantations of the Deep South. Here, the simple square designs became ornate mansions with large, Greek columns and front porches that extended across the facade. These porches were essential for ventilation and temperature control in the age before air conditioning. In the current south, you will see a much simpler, cleaner version of this style, sans the columned porches, known as "five, four, and a door." This refers to the symmetrical window placement.

Modern Victorian. Courtesy of David E. Wiggins, Architect.

Victorian

Made popular during the industrial revolution of the late 1800s, the Victorian style used lots of ornate wood trim that could be cheaply manufactured on the new machines of the era. It's a formal style with many smaller rooms, designed for privacy, and the conservative values of the day. The shapes and forms were many and varied with copious use of turrets and intersecting roof lines. Also known as Queen Anne and related to Mansard, Gothic, and Italianate styles common in the upper middle class suburbs of industrial cities; popular in the Northern and Midwestern States.

Cape Cod. Courtesy of Alan Mascord Design Associates.

Cape Cod

The Cape Cod is a popular, simple design common to New England and northern states. It's generally based on a story and a half design, with the second story embedded within a steep roof pitch, with symmetrical gables along the front to allow in light and ventilation. The steep roof pitch sheds snow easily and makes space for several bedrooms in the upper level. This design can be quite cost effective in its simpler forms, since the framing used for the roofline encompasses the second story. Related styles include the Salt Box (story and a half with large rear dormer) and simpler Georgian designs. Its compact footprint works well on smaller lots, either on slab or full basements. When wrapped with large porches on the ground level, these styles have become popular farmhouses in the South and Midwest.

Contemporary

This style covers a broad spectrum of unique designs, from the Art Deco period of the twenties to the modernistic revival of the seventies. Contemporary designs are in the midst of a renaissance, popular with younger families because of their modern and (sometimes) energy efficient features. The green housing boom has embraced this design due to the reduction of materials necessary for construction. They feature simple box styles and flat roofs that easily accommodate solar panels or green roofs and can often be assembled in factories and shipped to the site.

Contemporary. Courtesy of Alan Mascord Design Associates.

Ranch. Courtesy of Alan Mascord Design Associates.

Ranch

Not really an architectural style, per se, ranch homes became popular in the fifties with the boom in automobiles. Families migrated to the suburbs where land was cheap and big lots prevailed, allowing for sprawling, one-story designs with carports and garages. Although not appropriate for smaller lots, ranches can become quite space efficient when placed on full basement foundations. Since basement construction technology has improved in recent years, the basement space becomes fully livable and a source of low cost square footage. When combined with contemporary designs and flat roofs, the modern "two story" ranch has become a popular green building option.

Spanish Mission. Courtesy of Alan Mascord Design Associates.

Spanish Adobe

Also known as Spanish Colonial, Spanish Mission or Pueblo, this style sports clay tile roofs, adobe or stucco walls, and many Spanish influences. Popular in California, the Southwest, and Florida, this style is well suited to hot or dry environments. The thick adobe or concrete walls help to even out the broad temperature swings of these climates. This style has been modernized by recent technology. Insulated concrete foundations and rammed earth walls have made adobe style homes a popular choice for energy efficient construction.

Mediterranean. Courtesy of David E. Wiggins, Architect.

Mediterranean. Courtesy of David E. Wiggins, Architect.

Mediterranean

Also known as Italian or Tuscan design, Mediterranean style is similar to Spanish Adobe but is a bit more rustic, relying more on regional materials and building techniques. In traditional Mediterranean homes stone was covered with stucco, creating a great thermal mass for cooling in hot climates and insulation in cold weather. As the stucco aged, it revealed more and more stone. Modern architectural design captures this aging effect by using a mixture of stone and stucco from the beginning.

David E. Wiggins, Architect, is an authority on this style and elaborates, "Italian design incorporates outdoor elements into the architecture, creating courtyards and loggias, or breezeways. Roofs are clay tile or exposed mud for an even more authentic look, with low pitches and exposed eaves. The rustic appearance is reinforced through the use of corbels, rafter tails, rough sawn window and door lintels, and exposed beams. Windows are typically small to maximize the distinctive mass of the wall they penetrate. French doors that open to covered outdoor spaces add additional openness to the façade. Wrought iron is used to decorate windows, balconies, and terraces. Towers add an interesting architectural form in just the right place. Cut stone columns or rough sawn timbers are commonly used as vertical supports."

Living Small

House design and lifestyles seem to move in cycles, and the recent trend has moved away from the sprawling McMansions of the nineties to smaller, more efficient designs. Families are learning that a quality home comes from good planning and design, not sheer size. The recent housing collapse and the rise of energy costs has influenced this trend. You will find that a little ingenuity can stretch your construction dollar while shrinking your cost of living.

Smaller, well designed homes have many advantages:
- Lower property taxes
- Less area to clean and maintain
- Lower energy costs
- Simpler, smaller construction projects that are far easier to manage
- Lower construction costs, of course

Dennis Fukai is an Architect and expert in small house design. His books Living Small, How a House is Built, Building Simple, and Being Sustainable, are excellent resources for gaining an appreciation of the simpler, smaller lifestyle. Dennis points out, "As the size of a house grows, cost and complexity increases exponentially. More hallways are required to allow ingress and egress. Hallways can sometimes make up over 30% of a home's square footage. Larger homes not only consume more land, but much more of our natural resources. More than 600 trees must be cut down in order to build the average American home. This drain on resources continues over the life of the house, through maintenance and energy use."

Living small is difficult because builders don't like to build small. They make the most profit on the larger houses. The small house market is left to tract builders and rehabs of older, small cottage homes. By being your own contractor, you can buck this trend and achieve tremendous savings through the creative use of design and materials.

Advice for Living Large on a Small Budget

Much of the cost savings of your construction project start right during the design phase. By using these cost reduction strategies, you can save a tremendous amount of money before you drive the first nail. Make sure to study the chapter on Saving Money for more cost saving strategies that can be incorporated into your design. If you are customizing a stock plan, make sure to tell your designer or architect about these strategies so that they can be incorporated into the blueprints.

- Eliminate as many hallways and walls as possible. Hallways waste building materials and add to heated space while making a house seem smaller. When a hallway cannot be avoided, turn the long expanse of blank wall into shelves and cabinets. Double wall framing, with studs two feet on center, can be used as a framework for floor-to-ceiling cabinets and shelving,

while adding only a foot or so to the framed width of the hallway.
- Embed extra storage in empty walls. Blank kitchen walls can be framed with 2×8s and filled with shelving for canned goods between the studs. Put that empty space between interior wall studs to good use.
- Use an open area design. Combine the living, kitchen and dining rooms into one large entertainment area. This one area will make a smaller home seem far more spacious and will cut down on building materials.
- Two story and split level homes are generally less expensive to build than ranch style houses of the same square footage because there is less roof and foundation.
- Use closets as sound buffers between rooms.
- Avoid house plans that have many different sizes of windows. During purchasing, delivery and installation, mismatches can occur.
- Use the space under stairs for extra storage, a laundry room, or a play niche for children.
- Go with a square, rather than a rectangular or L-shaped home design. With the exception of circles, squares provide the greatest interior area for the area of exterior wall.
- Reduce corners and angles. Corners require several extra studs per corner and take longer to frame. Corners make roof lines and resultant framing far more complex, increasing labor and material costs. Instead, go with clean, simple building lines, and add back texture and interest through window and eave trim and varying siding materials.
- Create multipurpose rooms that eliminate additional walls. Turn the laundry room into a laundry, sewing, crafts and office room. Build fold-up Murphy beds into guest bedrooms so the room can be used as an additional office or playroom whenever guests are not present.
- Eliminate formal dining and living rooms, or convert them into multipurpose libraries and offices.

Evaluating House Plans

The pleasure of living in your new home will be greatly affected by its interior layout and traffic flow. When evaluating plans, examine the relationship of areas to one other, their traffic circulation, privacy, and room size.

Traffic Patterns

Study the circulation and traffic patterns of each room. A floor plan with good circulation should direct traffic flow to one side of the room rather than through its center. Some circulation problems can be improved simply by moving doors to the corners of rooms.

Kitchen Area

The location of the kitchen in relation to other areas of the house is critical. It should have direct access to the dining area and should be accessible to the garage or driveway for ease in unloading groceries. Being near the utility room is also convenient if you plan to have work in progress in the kitchen and utility room simultaneously. Ingress and egress traffic should not pass through the kitchen work area.

Make sure your plan concentrates the main kitchen appliances (fridge, stove, sink) in a compact traffic triangle. See the Designing Kitchens and Baths chapter for more detailed information.

Private Areas

To ensure privacy, the bedroom and bathroom area should be separated visually and acoustically from the living and work areas of the house.

Make sure you can reach the bathrooms without going through any other room and that at least one bathroom is accessible to the work and relaxation areas. One of the basic rules of privacy is to avoid traffic through one bedroom to another. Check the size of the bedrooms. They should have a minimum floor area of 125 square feet for a double bed and 150 square feet for twin beds.

Living Areas

The living areas include the dining room, living room, and den. In many older designs, these areas are broken into individual rooms. Modern designs combine separate rooms into single, multipurpose rooms. The den area is usually located at the front of the house, but rooms at the side or rear may be desirable, particularly if this provides a view into a landscaped yard. The main entrance should usually be located at or near the living room, if there is one. There should be a coat closet near this entrance and a passage into the work area without passing though the living room.

Designing Green

Energy efficient, green building has become increasingly popular in recent years, and you'll probably want to incorporate some of these features into your design. The energy efficiency of a home can be greatly improved with a little advance planning. Listed below are a few energy considerations:

- Orient your building site so that the exterior walls with the smallest window area face north. Southern exposure to sunlight will allow more solar heat gain during the day. Reverse this orientation when building in hot climates where heat gain is an issue.
- Combine a good orientation with intelligent landscaping to achieve considerable energy savings. Deciduous trees can provide shade in the summer months while letting sun through in the winter months.

Avoid planting trees too close to foundations, driveways, patios or walkways in order to prevent cracking.
- A window's insulating ability is much lower than that of a typical wall. Reducing the amount of window coverage will increase insulating efficiency. This can be done by reducing the number of windows (which also saves money) or the size of each window. Wooden windows, although more expensive, tend to insulate better than metal ones, although vinyl clad wood windows are gaining in popularity. Choose double or triple pane windows or use storm windows for the best insulation possible. Consider skylights or solar tubes for additional lighting in interior rooms.
- Framing your exterior walls with 2×6s 24" on center will allow 6" batts of insulation in exterior walls. However, framing doors and windows can be a problem with 6" walls. Make sure your building materials supplier stocks doors and windows designed for 6" wall depth. Another way to achieve a higher insulation value is to use thinner walls but an additional layer of foam sheathing on the exterior. Foam-in insulation can be used in thinner walls and has a higher R-value than fiberglass or cellulose. It also fills gaps and reduces air infiltration. Make sure the blueprints take into consideration extra thick walls or additional sheathing. See the Framing chapter and Saving Money chapter for more construction alternatives.
- The two most efficient heating systems are gas heat and heat pumps. Gas heaters are cheaper to install and are simpler to maintain; however, heat pumps use less energy to operate. As gas prices rise, heat pumps become an even greater bargain. Conventional air source heat pumps lose their efficiency when winter temperatures drop below 30°. In this type of climate, gas heat may be better. A geothermal heat pump is a modern alternative that captures heat from the ground to keep heat pumps efficient, even in the coldest weather.

House Plans

How many copies? Regardless of how you arrived at a design, you will need plenty of copies of your plan. Plans refer to a standard set of blueprints showing basic elevations of the home and a layout of each floor. Special detail plans will be covered later in this section. The following is a representative list of the copies you may likely need:

Personal Copies	2
Bank Copy	1
County Inspector	1
Permits	1
HVAC	1
Plumbing	1
Electrical	1
Framing	1
Total	9

Sound crazy? Copies are cheap, after the first one. Don't scrimp here. Get what you need. It's always nice to have that extra clean copy when your originals become ragged from use. You cannot directly make a blueprint from a blueprint. Blueprints are slowly being replaced by digital prints. These are easier to copy, but require large format copiers. If your plan is available in a digital CAD system, copies are very easy to obtain. For some purposes, such as permitting and loan approval, electronic files can be submitted in PDF format.

General Advice

When building your first home, consider using a standard plan, preferably a popular one designed by a reputable architect or plan company. They are usually popular for a reason. Builders love to build the same plan over and over. They learn from their mistakes and are able to refine the construction process. Their cost estimates will also be more accurate. By choosing a popular design, you benefit from the lessons learned by others.

Critically review the kitchen and kitchen cabinet layout. Cabinet layouts are normally there just to meet minimum requirements. You can probably do better. The kitchen design itself may not be the best either.

For an additional charge, your plan supplier may be able to provide you with a material takeoff. But DO NOT assume it to be accurate.

If you change interior partitions, be careful of any impact on load-bearing walls. Consult your architect or designer if you are in doubt. If you want to finish the basement, opt for a 9' basement ceiling to allow for a dropped ceiling and/or HVAC plenum. Consider open-truss floor joists so that wiring and HVAC can be hidden in the ceiling.

Plan Contents

How do you know if your plans are complete? At a minimum, a set of plans should include the following diagrams:

- Front elevation
- Side elevations (one for each side of home)
- Rear elevation
- Top view of each floor (floor plan)
- Detailed layout of kitchen cabinets
- Detailed foundation, footings and framing specifications
- Building and wall sections
- Detailed roof plan

Special Detail Plans

Although you have a set of blueprints, they may not include the following:

- Mechanical layout (HVAC and ductwork)
- Electrical layout

These are normally done by the respective subs. Your blueprints may also include:

- Detail cabinetry layout
- Trim detail layout
- Detail tile plan
- Window and door schedule

DETAIL CABINET LAYOUT – This should include a final layout of each specific cabinet you plan to have in the kitchen, including style, finish and placement. All cabinetry in bath areas, such as vanities, must also be spelled out.

TRIM DETAIL LAYOUT – This should include a room-by-room schedule of what trim is used where. Base, chair rail, crown and other decorative molding as well as paneling should be described in detail. Specific material finishes should also be covered. Trim detail layouts are not usually included with most standard plans.

DETAIL TILE PLAN – This should include diagrams showing areas to be covered with particular types of tile. All tiled shower stalls, bathtub areas, tiled walls and floors should be covered. Also, all fixtures such as mirrors, soap dishes, safety rails, towel racks, tissue roll holders, toothbrush and tumbler holders, medicine cabinets and related fixtures. Grout types and colors also need to be determined.

WINDOW AND DOOR SCHEDULE – Itemized list of all window types and dimensions. Make sure that if you plan to have shutters you have adequate room between window sets for the shutters (normally 12" or 15"). Also make sure you have symmetrical spacing between windows where applicable. Check before construction and during the construction process.

DETAILED ROOF PLAN – This plan is helpful, especially with complicated hip roofs. It consists of a framing diagram showing opening dimensions, hips, valleys, ridges, roof pitch, ridge venting and skylights.

Plan Changes

During your plan selection, also consider special changes such as zoned heating/AC, passive solar, special wall partitions, window modifications, special cabinetry, etc. Your original drawings should provide the basis for any details, refinements or other changes you wish to make. Be careful when planning changes. Some changes can have unforeseen effects. For example, changing the height of a ceiling on the first floor of a two-story home will change more than just studs framing; it will also add several risers and treads to every staircase, which will require more space for the stairs to spread out horizontally. When you decide to change a plan, mark the change on the plan with a red permanent marker. (This will keep the change from smearing if the plans get wet.) Make sure when making changes on plans that you mark ALL the sets of plans to eliminate costly problems later. Many lenders require that houses be built exactly as the plans indicate.

Building Information Models

Remember the old adage that we mentioned at the beginning of this chapter? Measure twice, cut once. The time you invest in getting to know your home design before you begin construction, will be the most productive time spent. Modern computer technology has given us powerful tools to visualize and refine the design before mistakes are made in the real world.

This technology, as it applies to construction, is called a Building Information Model and grew out of the CAD-CAM advances of the seventies (computer aided design — computer aided manufacturing). With ever more powerful computers, designers and engineers can build a virtual model of a construction project on the computer and attach manufacturing information to each building object. They can check for spatial interaction between parts, calculate the exact nature and quantity of materials needed, analyze lighting, and in the most sophisticated systems, fabricate parts directly from the computer.

These systems have become so powerful that the latest commercial airliners from Boeing and Airbus were designed and tested exclusively in virtual space before the first part was ever manufactured. Less powerful versions of these programs have become popular in the commercial construction fields. And now, thanks to ever more powerful computers, you too, can harness the power of building information models.

The online search company, Google, has released a powerful CAD drawing tool called SketchUp. It has vastly simplified the task of creating 3D objects on the computer and is accessible to all aspiring designers. Best of all, their non-professional version of SketchUp is free. With SketchUp, you can build a complete three dimensional model of your home on the computer — to scale.

Virtual modeling will allow you to walk through your design to see precisely how the home will appear when completed. Suddenly, dimensional and spatial errors become obvious, long before you have committed yourself to the actual construction project. SketchUp allows you to orient your house geographically and study the play of light and shadow throughout the home at differing times of the day and year (see The Lot chapter for an example of this). Reports can be generated that list the quantity of components used in the drawing. Why "guesstimate" the number of studs and rafters in a home, when you can know precisely how many were used within the model?

Dennis Fukai's books provide excellent tutorials for using SketchUp in the design of your home. Dennis has this to say, "We use our books to show how SketchUp can visually communicate the means and methods of an assembly as a series of distinct events. Though it's a great design tool, SketchUp is more than a pretty face. It plays an important role as an information tool, by modeling and communicating "time" as a sequence of events. Different scenes can be animated to show these events in a visual manner. This is important when discussing change orders and clarifications because it sets up a visual understanding of a problem from a common point of view."

Is it easy? Well, that's a question for debate. Compared to earlier CAD programs, SketchUp has reached a new level of usability. But make no mistake — it will take time and dedication to learn enough to successfully construct an entire house. Is it worth the time and effort? This author thinks so. Why? Because the act of modeling the house will teach you exactly how design elements and your decisions fit together during the construction of the real thing. It may take months to build out a complete construction model on a computer, but during that time you will become intimately familiar with all the nuances of your future dream home. This familiarity and expertise will impress the sub-contractors that work on the job and will help to head off problems while solutions are cheap.

What if you simply don't have the time or inkling to learned 3D CAD? One excellent compromise is to work with a designer or architect who is familiar with SketchUp. Due to its flexibility and power, many designers are rendering all their latest plans in SketchUp. Check with your designer of choice to see if SketchUp models are available for your perusal and analysis. Ask for "walkthroughs" of the design. Two excellent designers who work with SketchUp models are Alan Mascord and David Wiggins. See their contact information at the end of this chapter.

Designers and Architects Working in SketchUp 3D

Contact these designers and architects for more information on home designs and the Google SketchUp Pro design software.

Dennis Fukai, PhD, Architect
16708 SW 132 Lane, Archer, FL 32618
Website: www.Insitebuilders.com
Books: Building SIMPLE: Building An Information Model,
  Being SUSTAINABLE: Building Systems Performance,
Living SMALL: The Life of Small Houses, How A House
  Is Built: with 3D Construction Models

Alan Mascord Design Associates, Inc.
1305 NW 18th Avenue, Portland, OR 97209
Website: www.mascord.com
Books: Mascord Living Spaces, Mascord Efficient Living,
  Mascord Home Collection 2007

David E. Wiggins, Architect
2725 Highland Trail, Leander, Texas78641
Website: www.wigginsarchitect.com

Standard Dimensional Specifications

General
- Floor to ceiling: 7'- 6" minimum with no beams projecting down more than 6"
- Smoke detectors in each sleeping area or in the hallway located centrally to the sleeping areas.

Bedroom
- No dimension less than 7' in any direction
- Two means of escape in case of fire: one door, and a window minimum 6 sq. ft. area and no more than 44" above the floor.
- Closet in all bedrooms
- Easy access to a bath
- Space for bed and nightstand (double: 54"× 72", queen: 60"× 78", king: 72"× 78")
- Space on another wall for a 6' dresser
- At least one window
- Should not have to go through a bedroom to get to any other room
- Bed to closet: 36" (minimum)
- Bed to dresser 36" (minimum)
- Minimal space for dressing: 42"
- Minimal space between double beds: 22"
- Minimal space between bed and wall: 12"
- Minimal space between bed and dresser: 6"

Closet
- Depth: 24"
- Pole height: 5' to 6'
- Pole depth: 1' from closet rear wall
- Unsupported clothes rack (4' maximum)
- Door clearance at bottom: 1" from finished floor for ventilation)
- Shelving depth: 6"

Hallways
- Minimum width: 36", but 42" to 48" preferred
- Minimize overall hallway length

Dining Room
- 8-person dining room: 10'× 12' minimum
- direct access to kitchen
- closed door access for formal (consider double swing doors)
- should not have to walk through dining room to access any other room
- one large free wall for a hutch
- consider space for extra furniture and guest seating

Kitchens and Baths
See the chapter on Designing Kitchens and Baths.

DESIGNING KITCHENS AND BATHS

Introduction

Kitchens and baths have historically been the focal point of most households. Our fondest childhood memories often center around the aromas and tastes experienced in the kitchen. Our most memorable arguments probably took place at the bathroom door. The old adage that the "kitchen and bath sells the house" is more true today than ever. That's why you will want to spend extra time in planning the design and layout of these critical areas.

Professional Kitchen and Bath Services

Planning a kitchen can be challenging and may require the help of professionals to get the job done right. Unless you have a high confidence in your style sense, choosing a professional designer can save many hours of frustration and expense. You will probably deal with at least one of the following services:

Architect or Designer

Architects and designers will usually include a kitchen and bath design if requested, drawing the elevations and in some cases providing a rendering. However, many architects can be hired solely for the kitchen design. This can be expensive and isn't usually worth it unless your kitchen designing is quite challenging.

Kitchen and Bath Designer

This may be an independent cabinet dealer, or an in-house designer at a material supply store. They usually charge a flat fee for designing the kitchen and baths or, if they are selling you a cabinet set, they may include the design fee in the cost of the cabinets. Many suppliers now have in-house design facilities with computerized design services.

A kitchen and bath designer can be very helpful for designing stunning layouts, but may not be knowledgeable about the technical issues of your home project. You will still need to hire installation contractors to complete the kitchen project or depend on your plan provider. Look for kitchen and bath designers who have been certified by the NKBA – National Kitchen and Bath Association.

Interior Designers

Sometimes interior designers get involved in the design of kitchens and baths and often provide a comprehensive home plan with paint and wallcovering suggestions that will blend seamlessly with kitchen and bath choices. Look for designers who have been certified by the ASID — American Society of Interior Designers.

Kitchen Design

Now more than ever, the kitchen has evolved into a social center, work center and entertainment hub for families. Make sure you analyze how these various functions will integrate into your kitchen design. Well before starting construction; evaluate how you and your family plan to use the space. Besides the usual — cooking and cleaning — do you dine in the kitchen, congregate around the table and converse, chat on the phone, study, pay bills, write? When you entertain, do you like to have your guests visit with you while you cook? Or do you prefer the kitchen to be isolated so the mess from food preparation can be hidden away? Consider the following factors before starting your design:

- Budget
- Style and function
- House floor plan and traffic patterns
- Lighting and color
- Available counter space
- Appliances

Design Considerations

Once you know the budget and scale of the project, study the existing layout and look for improvements with the least cost and effort.

Traffic Patterns

By graphing the work triangle on your kitchen plan, poor traffic patterns will immediately become obvious. Draw arrows showing the usual traffic flow of family members and guests through the kitchen and out again. Think about access from the garage or carport. If you're coming in from the car with an armload of groceries, you'll want the kitchen nearby with a large surface on which to set things.

The four generally recognized arrangements for kitchens are the "U", "L", galley and one-wall types. Variations include the peninsula and island types. The arrangement you choose will depend on the space available, its shape, and the location of the doors. The smallest work triangles are the "U" and galley layouts. The one-wall arrangement works well where space is limited and the "L" arrangement fits best in a square kitchen with a dining table.

Doorways

Doorways should be located to direct traffic away from the work triangle. Doorways in corners should be avoided and door swings should not interfere with the use of appliances, cabinets or other doors. If door swings interfere with traffic in a hall or other activity area, consider using a sliding or folding door. Sliding doors can be expensive, but may be worth it.

Clearances

As you layout the design, pay attention to proper clearances and design options. The National Kitchen & Bath Association has developed a comprehensive set of kitchen guidelines and specifications for modern kitchens. These are included in the kitchen specifications section as NKBA specifications.

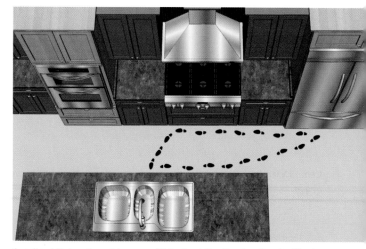

A galley kitchen layout is very efficient, while allowing thru traffic, but watch out for collisions between the chef and eager children!

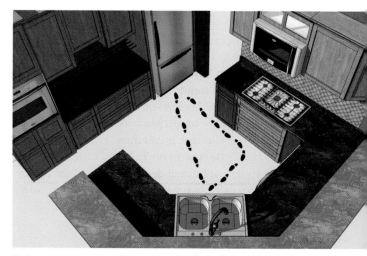

U-shaped kitchens provide a compact, efficient work triangle, but can have storage issues in the corners of the lower cabinets.

Central islands and peninsulas can recapture wasted space and serve as a buffer zone between meal preparation and dining. An extended shelf can be used as an eating surface or bar. An island can also centralize the stove or sink, but make sure to consider plumbing and gas utilities during planning.

Choosing a Style

The ambience and style of a kitchen is a very personal choice, but it will be influenced by other factors such as the architecture of your home and the lifestyle you wish to lead. You want to choose a kitchen style that complements and unifies the other visual elements of the home.

Stick with Classic Styling

Recent architectural trends may influence the kitchen design, but be careful. The latest trends can go out of style quickly. With the cost of kitchen makeovers, don't lock yourself into a design that might appear dated in a few years. The classic designs never fade from fashion and insure that your new kitchen design will stand the test of time. If you start with a time tested layout, you can add styling elements such as windows, trim, light fixtures, cabinet hardware, paint and countertops to impart that custom feel.

Popular kitchen styles
- Colonial
- Contemporary
- Eclectic
- English country
- European: Tuscan, French Country
- Shaker style
- Victorian

Match the Kitchen to Your Lifestyle

Do you like formal dining? Or do you prefer socializing with guests in the kitchen? Open walls between the kitchen and dining room will allow greater interaction. Mud rooms, pantries, and foyer closets can be integrated into the kitchen design to gain much needed storage. If you have children, consider emphasizing the breakfast nook and providing durable work areas where children can play or draw while meals are being prepared.

Kitchen Functions

During design, split the kitchen into work zones and consider each typical function and its impact on the overall layout. Typical kitchen functions include:

- Food storage
- Food preparation
- Cleaning
- Eating
- Organizing and paperwork
- Crafts

Lighting and Airflow

Kitchen windows should be large enough to make the kitchen a light, cheerful place. They also serve as ventilation for moisture, fumes, and cooking odors. The current trend toward indoor-outdoor living has fostered the "patio kitchen", with large windows, which extend to the outside and provide a "pass-through" or an outdoor eating counter. This is particularly useful in warmer climates, but it also has merit for summer living. Window lighting can be supplemented with careful placement of accent and track lights, to concentrate illumination in critical areas. Under-counter lighting using halogen or LED technology places the light right where you need it most: on the countertop.

Peninsulas can provide separation between the kitchen and other living areas, while providing a space efficient eating area.

Color and Texture

Color, contrast, and texture will dictate the atmosphere and brightness of the kitchen. Use lighter colored paints, cabinets, or flooring to open up a small kitchen area. Colors in the blue family are considered cool and tend to

Large windows, good lighting and lighter-wood cabinets will visually open up a kitchen.

Drawer refrigerators can be placed in kitchens, dens and leisure rooms.

Panel inserts can match the refrigerator to the cabinet style.

enlarge a room. Colors in the red/yellow/orange family are warm and tend to make a room smaller and "cozier."

If your kitchen is open and flooded with light, then darker colors can provide contrast. Darker colors also provide depth in larger spaces. If you use dark cabinets, consider lighter countertops and walls for contrast. If the kitchen is small, try to avoid dark colors or it will appear even smaller.

Think about the color scheme you have chosen and the overall "feeling" you are trying to create. Do you want a warm and cozy kitchen? Do you want a sterile, efficient appearance? It is well known that colors can be used to achieve just about any effect you wish.

Appliances

Since appliances can make up over half the cost of your entire kitchen, it's important that you choose units that match your aesthetic and functional goals. Luckily, even the low cost appliances of today work efficiently. You'll have an infinite range of prices, styles and features to choose from. But don't overlook the bargains. They can stretch the budget and work as efficiently as the "premiere" brands. Expensive kitchen appliances are much like high priced cars: Part of what you pay for is their prestige and brand. Visit appliance retailers and study the options carefully. You'll be spending a lot of time using these machines (assuming you like to cook) and even the tiniest details can make a big difference in functionality.

For instance, are the stove tops easy to clean? Do the grills pop out easily? Are the cooktop knobs located on the cooking surface where they can pick up dirt and stains? Will you burn your hands on hot pots when trying to adjust the burners? Or are the knobs located on the front of the unit where you are more likely to bump into them? Everyone is shaped differently and works differently in the kitchen, so make sure your appliances fit you and not the other way around.

Appliance Styling

Developing a unified style in the kitchen means coordinating all the different parts, from paint, to cabinets, to flooring, to appliances. A bright stainless steel refrigerator may look fine in a contemporary kitchen, but would stick out like a sore thumb in a rustic Tuscan kitchen. Most manufacturers have caught on to these styling issues and now offer interchangeable panels for their appliances so that you can alter their appearance in the same way you would choose a color and option scheme for a car. Many refrigerators offer reversible doors and options to allow you to match the door front to your cabinetry.

Counter Space and Storage

One major goal in any kitchen design is to provide ample storage space. A modern kitchen should have at least 6 square feet of storage space in the wall cabinets for each resident. An additional 12 square feet should be added for miscellaneous storage and entertaining. Since the base cabinets usually occupy all the space under the wall cabinets, triple the wall cabinet space to arrive at total usable storage space.

Counter space should be located in strategic areas. There should be at least 3' of counter space between the sink and the refrigerator for food preparation and to make room for a dishwasher under the counter (2'). A minimum of 2' of counter space should surround the stove top to make room for pots or food preparation. Try to leave about 2' of space next to the refrigerator on the side where the door opens as a transition area for food going in and out.

Cabinets

Kitchen cabinets can be custom made or purchased in units as stock items. Stock cabinets are usually purchased in widths varying in 3" increments between 12" and 48".

Although custom cabinets are nice, the extra expense and time involved is rarely worth it. There are abundant high quality prefab cabinets available today. When shopping for cabinets, pay attention to space-saving devices such as revolving corner units, drop down shelves and slide-out shelves. These options can significantly increase the usable space in your cabinets.

Visit a few kitchen cabinet stores or kitchen design stores to become acquainted with the many types of available cabinets. Many American companies are turning out impressive reproductions of fine, expensive European models at a fraction of the cost.

Counter Tops

Counter tops typically should have a standard depth of 24" and may have a 3" or 4" splash block against the wall. Counter tops are priced and installed by the foot, while special charges may be incurred to cut openings for stoves, sinks and fancy edgework.

Floor to ceiling cabinets and shelving over openings vastly increase the storage possibilities in this eclectic design.

Basic cabinetry: Wall cabinets, base cabinets and an island make this an affordable but great-looking and functional kitchen.

Bathroom Design

Next to the kitchen, bathrooms make up the largest portion of a home's interior cost. The biggest challenge is to achieve your goals while staying within budget. Since bathrooms are difficult to retrofit later, due to their extensive plumbing and wiring requirements, it's important to get this right the first time. You can, however, plan for the future by "stubbing out" future bathrooms with plumbing and waste lines as long as you have your floor plan finalized.

Start by assessing your needs, as you did with the kitchen.

- Will the children share a bathroom? If so, can it be divided into zones to increase usability and privacy?
- Will there be a dedicated powder room for visitors? Should it include a shower so it can be used as an auxiliary bathroom for overnight guests?
- Can the bathrooms be grouped together to reduce rough-in costs?
- Do family members or frequent guests have a disability?
- Is there adequate storage space for linens and towels nearby?
- Will you be using water saving toilets?
- Do you want private alcoves for toilets? If you're short on room, can a half-height wall substitute?
- Is there enough space? Sometimes bathroom floor plans just aren't quite large enough to accommodate fancy additions to a bathroom design. Instead of

making an expensive change to your blueprints, consider a cantilever. Cantilevers can add an all-important 2' or more to the bathroom area. This can be enough to squeeze in amenities like a linen closet or large whirlpool tub.

Choosing a Style

Like the kitchen, you should choose a bathroom style that reflects your lifestyle. You will have far more creative freedom here because each bathroom is a separate living space and can reflect wildly different styles from one bathroom to the next. However, the styling and appliances in bathrooms can be very expensive. Consider how many guests will be viewing your "design masterpiece." Save the elaborate décor for the powder room. Since this bath has guest access, it should generally reflect the styling used in the rest of the house. Master baths will seldom be viewed by guests. You and your spouse are the main audience, so make it the perfect match for your taste and preference.

Don't go overboard with bold design. Your choice of floor tiles, vanities, lighting, faucets and accessories will be more permanent than most interior finishes. If you grow tired of the design, or wish to sell the house later, these items will be far more expensive to replace than a new paint job or wallpaper.

A combination of function and style makes for a versatile bathroom.

Common Bath Styles
Art deco
Arts and crafts
Contemporary
Country
Shaker
Traditional
Victorian

Handicapped Access
For people with disabilities, the bathroom can be a maze of frustration and embarrassment. Even if your family is in good health, consider relatives who might come to visit. As we age, we all begin to lose mobility. If you intend to stay in your dream home for years this eventuality should be considered NOW when it is convenient to make adjustments. More information on design choices can be obtained on the ADA.gov website. (Americans with Disabilities Act)

Improved access can include:
- Extra height toilets, known as ADA toilets. They have the added benefit of better flushing action and are more comfortable for most adults.
- Grab bars near toilets and shower
- Wider doors and access lanes for wheelchairs
- Faucets with levers, or with sensors that turn water on and off automatically
- Lower height sinks and vanities with under-sink access for wheelchairs.
- Shower heads that can be adjusted up and down or removed from the wall for handheld use.

Consult with your architect or designer for additional specifications and dimensions for handicapped access. Many of these features will not only be welcomed by disabled visitors, but can be useful for anyone recovering from an injury or simply suffering from early morning grogginess.

Plumbing Walls
Avoid placing plumbing and venting in the walls separating the master bath from the bedroom, or the powder room from the main entertainment area. The noise from pipes and the sound of whooshing water from toilets can be quite loud and disturbing. Whenever possible, place the plumbing wall between the vanity and toilet area or between the bath and closet.

While designing your bathroom floor plan, try to think in three dimensions and imagine the layout and routing of plumbing and waste pipes. Make sure the "utility wall" will have a clear path to the roof for venting unless you plan to use air admittance valves. Never place the plumbing and drain lines in outside walls, as this increases the chance of freezing pipes in cold weather.

Other Utilities
The utilities in the bathroom are interrelated and pose several challenges due to the humid environment.

Lighting
Lighting in the bathroom can be tricky. On one hand, you will want a lot of natural light to open up a small space visually. But you'll also want privacy. Consider placing windows high on the wall or use skylights. Textured glass blocks are also a good option for letting in light while maintaining privacy. Try to avoid curtains which invite mold in humid spaces, and plan for triple pane windows to help reduce condensation on the windowpane.

Good task lighting at the vanity is very important. This will illuminate your face in the mirror when applying makeup or grooming. The best lighting comes from both sides of the mirror and illuminates the face evenly, with no shadows. Since incandescent bulbs are being phased out nationwide, make sure to choose a light source that will provide pleasing natural color. Some compact fluorescent bulbs emit an ugly greenish light. Natural light fluorescents and LED's make the best choices. If your medicine cabinet doubles as the vanity mirror, make sure the lights project past the frame of the cabinet, or shadows may occur.

Old world style adds atmosphere and romance to bathroom design.

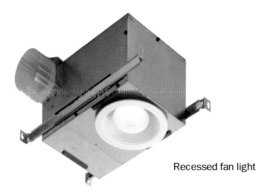

Recessed fan light

Delta hands-free faucet provides automatic on/off flow and electronic temperature control.

Ventilation

Bathroom ventilation should be considered a requirement, not a luxury. Humidity can take its toll on wall coverings, fabrics, and towels, leading to that "musty" smell. In energy efficient, tightly sealed homes, moisture control is critical in baths and kitchens. Mold can accumulate quickly. Ventilation units should operate on a timer so that they can run for several minutes after the bath has been used. The vents must exit to the outside. Do not vent into a crawl space or attic. You will only be transferring your moisture problems from one area to another. If your home has a HRV (heat recovery ventilation) system, consider tying your bathroom and kitchen vents into it to increase energy efficiency. Place the ventilation unit above or near the shower area to capture the most humidity.

Other Design Considerations

- Pedestal sinks are very stylish, but they lack the storage space of a vanity. If you choose to go with the pedestal style, make sure to expand the size of storage in other areas, such as the linen closet and medicine cabinet. If your medicine cabinet has sufficient depth, consider building in an electrical outlet and alcove for a hair dryer.
- Modern water saving toilets can flush effectively with as little as 1.28 gallons. Dual-flush toilets use even less. Plumbers however, are accustomed to installing older toilets with greater flush volume. Make sure the drain pipes for water saving toilets have been sized properly and have a steeper pitch to aid in drainage.

- Plumbers charge by the connection, so reduce the number of sinks and toilets if possible, or if you stack your plumbing connections back to back, negotiate a reduced price per connection.
- Pocket doors make maximum use of space in cramped bathrooms. Consider using one to separate a guest bath from the powder room to achieve double duty.
- Toilet paper holders are a commonly overlooked design detail. Make sure your plan allows for a toilet roll location that is convenient to reach without the need for contortionist movements.
- Use compartmentalization to enhance privacy in family bathrooms. Separate the bathing, grooming, and toilet areas so they can be used simultaneously. Family bathrooms used by children need plenty of storage to combat clutter. Install large medicine cabinets with zones for each child. Use vanities with several drawers. Provide ample storage for a variety of towels and toiletries near the shower.
- Whirlpool tubs become very expensive in larger sizes. They can also hold a surprising amount of water, weighing 8 lbs. per gallon. If used often, they can negate the savings from water saving toilets. Floor framing should be structurally reinforced to handle the additional weight.
- Safety locks should be installed on all bathroom (and bedroom) doors. These locks have small access holes that allow the door to be unlocked with a nail or paper clip. Since the bathroom is the site of a surprising number of accidents in the home (slipping in the tub for instance) it's best to allow access by other members of the family in case of trouble.

Combo light and ventilation unit for the bath

Supplemental Heating

Bathrooms can be chilly places. You will likely be walking barefoot on hard, cold surfaces in a state of undress. Consider radiant heat under the floor for that comfy feeling in the morning on cold days. A combination vent and infrared heater near or above the shower can provide instant warmth in an area where the air will be moving up toward the vent, adding to the perceived chill factor. Towel warmers can provide an instant jolt of warmth right out of the shower and will help to dry towels more rapidly. Make sure all these supplemental venting and heating sources are wired to a timer so you don't have to remember to turn them off on the way out of the bathroom.

Related Spaces

The versatility of bathrooms has expanded in recent decades, leading to far larger and more expensive spaces than in the past. Its form and function has affected other related spaces as well.

Closets

The location of the closet can be critical to a practical bath plan. When placed in or near the bathroom zone, closets allow dressing and undressing without waking spouses who may have a different morning schedule. Walk-in closets attached to the bath area are also good locations for a makeup vanity. Makeup vanities that are too close to the shower can be impractical, as the added humidity makes applying and preserving makeup more difficult. Make sure to add a door to the closet area to reduce the migration of moisture from the bathroom and make sure to install adequate bathroom ventilation.

Exercise Room

What about an exercise room in the bath? This is a popular new trend, but does it really make sense? The floor and wall coverings in bathrooms are usually more expensive tile products. Adding the extra room for an exercise area can really drive up the cost of these expensive coverings. It will also be harder to expand the exercise area later, and who wants to risk dropping a dumbbell on a new tile floor?

Laundry Room

Placing a laundry area in the bath makes sense. The idea is to move the washer to the source of dirty clothes. Modern washer-dryer combinations are very space efficient and make good use of the nearby plumbing pipes and vents. The latest laundry offerings combine washer and dryer into one unit which is appropriate for smaller loads washed more frequently. This may not be advisable for large families, as you may not want the increased traffic into your master bath.

Glass block provides light and privacy in alternative bath areas.

Sauna

Saunas and steam rooms have become more popular in recent years, but they can add significantly to the cost of the master bath and take up considerable space. Consider installing a sauna/shower combo instead.

Stacked washer and dryer units are perfect for creating a small, convenient laundry room as an addition to the master bath.

Other Design Considerations

- Pedestal sinks are very stylish, but they lack the storage space of a vanity. If you choose to go with the pedestal style, make sure to expand the size of storage in other areas, such as the linen closet and medicine cabinet. If your medicine cabinet has sufficient depth, consider building in an electrical outlet and alcove for a hair dryer.
- Modern water saving toilets can flush effectively with as little as 1.28 gallons. Dual flush toilets use even less. Plumbers however, are accustomed to installing older toilets with greater flush volume. Make sure the drain pipes for water saving toilets have been sized properly and have a steeper pitch to aid in drainage.
- Plumbers charge by the connection, so reduce the number of sinks and toilets if possible, or if you stack your plumbing connections back to back, negotiate a reduced price per connection.
- Pocket doors make maximum use of space in cramped bathrooms. Consider using one to separate a guest bath from the powder room to achieve double duty.
- Toilet paper holders are a commonly overlooked design detail. Make sure your plan allows for a toilet roll location that is convenient to reach without the need for contortionist movements.
- Use compartmentalization to enhance privacy in family bathrooms. Separate the bathing, grooming, and toilet areas so they can be used simultaneously. Family bathrooms used by children need plenty of storage to combat clutter. Install large medicine cabinets with zones for each child. Use vanities with several drawers. Provide ample storage for a variety of towels and toiletries near the shower.
- Whirlpool tubs become very expensive in larger sizes. They can also hold a surprising amount of water, weighing 8 lbs. per gallon. If used often, they can negate the savings from water saving toilets. Floor framing should be structurally reinforced to handle the additional weight.
- Safety locks should be installed on all bathroom (and bedroom) doors. These locks have small access holes that allow the door to be unlocked with a nail or paper clip. Since the bathroom is the site of a surprising number of accidents in the home (slipping in the tub for instance) it's best to allow access by other members of the family in case of trouble.

SPECIFICATIONS

Note: Add the appropriate specifications to the sub-contractor agreement to avoid confusion, disagreements, and to insure the highest quality work. Customize as needed.

Refer to also: flooring specifications, tile specifications, electrical specifications, framing specifications

Kitchen
Cabinets – Specifications
- Bid is to provide and complete cabinetry and counter top installation as indicated on attached drawings.
- All cabinets are to be installed by qualified contractors.
- All cabinets are to be installed plumb and square and as indicated on attached drawings.
- All hardware such as hinges, pulls, bracing and supports to be included in bid.
- All necessary counter top cutouts and edging to be included in bid.
- All exposed corners on counter tops to be rounded or covered.
- All tile counter tops to be glazed tile only.
- All grout seams to be sealed.
- Install one double stainless steel sink (model/size).
- Special interior cabinet hardware includes: Lazy Susan, Broom rack, Pot racks, Sliding lid racks

Counter Tops – Specifications
- Bid is to manufacture and install the following cultured marble tops in ¾" solid color material. Sink bowls to be an integral part of top.
- Vanity A to be 24" deep, 55" wide, finished on left with backsplash; 19" oval sink cut with rounded edge with center of sink 22" from right side. Color is solid white. Spread faucet to be used.

NKBA Kitchen Specifications
The following kitchen specifications are provided by the National Kitchen and Bath Association.
- A clear walkway at least 32" wide must be provided at all entrances to the kitchen.
- No entry or appliance door may interfere with work center appliances and/or counter Space.
- Work aisles must be at least 42" wide, and passage ways must be at least 36" wide for a one-cook kitchen.
- In kitchens 150 square feet or less, at least 144" of wall cabinet frontage, with cabinets at least 12" deep and a minimum of 30" high (or equivalent), must be installed over counter tops. In kitchens over 150 square feet, 186" of wall cabinets must be included. Diagonal or pie cut wall cabinets count as a total of 24". Difficult to reach cabinets above the hood, oven or refrigerator

do not count unless specialized storage devices are installed within the case to improve accessibility.

- At least 60" of wall cabinet frontage with cabinets which are at least 12" deep and a minimum of 30" high (or equivalent) must be included within 72" of the primary sink centerline.
- In kitchens 150 square feet or less, at least 156" of base cabinet frontage, with cabinets at least 21" deep (or equivalent) must be part of the plan. In kitchens over 150 square feet, 192" of base cabinets must be included. Pie cut/lazy Susan cabinets count as a total of 30". The first 24" of a blind corner box do not count.
- In kitchens 150 square feet or less, at least 120" of drawer frontage or roll out shell frontage must be planned. Kitchens over 150 square feet require at least 165" of drawer/shelf frontage. (Measure cabinet width to determine frontage.)
- At least five storage items must be included in the kitchen to improve the accessibility and functionality of the plan. These items include, but are not limited to: wall cabinets with adjustable shelves, interior vertical dividers, pull out drawers, swing out pantries, or drawer/roll out space greater than the minimum.
- At least one functional corner storage unit must be included. (Rule does not apply to a kitchen without corner cabinet arrangements.)
- Between 15" and 18" of clearance must exist between the counter top and the bottom of wall cabinets.
- In kitchens 150 square feet or less, at least 132" of usable countertop frontage is required. For kitchens larger than 150 square feet, the countertop requirement increases to 198". Counter must be 16" deep to be counted; corner space does not count.
- No two primary work centers (the primary sink refrigerator, preparation center, cook top/range center), can be separated by a hull-height, full-depth tall tower, such as an oven cabinet, pantry cabinet, or refrigerator.
- There must be at least 24" of counter space to one side of the sink and 18" on the other side. (Measure only countertop frontage, do not count corner space.) The 18" and 24" counter space sections may be a continuous surface, or the total of two angled counter top sections. If a second sink is part of the plan, at least 3" of counter space must be on one side and 18" on the other side.
- At least 3" of counter space must be allowed from the edge of the sink to the inside corner of the countertop if more than 21" of counter space is available on the return. Or, at least 18" of counter space from the edge of the sink to the inside corner of the countertop if the return counter space is blocked by a full-height, full-depth cabinet or any appliance which is deeper than the counter top.
- At least two waste receptacles must be included in the plan, one for garbage and one for recyclable or other recycling facilities should be planned.
- The dishwasher must be positioned within 36" of one sink. Sufficient space (21" of standing room) must be allowed between the dishwasher and adjacent counters, other appliances and cabinets.
- At least 36" of continuous countertop is required for the preparation center, and must be located close to a water source.
- The plan should allow at least 15" of counter space on the latch side of a refrigerator or on either side of a side-by-side refrigerator. Or, at least 15" of landing space which is no more than 48" across from the refrigerator. (Measure the 48" walkway from the counter top adjacent to the refrigerator to the island countertop directly opposite.)
- For an open ended kitchen configuration, at least 9" of counter space is required on one side of the cooktop/range top and 15" on the other. For an enclosed configuration, at least 3" of clearance space must be planned at an end wall protected by flame retardant surfacing material, and 15" must be allowed on the other side of the appliance.
- The cooking surface can not be placed below an operable window unless the window is 3" or more behind the appliance, and/or more than 24" above it.
- There must be at least 15" of landing space next to or above the oven if the appliance door opens into a primary family traffic pattern. 15" of landing space which is no more than 48" across from the oven is acceptable if the appliance does not open into traffic area.
- At least 15" of landing space must be planned above, below, or adjacent to the microwave oven.
- The shelf on which the microwave is placed is to be between counter and eye level (36" to 54" off the floor).
- All cooking surface appliances are required to have a ventilation system, with a fan rated at 150 CFM minimum.
- At least 24" of clearance is needed between the cooking surface and a protected surface above. Or, at least 30" of clearance is needed between the cooking surface and an unprotected surface above.
- The work triangle should total less than 26'. The triangle is defined as the shortest walking distance between the refrigerator, primary cooking surface,

and primary food preparation sink. It is measured from the center front of each appliance. The work triangle may not intersect an island or peninsula cabinet by more than 12". No single leg of the triangle should be shorter than 4' nor longer than 9'.

- No major household traffic patterns should cross through the work triangle connecting the three primary centers (the primary sink, refrigerator, preparation center, cooktop/range center).
- A minimum of 12"× 24" counter stable space should planned for each seated diner.
- At least 36" of walkway space from a counter/ table to any wall or obstacle behind it is required if the area is to be used to pass behind a seated diner. Or, at least 24" of space from the counter/ table to any wall or obstacle behind it, is needed if the area will not be used as a walk space.
- At least 10% of the total square footage of the separate kitchen, or of a total living space which includes a kitchen should be appropriated for windows/skylights.
- Ground fault circuit interrupters must be specified on all receptacles that are within 6' of a water source in the kitchen. A fire extinguisher should be located near tile cook top. Smoke alarms should be included near the kitchen.

Bathroom

- Bid is to perform complete plumbing job for dwelling as described on attached drawings.
- Bid is to include all material and labor including all fixtures listed on the attached drawings.
- All materials and workmanship shall meet or exceed all requirements of the local plumbing code.

NKBA Bathroom Specifications

The following bathroom specs are provided by the National Kitchen and Bath Association.

- A clear walkway of at least 32" must be provided at all entrances to the bathroom.
- No doors may interfere with fixtures.
- Mechanical ventilation system must be included in the plan.
- Ground fault circuit interrupters specified on all receptacles. No switches within 60" of any water source. All light fixtures above tub/shower units are moisture-proof special purpose fixtures.
- If floor space exists between two fixtures at least 6" of space should be provided for cleaning.
- At least 21" of clear walkway space exists in front of lavatory.

- The minimum clearance from the lavatory centerline to any side wall is 12".
- The minimum clearance between two bowls in the lavatory center is 30", centerline to centerline.
- The minimum clearance from the center of the toilet to any obstruction, fixture or equipment on either side of toilet is 15".
- At least 21" of clear walkway space exists in front of toilet.
- Toilet paper holder is installed within reach of person seated on the toilet. Ideal location is slightly in front of the edge of toilet bowl, the center of which is 26" above the finished floor.
- The minimum clearance from the center of the bidet to any obstruction, fixture or equipment on either side of the bidet is 15".
- At least 21" of clear walkway space exists in front of bidet.
- Storage for soap and towels is installed within reach of person seated on the bidet.
- No more than one step leads to the tub. Step must be at least 10" deep, and must not exceed 7¼" in height.
- Bathtub faucets are accessible from outside the tub.
- Whirlpool motor access, if necessary, is included in plan.
- At least one grab bar is installed to facilitate bathtub or shower entry.
- Minimum usable shower interior dimension is 32" × 32".
- Bench or footrest is installed within shower enclosure.
- Minimum clear walkway of 21" exists in front of tub/shower.
- Shower door swings into bathroom.
- All shower heads are protected by pressure balance/ temperature regulator or temperature-limiting device.
- All flooring is of slip-resistant material.
- Adequate storage must be provided in plan, including: Counter/shelf space around lavatory, adequate grooming equipment storage, convenient shampoo/ soap storage in shower/tub area, and hanging space for bathroom linens.
- Adequate heating system must be provided.
- General and task lighting must be provided.

SAVING MONEY

Aside from the obvious design decisions made in the previous chapters, there are hundreds of other measures that can be taken to save money without compromising on quality. The real savings are hidden in the details, the collection of nickels and dimes that will all add up to thousands of dollars over the span of an entire construction project. This chapter takes the savings discussed in the design chapters to another level. Let's look at the many ways to save money before, during and after construction.

Design Choices That Save Money

As mentioned in the chapter on Designing Your Home, the greatest cost savings come from decisions made before construction even begins. The choice of house design, construction method, foundation type, siding and roofing materials, appliances, kitchen and bath designs, flooring, windows, insulation and HVAC will all have tremendous effects on the final construction cost. This section covers those construction decisions that go beyond the style and floor plan of the home.

Consider the long term implications of your choices. Professional builders are primarily concerned with the initial cost of their home "product." Their profits are made by keeping initial costs low, which often leads to substandard product choices.

As the future owner of your "self made home" you should be concerned not only about up-front costs, but also the cost of ownership down the line. Decisions that save money in the short term can cost much more during the life cycle of the home in maintenance and energy costs. Oftentimes, you will benefit by investing more up front — in insulation, energy saving appliances and energy efficient HVAC systems — which will increase savings down the road. Remember, the cost of living includes not only the mortgage payment, but energy and maintenance costs as well.

Choose the Right Foundation

Let's start at the bottom and work our way up. Foundations are a large percentage of the cost of construction. Choosing the proper foundation can save a lot of money. These foundations are listed in order from the least expensive to the most expensive. We will see later, however, that choosing the cheapest approach is not always the best strategy.

All Wood Foundation

An all-wood foundation can be installed at any time of the year and saves labor costs because only one sub-contractor (carpenter) is used. It is acceptable for crawl spaces and full basements. All-wood foundations have been used for fifty years and have excellent records of durability, livability and water tightness. They are easy to construct, and easy to insulate. Finishing an all-wood foundation later, is no more difficult than working with any wood framed structure, making it a good do-it-yourself project. They make a seamless transition between below-ground and above-ground portions of a basement.

So why are they not more popular? There are lingering doubts by many about their resistance to rotting and termites, even though studies have shown high durability and performance when properly installed. This public doubt may make the home more difficult to sell later. It's also difficult to find subcontractors who are experienced at installing them properly. However, if you plan to live in the home for many years, an all-wood foundation can be an excellent cost saving strategy.

Monolithic Slab

A monolithic slab is the least expensive concrete foundation and a good choice for quality and savings. Work can be completed quickly by a small team of subcontractors for additional savings. Monolithic slabs work best on level lots. Care must be taken to position plumbing accurately before pouring concrete. Check the plumbing fixtures during the pour to make sure none have shifted out of position.

Disadvantages: Any plumbing changes, repairs, or additions later will require a jackhammer. You will also lose the space advantage of a full basement. Termites are another concern. The height of monolithic slabs is usually less than eight inches above ground, making it easier for termites to gain access to the house through breaches around forms and plumbing or through mud tunnels built by the termites up the side of the slab. Make sure to install termite shields around plumbing and remove all wooden forms. The slab's shallow depth also increases the chance of "frost heave" so make sure the slab has been insulated properly.

Cinder Block

Cinder block is somewhat cheaper than concrete but requires the use of several different sub-contractors to install. This causes a longer completion period and more scheduling issues. It's not as strong as concrete and more vulnerable to leaks, making it a poor choice for full basements. Cinder block is still the foundation of choice for crawl spaces. Good ventilation and moisture control is essential.

Poured Concrete

Poured concrete is the standard choice for full basements due to its high strength and resistance to leakage. It is more expensive, but time and money can be saved because only one sub-contractor is used. That contractor will place the concrete forms and insure that they match the dimensions of the plan precisely. While expensive if used simply as a foundation, poured concrete is very cost effective when used as additional living space. This will require good drainage control and insulation to control moisture.

When installed properly, modern basements can be just as livable as any other part of the structure.

Precast Concrete Foundations

These are foundation units that are poured inside a factory and delivered to the site in completed form. Promoted by firms like Superior Walls, they offer an excellent alternative to poured foundations. The panels are cast using 5,000 psi concrete, rather than the standard 3,000 psi concrete. This denser concrete is naturally waterproof and will not allow moisture to migrate through to the interior. Insulation is usually cast within the panel, eliminating the need for installation later. Precast panels are more expensive than poured foundations, but they can be installed quickly, even in inclement weather, and can often pay for their extra cost through labor reduction in other areas of the project.

ICF's – Insulated Concrete Foundations

This new technology uses insulated foam as the form for the poured concrete wall. The form is left in place after pouring and provides excellent insulation and moisture control. The forms themselves are extremely easy to construct, making them a good do-it-yourself option. Any one who has built with plastic blocks as a child will instantly feel at home with the assembly process. The actual pouring of the concrete into the forms can be tricky and the forms must be reinforced during the pour. While more expensive than any other type of foundation, the extra stability and insulation can reduce energy costs significantly.

Use Room Dimensions to Minimize Waste

When designing the floor plan, make sure to take advantage of the natural dimensions presented by the materials you will use for construction. Carpet normally comes in 12' rolls. In estimating your job, a carpet sub will charge you for any and all scrap necessary to carpet designated areas. When possible, plan rooms to be multiples of 12 feet. This reduces waste significantly and also eliminates carpet seams that require extra labor and have the potential to come loose later.

When combined with modular construction and the "optimum value engineering" concepts discussed later, dimensional planning can save up to 30% on building materials. These techniques aim to reduce excess waste in residential construction.

Drywall comes in several lengths, from eight feet to twelve feet. Create a drywall plan and order a mixture of sheet lengths that will reduce the number of joints that must be covered with drywall compound. This can be especially valuable on ceilings, where subs usually charge extra for finishing work. For instance, 12' drywall boards will cover the entire ceiling of a bedroom that is less than twelve feet in length without the need for butt joints, which cost extra to finish.

Use Modular Units of Construction

Blueprints are often designed with esthetic purposes in mind. That's okay, but remember that many materials, lumber especially, are designed around 16", 24" and 48" units. To a reasonable extent, plan your home to be composed of 24" or 48" units, eliminating waste and cutting time. Refer to the framing section, which describes this approach in more detail.

Plan for Extras Later

Almost anyone would like to have all or some of the following items in their home:

- Stone Fireplace
- Built-in microwave
- Intercom system
- Central vacuum system
- Electronic alarm/security systems
- Skylights
- Finished basements or bonus room(s)
- Patio or deck
- Swimming pool and hot tub
- Electric garage door opener
- Crown and other fancy molding
- Wallcoverings
- Sidewalk or other paved areas
- Expensive landscaping
- Den paneling and built-in bookcases
- Wet bars
- Paved driveway/garage door
- Gas grill
- Solar panels and solar water heating

But these items, and many others, are not essential for obtaining a Certificate of Occupancy. With a little advance planning, you can allow for the addition of these items later, when budgets permit. The trick is to prepare for extras by doing work that is difficult or impossible to add later. If you want a gas grill, run the gas line now and cap it. Gas lines, plumbing and wiring are fairly inexpensive to install during construction and will wait for you until you are ready to add those extras.

Want a zero clearance fireplace in the den? Have the framer install the header and chase for the fireplace you desire, but don't install it. Simply cover the area with drywall. When you can afford the fireplace later, the framing will be waiting for you. A few raps with a sledgehammer will expose your advance planning.

Want a central vacuum system? Run the PVC pipes before finishing the walls, but don't buy the expensive vacuum unit until you can afford it, or until a great bargain comes along. Run wiring for cable and Ethernet hookups throughout the house.

What about a central security system? Run the wires to each window and door during the framing phase. A security system can be installed years later, but the existing wiring will vastly reduce the cost of installation. Frame the roof for future skylights. Have the expensive licensed plumber and electrician run "stub-outs" in an unfinished basement to make your job far easier when you get around to finishing the basement later.

Can't afford that fancy stacked stone façade on the house? Use textured plywood siding instead (which can double as structural sheathing). Just make sure you've added a brick ledge to your foundation. The textured plywood is sufficient for a Certificate of Occupancy. Later, when time and money is available, you can easily hire a mason to add the stone veneer.

Wet bars, paneling, patios, decks, landscaping, and bookcases can become do-it-yourself projects months or years after your family has lived in the home. Bargain hunting is a great way to save money. Unfortunately, it takes time, one thing you won't have a lot of, during construction. Save the expensive luxury items for later when you have the time and resources to find bargains.

Build Closer to the Street
Moving the house closer to the street will reduce the cost of the driveway and front yard landscaping. The additional benefit will be a bigger backyard — an added plus with modern houses. Just make sure that you meet the minimum setback requirements.

Reduce the Footprint of the House
The smaller the footprint of the house, the smaller (and less expensive) the foundation. How do you reduce a house's footprint without reducing its size? By using cantilevers. Bay windows, breakfast areas and fireplaces can all be extended out from the house 2'- 4' or more. These cantilevers also serve to break up the straight lines of the house — adding style and texture.

Eliminate Walls
Older homes were often designed with many enclosed rooms. This was partly due to the need for privacy when families were generally larger than they are today. The smaller rooms also simplified the structural design of framing timbers and their allowable spans.

Modern construction techniques that use steel beams, laminated beams, truss joists, and truss roofs allow for longer spans and more open spaces. Open designs have become very popular, with good reason. Interior walls cost money. Every wall partition adds to the budget with framing material, labor, trim, doors, drywall finishing, electrical, painting, etc. Every unnecessary wall that is eliminated reduces costs and gives the floor plan the appearance of being much larger. Every separate room requires its own traffic pattern. Combined rooms share this foot traffic

area and can be functionally smaller than two separate areas. The heightened sense of space and openness will prevent smaller homes from seeming cramped. Modern families are better able to congregate around entertainment centers like large screen TV's. Heating and cooling needs are simplified. The "air" in an open space is the cheapest building material around.

Build Only What You Need
Can't afford to build the home of your dreams now, but you don't want to move again later? Plan ahead for expansion when you place the home on the lot. The largest single cost in a new home is the finishing work. You can leave entire areas of the house unfinished and complete them as needed. The finishing of attics and basements become excellent projects for the do-it-yourselfer. The second story of a "story-and-a-half" design can house the bedrooms of future children or guests. Make sure the stair access, plumbing, electrical rough-outs, and insulation are in place. These are expensive to add later.

A basement is the perfect expansion area for an office, den, or workshop. Make sure you add windows for natural lighting and outdoor access to improve the quality of finishing work you will complete later. Make sure rough-in plumbing is in place and that the basement floor is high enough to allow proper sewer drainage. If not, "flush-up" toilets may be required. Design a nine-foot basement ceiling to leave space for drop ceilings. Use open-web trusses for the basement ceiling. Trusses can span large distances without support and the open web allows the plumbing, electrical, and HVAC to be placed within the truss itself.

Costs Rise Exponentially in Larger Houses
Larger homes add complexity and with it, expense. The additional size requires larger HVAC units, more powerful electrical service panels, larger water heaters, etc. Other factors contribute to spiraling costs in larger houses:

- Larger homes typically require larger rooms with correspondingly longer spans for floor and ceiling joists. Longer pieces of lumber cost more per board foot than shorter pieces. Longer spans may also require thicker lumber in order to carry the weight over the longer spans.
- Larger homes tend to have higher ceilings — 9', 10', 12' and up. Longer studs cost more per square foot. Taller rooms use more drywall, paint, trim, etc.
- Larger homes tend to have more structural weight, which often means thicker joists, steel support and I-beams; all of which add to the cost.
- Larger homes have larger and fancier doors, windows and trim.
- Larger kitchens mean more and fancier cabinetry.

Economize Materials

Since framing is hidden, appearance is not an issue — only workmanship and safety. The Department of Housing and Urban Development (HUD) has worked closely with all the major building code associations and the NAHB Research Foundation to develop a series of construction methods that can reduce construction material by over 15%. It is called Optimum Value Engineering (OVE). You can use all or part of these suggestions to reduce costs considerably.

General

Locate partitions to intersect exterior wall studs. This will save additional framing material at the intersection and simplify insulation.

Use the empty space in walls to your advantage. The space between studs can be used for pantries, shelves, and medicine cabinets. The wasted space inside interior walls can be astounding. Put it to good use.

Coordinate window and door openings with normal stud locations to minimize framing. Windows and doors can often be moved several inches without upsetting the symmetry of the design.

Plan for straight run stairs. Orient stairwell openings parallel to floor framing so as to disrupt as few floor joists as possible.

Locate attic/crawl space access doors between framing members in closet, hallway, and other appropriate areas. A 24" o.c. spacing of framing members provides ample access.

Use in-line framing. Use the same spacing (24" o.c.) for all roof, exterior wall and floor framing and align the framing members, one under the other, to provide structural strength. This will also reduce the need for double top plates on walls.

Consider centralized "back-to-back" plumbing. Cluster bathroom, utility room, and kitchen fixtures as close together as possible to reduce plumbing and allow sharing of vent stacks. In two story plans, arrange upper level plumbing over lower level plumbing so they can connect to the same stack. Concentrate plumbing in the same wall, to minimize framing for plumbing.

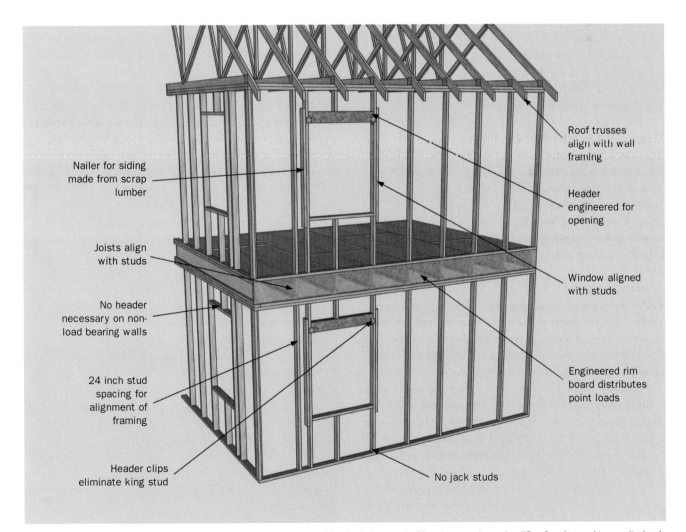

Nailer for siding made from scrap lumber

Joists align with studs

No header necessary on non-load bearing walls

24 inch stud spacing for alignment of framing

Header clips eliminate king stud

Roof trusses align with wall framing

Header engineered for opening

Window aligned with studs

Engineered rim board distributes point loads

No jack studs

OVE, or Optimum Value Engineering can save money and cut waste. Vertical alignment of framing members simplifies framing and transmits loads directly down through structural members. This allows for 2' o.c. framing and leaves more room for insulation, increasing energy efficiency.

Lay out plumbing to avoid structural members in the floor and roof. Do not place plumbing in exterior walls because of structural and insulation difficulties.

Foundations

Consider a crushed rock footing under poured or precast basement walls. In all but the most unstable soils, the foundation simply serves to even things out for basement or crawl space walls. Many footings are totally unnecessary, and tend to block the drainage of water around the bottom of basement walls. Consider filling the foundation trench with crushed stone instead. It saves concrete, and incorporates the drainage system into the foundation itself. However, make sure the soil's bearing capacity has been checked by a soil engineer. Many types of foundation walls, like treated wood foundations or precast concrete panels, are designed to rest directly on level gravel.

Size the basement walls to match needs. Shallow or above ground basements can be built with thinner walls and less material. The thickness and strength of a basement wall is designed to hold back soil and water pressure. Consider a raised basement design with only 4 feet of the wall underground. This not only allows for basement windows, it reduces the need for thicker walls to hold back soil, reducing significantly the amount of concrete needed.

Use concrete basement walls below ground only. On sloping lots, much of the basement will be above ground. In these areas, switch to traditional lumber framing, which is far cheaper per square foot than poured concrete. This may require "stepping" the concrete walls. Windows and doors can be installed far more easily in these wood framed walls.

Add fly ash to your concrete mix. This is a byproduct of coal fired electrical plants. Fly ash not only reduces the cost of the concrete, its mineral content increases the concrete's structural strength. It also makes the concrete easier to work, reducing labor.

Floor Construction

Use built-up wood beams. If a center beam is required to support floor joists, construct a beam from standard 2× lumber rather than more expensive steel or laminated wood beams. The joists can be fastened directly to the wood beam.

Space floor joists 24" o.c. to correspond with the two foot planning module. This will allow wall studs placed 24" o.c. to be aligned directly above each joist. Often, the same size floor joist used for 16" o.c. can be used if the plywood floor is screwed and glued.

Use engineered floor trusses instead of standard lumber. These trusses will be engineered for the span of the floor. They can span greater distances and work well when spaced 24" o.c. Most floor trusses have a wider surface area for the sub-floor adhesive. While they may be more expensive

than solid lumber, their open web design can cut labor and material. The open webs allow space for running plumbing, wiring, and HVAC vents, reducing the labor of utility sub-contractors significantly and eliminates the need to box-in a utility duct. Make sure to bargain with subs for decreased rates on labor costs when using trusses.

Use single, long I-joists over floor beams. I-joists can be fabricated to cover the entire width of the floor, even if span tables require a center beam. The single I-joist requires less labor to install and will be stiffer than two I-joists aligned end-to-end.

Use single, long I-joists over floor beams. I-joists can be fabricated to cover the entire width of the floor, even if span tables require a center beam. The single I-joist requires less labor to install and will be stiffer than two I-joists aligned end-to-end.

Exterior Wall

Space studs 24" o.c. to reduce the amount of framing lumber and labor in conjunction with the 2' planning module. A 24" stud spacing is acceptable in most areas of the country for one story construction and for the second story of two story construction. If you are building a two story house and want to preserve the 2' planning module, use 2×6 lumber on the first floor and 2×4 lumber on the second. This also allows space for more insulation on the first floor.

Use 2-stud corners. A 3-stud corner post is not required and is used only to provide a backing for the drywall. By using metal drywall clips or wood cleats, the third stud can be eliminated. It also leaves more space for insulation.

Eliminate partition posts. A partition post is not required where an interior partition intersects an exterior wall. It serves only to back up the drywall. Drywall clips can be used here as well.

Engineered floor trusses can span long distances and provide an open framework to run plumbing, electrical, and HVAC.

Single, long I-joists resting on a center beam are stronger than two shorter joists. Note blocking for stability.

Automated screw guns have made "screwed and glued" floors more popular.

Eliminate mid-height blocking. Blocking is not required in exterior walls as a fire stop or structural bracing. Standard platform framing provides all the fire stopping that is necessary. The lack of blocking also makes the application of wall insulation much easier. Old habits die hard and some framers must be talked out of using blocking.

Eliminate structural headers in non-bearing walls. Structural headers are used to support loads over openings in load-bearing walls. However, in practice, they are often used over openings in walls that are not actually load-bearing. If the house uses roof trusses, only the walls that support the truss need headers over openings. The walls under the gable ends do not need headers.

Use tuned headers on load bearing walls. Often, headers are over engineered for openings. Often, a single 2×8 or 2×10 is all that is necessary. Oriented strand headers are specifically engineered for certain opening sizes. Single depth headers allow space for insulation to serve as a thermal break.

Use 22½"-wide windows fitted between studs spaced 24" o.c. and aligned with other framing members. Since the window fits between the studs, no header is necessary. Several windows can be placed side by side to provide the effect of a larger window.

Use a glued-and-screwed plywood header. Where a header is required, use a plywood box header. Make a box header by gluing and screwing a plywood face over the framing above an opening. The plywood face can be used in place of the exterior sheathing in the space above the door. Make sure to use a plywood thickness that matches the sheathing thickness.

Use single layer panel siding. Siding products that do not require the additional support of sheathing may be applied directly to the studs. However siding products with open joints or that might otherwise allow wind or rain to gain access to the interior should be installed over a suitable non-vapor barrier building paper, such as 15 lb. felt or house wrap.

Minimize exterior trim. However, exterior trim often serves a valuable function by covering construction joints and providing tolerances at floor, wall and roof intersections, windows, corners, etc.

Eliminate interior window trim altogether by wrapping the drywall around the opening. Besides saving money on trim, this treatment creates clean lines around windows.

Roof Construction

Use engineered roof trusses spaced 24" o.c. in conjunction with aligned modules. These trusses will be specifically engineered for the house and often will span the entire interior space without load-bearing interior partitions. This saves headers in the interior walls. There are several truss designs available, including ones that provide an attic space.

Use pre-fab gable end trusses. These are similar to the regular roof trusses but with vertical members for nailing the gable siding.

Use ⅜" Group 1 plywood roof sheathing or OSB with ply clips to support the edges of the plywood between the trusses.

Eliminate the rake overhang on a gable roof. The rake overhang is a costly detail and serves no real function except as partial shade for the windows. In warmer climates, it is best to leave overhangs in place.

Use an open soffit overhang. If you use an overhang, use an open soffit detail with blocking to fill the opening between the trusses at the wall. A dark stain used in the open soffit will conceal shingle nails or staples which penetrate the wood sheathing.

Avoid complex hip or mansard roofs. While popular, hip roofs can be a nightmare to frame. Virtually all the rafters must be cut with unique and complex multiple angles. This complexity grows exponentially when multiple corners and valleys are added to the roof line. Straight gabled roofs are far simpler and more appropriate for truss framing. Gabled roofs also have the advantage of increased attic space or room for extra living areas.

Interior Walls

Use light gauge steel studs for non-load bearing walls if they offer a cost advantage. Steel studs install rapidly and are very common in commercial construction. Drywall must be installed with screws, however.

Use single frame openings in non-load bearing partitions. It is not necessary to double the studs at either side of interior door openings or install headers or cripples over the opening.

Reduce blocking. Non-load bearing partitions that are parallel to floor framing do not require blocking or other

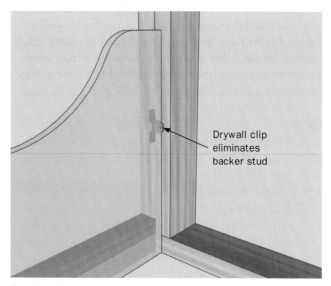

OVE two-stud corner

special support in the floor where ⅝" or thicker plywood flooring is used. Where partitions are parallel with ceiling joists or roof framing overhead, use precut 2× blocks spaced at 24" o.c. between the framing members to secure the top of the partition and to provide drywall backup.

Wrapping drywall around the interior opening of a window not only saves the cost of trim and sill, but creates a clean, modern line.

Interior load bearing walls should be aligned with the framing members in the floor and roof structures just like the exterior walls.

Gang doors together to allow more accessible wall area for furniture.

Reduce the number of windows. They're expensive and reduce the insulation quality of the wall. A few well-positioned windows can let in lots of light while keeping costs reasonable.

Simplify closet framing by providing a full width opening at the front, dimensioned to receive a standard width bi-fold or sliding door. As with windows, wrap drywall around the opening to eliminate trim.

Use 4×12 foot sheets of drywall installed across the framing on walls and ceilings. Drywall ½" thick is approved for application over framing 24" o.c. Blocking is not required behind tapered drywall edges.

Place drywall butt joints at windows and doors. Butt joints require more labor and several extra coats of drywall compound when compared to tapered joints, since the joint must be feathered. Place these joints above doorways or at windows to reduce the size of the butt joint.

Eliminate the bulkhead over kitchen cabinets. This is a costly detail and is difficult to insulate properly. Kitchen cabinets may be hung on the wall at the proper height with

(Image caption for img_2) Drywall clip eliminates backer stud

an open top. This area becomes additional storage space. Or you can use full height cabinets to enclose the entire area.

Use metal drywall clips to eliminate studs and blocking for drywall back-up.

Plumbing

Cluster plumbing. This is a cost saving design concept also referred to as back-to-back plumbing. The basic principle is to arrange typical plumbing groupings such as baths, kitchen, and laundry on a common wall or, in multiple stories, vertically so that all fixtures can be attached to a common stack. This minimizes drain, water pipe, waste, and vent materials and reduces installation labor. Make sure to bargain with the plumbing subcontractor, who often charges by the connection. Point out that your design will use considerably less material and labor and ask for a substantial discount. Locate the water heater near this grouping. This will greatly reduce the time it takes for hot water to reach the faucet.

Frame shared wet walls with 2×6 or 2×8 lumber. A single 2×8 wall, 2' o.c. will be cheaper than two 2×4 walls back-to-back and will allow extra room for venting pipe.

Use pre-fab shower and bath modules. These pre-fab units eliminate the cost and installation of ceramic tile, a significant expense. Other advantages include eliminating leaks and callbacks associated with tile installation, one of the trickiest installation steps. The material is also easier to clean.

Combine the shower and walk-in tub into one area. Many modern bathrooms contain not only a stand up shower, but a large tub in two separate areas. How often does the tub get used? Do you then rinse off in the shower? Why not combine these two areas? Enclose the tub area with glass and place the shower stall in a space directly in front of the tub. This creates a "bathing zone" which saves space and increases the "elbow room" in the shower area. When leaving the tub, you will be stepping directly into the shower area, convenient for rinsing off. There is no worry about dripping water on the floor. If you lack the budget, omit the tub, but leave room for it. You will have a mansion sized shower stall at a fraction of the cost. A fancy whirlpool tub can be added later.

Reduce venting. Most homes have one or more 3" vent stacks plus several 1½" and 2" fixture vents. The vent's primary function is to equalize atmospheric pressure in the plumbing system in order to preserve the trap seal. The trap seal is the "S" shaped pipe that prevents sewage gas from leaking into the house. Research shows that a single vent is adequate for several shared fixtures.

Use stack venting. Another simplified system is referred to as stack venting, which is recognized by most plumbing codes. In this system, each fixture drain is individually connected to the plumbing stack, eliminating the need for individual fixture vents (see Plumbing chapter).

Substitute air admittance valves for traditional venting.

Air admittance valve

Traditional plumbing vents have to extend through the roof. That requires more framing and the installation of vent boots on the roof that are prone to leakage years later. The plumber must also install the vents before the roof can be applied and the house dried in. Air admittance valves are installed under counter and replace expensive vent stacks. They simplify framing and plumbing and reduce costs significantly. Note that plumbing codes require at least one "through-the-roof" vent regardless.

Use CPVC (chlorinated poly vinyl chloride) plastic piping in place of copper. CPVC costs 20% to 40% less than copper. It installs quickly and easily and reduces the danger of fire from soldering the copper joints. The installation is so simple that you might consider doing this job yourself. The only disadvantage is its large expansion and contraction ratio when hot water is running through. This can lead to popping and cracking noises in the wall when the hot water is turned on. To prevent this, do not anchor the plumbing to the wall. Leave additional area around the pipes to allow them to expand and contract freely. PEX plumbing offers a cost effective and superior solution.

Use PEX tubing with home run plumbing. PEX is a proven flexible piping product that has been used in Europe for 50 years (see Plumbing chapter for more information). Its flexibility makes it far easier to install, with less joints, reducing labor considerably. Make sure to bargain a reduced price with plumbers for the labor savings.

HVAC

Select the most efficient HVAC system. There are two costs to consider here: the installation cost, and the operating cost. Choose the system that best fits the climate in your area. For instance, gas heaters are inexpensive to install and operate at a moderate cost. Heat pumps are more expensive to install, but need very little electricity to operate. However, they lose efficiency as temperatures drop, so they don't work as well in northern climates with low winter temperatures. Generally speaking, heat pumps work best in southern climates, and gas furnaces work best in northern climates. You should also consider the relative costs of electricity and gas in your area when making your decision.

Size the equipment to the house. Make sure the BTU rating of the HVAC unit matches the size and insulation efficiency of your house. A system too large will turn off and on much too often and will not operate efficiently. A system too small will be on all the time and will not be able to handle temperature extremes. If you insulate the house well, a smaller unit is required, saving money on installation and operation.

Locate heating and cooling equipment centrally to minimize duct runs and sizes and to provide good distribution.

Downsize duct systems. If the house is well insulated, with double or triple pane windows, you can specify smaller ductwork runs that end near the center of the house rather than underneath windows, which is the traditional location. In the past, vents were located near the areas of least insulation such as windows. The idea was to locate the conditioned air near the warmest or coolest areas of the room. With improved windows and insulation, this is no longer necessary. Shorter duct runs not only save materials and labor, but reduce the possibility of leakage and energy loss.

Locate the HVAC in the center of the house. Ducting is more efficient and less expensive when it radiates out from a central point and is connected directly to the HVAC unit.

Use high efficiency units with a high EER (energy efficiency rating). These units may be a little more expensive to install, but they pay back the expense in savings on monthly utility bills.

Alternatives to Stick-built Homes

Advances in factory automation have made modular and panelized homes a low-cost, viable alternative to traditional "stick-built" homes. They are not only cheaper, but better built. With much of the work done in the factory, there is less work to be done on site. For a self contractor, this can be a big advantage. The project goes faster and smoother and there's less chance of delays and change orders. You save money in several ways:

- These pre-constructed homes go up faster, shortening the overall project duration and saving you interest payments.
- There are no material surprises. All the materials of the structure will be included in the sales price. You don't have to worry about inaccurate estimates. Material can't "mysteriously" disappear from the building site.
- Usually, the equivalent manufactured home is less expensive than stick built. This is because the factory can use materials more efficiently. Also, time and labor is saved on site. There is less opportunity for cost overruns. Modular manufacturers have bargaining

power when placing huge purchase orders with building material suppliers.

Don't confuse modular or panelized homes with "manufactured housing," otherwise known as mobile homes. Modular construction must follow the same exact set of building codes as traditional stick-built construction. Mobile homes, on the other hand, have their own unique set of building codes.

It's a shame that most people want their house built the "good old-fashioned way." Homes built in the factory are made from high quality, straight lumber. Nailing and framing is done in jigs to a precise accuracy unmatched in the field. These precise dimensions lead to a more air-tight and well insulated home. There are several types of factory built and alternative construction techniques available to homeowners.

Modular Homes

Modular homes are completely constructed in the factory in "modules" that are shipped to the building site and connected together. Virtually all the construction except for some finish work is done at the factory. They go up fast; sometimes "dried in" in less than a day. The individual modules usually have the wiring, plumbing, and interior wall finishes already installed.

The obvious limitation with modular construction is the size of the module. Since it must be shipped to the site via public highways, the dimensions of each individual module can not exceed 12'-14' in width. In the industry's infancy, houses built from modules had that "blocky", square look. Modern advances in computer design and factory fabrication has changed this however. Now, custom home designs can be produced in the factory that are indistinguishable from site built homes, once they are assembled. Modular factories have come up with innovative ways to work around the dimensional limitations of their modules.

Caution: You must make sure that your foundation is as accurate as possible. There's nothing more upsetting than to get your new home to the site and realize your foundation is off by a foot.

Panelized Construction

This is a hybrid between modular and stick built construction. The wall "panels" are constructed on a jig in the factory. The windows and doors, siding, and insulation can be installed. The panels are then stacked in a truck along with the rest of the house and roof trusses. Since the "air" is eliminated, the entire house can be shipped on one or two trucks. These houses are not limited by modular shipping dimensions, so any house plan can be converted to a panelized house. In fact, many factories specialize in taking your plan, pricing it, and constructing a custom design. Panelized homes are just as cost effective as modular.

Once the panels reach the site, they are unloaded, and assembled on the foundation. Small variations in the foundation can be fixed by a little sitework. Once the walls are up, the roof trusses are installed, covered with sheathing and the siding is completed. Small pieces of siding will be installed to cover the seams between the panels. These siding pieces will be staggered so that, once they are installed, it will be impossible to distinguish it from a normally constructed home. Panelized houses go up as fast as modular housing, sometimes two days or less. The panels often do not require a crane. They can be handled by small work crews. It is truly amazing to see an entire house go up in a couple of days, complete with windows and doors and fully dried in. This allows you to cheat the weather and get inside quickly, where the finishing work can be completed in warm and dry surroundings. Your building materials can be locked inside to protect them from theft.

Make sure you find a building crew that is familiar with panelized construction. A rookie framing crew can take as long to construct a panelized house as they would to stick-build it. There goes yours savings. Many panelized dealers offer a list of framers familiar with the intricacies of panelized assembly.

Kit Homes

Kit homes are not pre-constructed, but they offer the advantage of having all the necessary materials cataloged and priced. Kit homes are especially popular for certain types of tricky construction or homes made from special materials. Examples include log homes, post and beam construction, cedar homes, steel construction, and solar homes. The materials may come pre-cut, as in a post and beam home, where large beams are connected together with mortise and tenon joints. Other kits, such as steel framed homes, include all the materials, but you will still have to hire a framer to cut and assemble them. Still, you will know that all the necessary materials are available for a fixed price.

SIP's

Structural Insulated Panels are made from a sandwich of two OSB (oriented strand board) panels with a "filling" of expanded polystyrene insulation in the middle. SIP panels are incredibly strong, simple to put up by experienced contractors, and offer excellent insulation and air tightness. They are more expensive than other construction methods, but sometimes the extra cost can be recovered through simpler and faster construction and the downsizing of other costs, such as the HVAC unit. SIP's are commonly used by themselves or as a "skin" over a post and beam home. The extra cost you pay up front may be recovered over time through reduced energy costs.

ICF'S

Insulated concrete foundations, as the name implies are most often used just for foundations, but can also be used for the entire house structure. They create extremely strong, energy efficient homes. However, costs are 20% higher than other types of construction. So why mention them under a section titled "Saving Money?" As we will see in a moment, higher cost structures have their cost advantages in certain circumstances. If you plan to build a home in tornado or hurricane country, these structures can save a significant amount on insurance costs, not to mention energy savings. Remember to consider the "life cycle" cost of your home.

Rammed Earth

This technology is as old as man himself and is seeing a renaissance. This can be the cheapest construction method of all and consists of clay, often harvested from the building site, mixed with sand and concrete. It is mechanically rammed into a form and results in a strong and energy efficient home. The tradeoff comes from the large amount of labor involved. However, if you can provide much of that labor as "sweat equity," then this may be the construction bargain for you. If your site lacks good clay soil, you might want to pass on this option.

Steel Framing

Steel framed homes have been around for awhile, using steel studs common in the commercial industry. There have been recent advances in fabrication methods that make this option intriguing. One company fabricates the steel studs on site from large rolls of steel sheet. A computer controls the size, order, and number of studs according to the house plan and a team assembles the walls, floors, and roof as the machine spits out each piece. This method has the ability to be quite cost effective and generates a home that is strong and termite proof. You may want to research availability of this service in your area.

MORE IS OFTEN LESS

As you can see, there are hundreds of ways to trim costs. After reading through all of them a couple of times, step back and try to see the forest rather than the trees. Your entire construction project is an organic being, a carefully balanced chaotic system. Changes you make in one area can ripple through the rest of your project. This fact is often ignored by builders who often find themselves consumed in the minutia of the moment. They fail to think about the future ramifications of present day decisions. You may find some construction choices to be more expensive than others on their face, but they may end up saving money elsewhere through the "trickle down" effect.

Here's an example: you may choose to use SIP wall construction and triple glazed windows. This will cost at least 20% more than conventional windows and framing, but let's look further into the distance. SIP's eliminate the insulation sub-contractor altogether. Sheathing will not be needed on the exterior, since the SIP provides that function. Extra bracing will not be needed at corners because of the strength of SIP walls. Energy efficiency will be greatly increased, so you can downsize your HVAC significantly, saving money. Air ducts do not need to run to windows because of their high insulation and air tightness. SIP roofs allow for more storage space in upstairs bedrooms. Future energy bills will be lower. The house will go up faster, so interest on the construction loan will be reduced. Do these savings outweigh the initial premium of SIP's and windows? Only a thorough analysis can answer that question.

On the other hand, your electrician may charge more to run wires in SIP walls. The extra air tightness of the home may require the purchase of a heat exchanging air ventilation system. Some sub-contractors may be unfamiliar with SIP construction and could make more mistakes than usual.

The moral of the story is this: You must consider all side effects of every decision you make. Sometimes you can save money by spending more. Sometimes you can't. This is another good reason to consider constructing your home in virtual space and using a building information model.

Sound like a lot of work? It's easier than combating problems on the job site if you rush into the project. Think of this also: professional builders seldom take the time to analyze their construction projects with due diligence. That means there is a lot of room for improvement and cost reduction when you take the time to study not only the trees, but the forest. Take that extra time for analysis and you will be far ahead of the game.

SAVING MONEY DURING PRE-CONSTRUCTION
Once an efficient plan design and foundation has been chosen, there are more tweaks that can be used to reduce costs during the pre-construction phase. Many of these cost reduction methods will alter details of your blueprints, so make sure you incorporate them into the design before you have the blueprints produced. Give your designer or architect a list of these options during the customization of your plan.

Shop for Prices Aggressively
This means not accepting bids or prices on face value. Use your cost take-off to help keep prices down. Put out bids to several material suppliers and contractors at the same time. You may be surprised at the range of prices. There are now many builder surplus operations across the country. Search them for bargains. Everyone is out to earn a living; and they will try to get the best price possible. If you say the price is right, it's a done deal. Don't be too anxious to give in. The more you save over your estimate

early in the game, the more "cushion" you'll have later for unexpected expenses.

Coordinate Sub-Contractors
The task of timing and coordinating subs can be difficult but can yield savings by reducing construction time. In some instances, promoting cooperation between subs can make their jobs easier and give you a reason to bargain for a lower price. One example would be coordinating the painter and trim carpenter. If trim and walls are to be of a different color, then have the painter put on his primer coats before trim and doors are installed. Primer coats can also be put on trim and doors before they are set. This will save the painter considerable time and effort and will speed up the construction process.

Contractor coordination begins early. Create a timeline and critical path for your project well in advance of the start of construction. Track your schedule and keep it updated. Use the MasterPlan in this book as a starting point. Bring the sub-contractors together ahead of time, if possible, and discuss the coordination. When subs get to know each other and know that their schedules are inter-dependent, they are more likely to stay on schedule. The framer is not just putting you behind with a delay, he's messing up Joe the plumber's schedule too. Make sure he knows it, and knows Joe the plumber personally.

Bargain with Subs
Provide highly detailed drawings and specifications to subcontractors. When you have gone to the trouble to optimize materials and labor in your design, try bargaining with the subcontractor for an additional discount. Try to work out all your construction details before the first nail is driven. Plan the placement of electrical outlets and plumbing to avoid collisions with wall and floor framing. When you specify the project thoroughly, you make the sub-contractor's job easier. You also make it easier for the sub to estimate his costs and labor. Adding complete speci-fications to your contract is essential in order to guarantee quality work at a low cost.

Don't be afraid to question the sub-contractor's estimate. Many subs use simple formulas to keep job estimates simple. For instance, plumbers charge by the connection, electricians by the outlet. If your planning and material choices make the sub-contractor's job easier, point that out and expect a discount on the standard rule-of-thumb estimate.

SAVING MONEY DURING CONSTRUCTION

Do Work Yourself
This can be lots of fun. You'll feel (and be) helpful, while saving even more money. In most cases, every hour you spend productively on site can save you $10 to $15. Great! But be careful. There are two main reasons why you shouldn't get carried away with doing too much of the work yourself:

- There are many jobs you shouldn't even consider doing. Remember, where special skills are needed, skilled tradesmen serve an important purpose: They accomplish professional results and save you from injury, expensive mistakes, frustration and wasted time. Certain jobs may only appear easy. You can ruin your health sanding hardwood floors or drywall. You can ruin a den trying to put up raised paneling and crown molding.
- Work can distract from your more important job of overseeing the work of others. Your most valuable skill is that of boss — scheduler, inspector, coordinator and referee. Saving $50 performing a small chore could cost you even more than that in terms of problems caused by oversight elsewhere on the job. Doing lots of work yourself invariably extends the project timetable. Let the pros do the job quickly.

Jobs to Consider Doing Yourself
- Paint, sand and fill
- Wallpaper
- Wiring: cable, phone, security, Ethernet. Note: a licensed electrician is still required for service hookup
- Install door and light fixtures
- Install mailbox
- Light landscaping
- Light trim work
- Clean-up
- Organizing materials and locating them where needed (stack bricks where masons will use them, etc.)

Jobs Better Left to Pros
- Framing
- Masonry
- Electrical panel
- HVAC
- Roofing
- Cabinetwork
- Plumbing
- All others not listed

Make Pre-fab Substitutions
Many items, such as fireplaces, can be purchased as pre-fabricated units. This saves you labor costs. When pre-fabricated items are used, make sure the sub knows of this and takes it into consideration in his bid. One-piece fiberglass shower and tub enclosures, attic staircases and pre-hung doors are other examples.

Downgrade If Necessary
Certain items can make a major impact on cost reduction without making a visible impact on your home. Consider hollow vs. solid-core doors, pre-fabricated cabinets vs. custom cabinets, pre-fabricated doors and windows vs. custom doors and windows to name a few of your options. Another example is trim. Trim that will be stained will show all imperfections, so it should be as high a grade as possible. But if you plan to paint the molding, use "finger-joint" grade. This is molding made of lots of little scrap pieces glued together. It costs less, but will serve the purpose.

Windows with true divided panes are more expensive than solid pane windows with "fake" wood pane inserts. In most cases the fake panes look just as good and have the added benefit of easier window cleaning and better insulation.

Brick provides a great look to the house but is very expensive. If you use brick, consider bricking only the front of the house — that part visible from the street. Use siding on the back side.

Recycle Whenever Possible
Used building materials can be a gold mine of opportunity and savings. They also contribute to green construction. Anything recycled saves the energy and expense of manufacturing a new part. Look for used bricks, which can lend character to a new home. Old timbers from farmhouses and demolished structures can be used for structural members and as a decorative touch. Weathered beams work wonders for ambiance in a den. Old lumber can be planed and used for flooring or cabinetry. Lumber from old structures is often superior to modern materials. They were likely harvested from old growth forests and have tighter, denser grain. If your lot contains an old house or structure that will be demolished, reuse as much of the material as possible. Old stone foundations can be used as stone veneer siding. Old concrete can be crushed and used as drainage material. Look for old wrought iron that can be used for fencing, deck banisters or gates. Use your imagination.

SAVING MONEY AFTER CONSTRUCTION
Here, savings are small, but every little bit helps:
- Clean up yourself
- Move in yourself as early as possible

Cleaning Up
Cleaning up yourself can save you money. Normally, a cleaning crew works on a home for hours before it goes on sale. This is work you can do yourself. Treat yourself

(and your family) to dinner out after finishing this one. Scraping windows with razor blades, paint touch-up and trash hauling will take days.

Rather than hauling off waste, consider recycling it. Modern grinding machines can turn drywall waste, concrete waste, and lumber into mulch and landscaping material. You save not only on dumping fees, but the cost of the landscaping material your waste will replace.

Moving In

Most likely, you will build in the same city (or nearby) where you live during the project. Moving yourself can save hundreds of dollars. You can save yourself a little rent if you move in early and "finish up" while you're at home. You should ask the local building inspector about minimum requirements for occupancy.

LEGAL AND FINANCIAL INFORMATION

Financial Issues

Obtaining financing for your home will be one of the most critical and demanding phases of your building project. You must be bookkeeper, financier and sales-man. The lender will want evidence that you know what you are doing and that you can finish your project within budget.

A Note from the Authors: The recent collapse of the construction and financial markets means that the way loans are approved will probably change drastically in upcoming years. One of the main reasons for the collapse was the lax attitude of some lending institutions toward loan requirements. They simply gave out too many loans to people who could not afford mortgage payments designed to increase over time.

It was a ticking time bomb that was destined to explode — and explode, it did. That policy is sure to change as banks reexamine their qualification procedures and return to the more conservative policies of the past. The advice given here is based on a very conservative model — one that was followed decades ago.

The authors have one major piece of advice: Forget what the banks will allow. Instead, ask yourself the amount of mortgage liability you can comfortably take on while leaving yourself extra funds for the unknown.

It's far better to take a conservative approach to mort-gages. Thankfully, if you are contracting your own home you have already engineered extra savings into the price of the home. Don't squander those savings by pushing the envelope in mortgage liability.

As the contractor and future owner of your home, you will be dealing with two loans: The construction loan, and the home mortgage. Each one has different requirements, lending sources, and durations. You will be wearing two different hats when you visit the bank. If you are lucky,

you will be able to obtain both loans from the same bank and the same loan officer. If you're not, then you may have to work with two different loan officers. Many mortgage companies do not handle construction loans and visa versa.

The first loan that must be addressed is the home mortgage. You must qualify for the long term financing before any bank will consider a construction loan. While choosing the appropriate plan for your dream home, you calculated the maximum loan you could afford. To satisfy the loan officer, you must go into greater detail.

THE HOME MORTGAGE

Pre-Qualification

Pre-qualification enables you and the lender to calculate how much home you can afford to build. This step takes the guesswork out of the whole process of knowing what you can afford, giving you the knowledge of knowing what your construction budget can be. The lender will have you fill out a Mortgage Credit Analysis Worksheet to calculate your purchasing power and your creditworthi-ness. The lender will depend on certain formulas to calculate the amount of loan you can afford:

- Mortgage Payment-to-Income Ratio (MR). The ratio of your proposed PITI (Principal, Interest, Taxes, and Insurance) payment to your gross monthly income. This varies from 28% to 30%, depending on the lender.
- Total Debt-to-Income Ratio (DR). The ratio of your total monthly debt payments to your gross monthly income. This varies from 36% to 41%.
- Loan to Value (LTV). The ratio of the total value of the house to the loan amount. Most conventional loans require an LTV of 90% or less. This means that the loan you request is less than 90% of the appraised value of the house. Exceptions to this include the government guaranteed loans.

The lender may use one or a combination of these formulas to determine a maximum monthly payment amount. Here's how it usually works:

- The lender will calculate the MR ratio first. This will determine the absolute maximum house payment you can make.
- The lender will then add in your other debt obligations and calculate the DR. If it is less than the DR maxi-mum ratio, then your maximum payment will the MR amount. If it exceeds the maximum DR percent-age allowed, this will drive down the maximum PITI payment you can make. By eliminating some of your other debt payments, (by paying off debts) you may be able to push the maximum back up to the MR ratio.
- Once your maximum payment is known, the lender will calculate the total loan amount that will generate this monthly payment. This amount will vary according

to the type of loan and interest rate you choose: Fixed rate, adjustable, graduated payment, etc.

- The resulting loan amount cannot exceed 90% of the house's value.

However, in the case of a homeowner building his or her own home, the bank may require a larger down payment. They will be nervous about your ability to complete the project. If you contribute more money to the project, they will feel more comfortable. If your handiwork falls short of producing a home with the expected appraisal value, the bank will have less of their money on the line. Your challenge will be to convince the lender that you can finish the job on time and on budget. This shouldn't be too much of a problem for you because of sweat equity.

Sweat Equity

This refers to the labor "sweat" and materials you contribute to the value of your house. Most lenders will consider sweat equity as part of your down payment. As the contractor for your house, this becomes your secret weapon. This allows you to build a home with a value much higher than the loan amount. This allows you to have a much lower loan to value (LTV) ratio. Most lenders should give you credit for value added through sweat equity.

Improving Your Chances

It's unfortunate that the banks seem to focus on one item: your monthly debt payments versus income. They don't consider your assets or savings to be quite as important. They are more interested in your credit payment history and your current debt obligations. There are several things you can do to improve your "score on paper" if you have additional funds:

- Pay off as many credit card debts as possible. This will incur no pre-payment penalties and will reduce your monthly debt payments.
- Try to keep your debt payments timely. Most credit companies will report a late payment to a credit bureau after it is more than one month behind.
- Obtain a credit report on yourself before visiting the lender. These are available from all the credit reporting companies. By studying your own credit history first, you can make adjustments or file a letter of explanation for certain problem areas. This letter will then be available to any bank that requested the report. Check out your credit score. The credit score is primarily based on a statistical analysis of a person's credit report information, typically from the three major American credit bureaus: Equifax, Experian, and TransUnion. Each of the three credit bureaus may have different information and there are many different credit scoring models, which means you may

have several different credit scores simultaneously. If you think your credit score is too low, contact the credit company and ask for an explanation of how the score was calculated.

- If you've had payment troubles with one credit company in particular, pay off that balance completely. This will show your determination to fix the problem permanently.

How Flexible Can the Lender Be?

In the past, banks were far more flexible than they should have been. The future? Who knows? One thing is certain: The loan officer will seldom have leeway in making judgment calls. That's why you'll find most lenders focused on the "paper details" instead of common sense. Use this knowledge to your advantage. Make sure you qualify on paper before contacting the loan officer. With his requirements met, the he will be free to help you structure a workable lending package.

Applying for the Mortgage

This mainly involves filling out lots of forms. Most lenders use a generic form called the Uniform Residential Loan Application. Make sure you bring all your background information with you (listed below) to expedite the process.

Things to Bring

- Bank and Credit statements
- Current Employer information: address, your supervisor, your title, and length of employment. A letter from your employer verifying these facts is useful. If you are self employed, you will need two or three years of professionally prepared financial statements or tax returns for your business.
- Neighborhood or Homeowners Association covenants and By-Laws
- Information on investments and pension plans.
- List of other assets. You may be asked to provide verification.
- Social Security Number
- Two years of tax returns and W-2's.

Additional Loan Costs

There are numerous costs and fees involved in originating and processing the loan. You need to add these costs to your down payment and points when calculating your monetary limits. RESPA (Real Estate Settlement Procedures Act) is a government law that requires the lender to provide you with an accurate estimate of these costs. He will also give you a copy of the governments Settlement Cost booklet, which explains your rights.

Loan Costs
- Cancellation fee
- City/County Taxes and Stamps
- Credit report

- Discount Points
- Escrow fees
- Flood Report
- Loan origination fee
- Loan processing fee
- Notary fees
- Plan Appraisal
- Recording fees
- Title insurance
- Title search

Prepaid Reserve Fees
- Homeowners insurance
- Mortgage insurance
- Property taxes

Things Lenders Like to See

- A good credit history. No bankruptcies, foreclosures, or late payments.
- A healthy net worth.
- A large down payment or substantial equity in the project. The more you have invested, the less likely you are to default on the mortgage.
- A standard or popular house plan. They want to know that the house will sell easily in the future.
- Free and clear title to the building lot with no loans due. Lenders don't like to be second in line for their money.
- Minimal debts payments. This shows a responsible attitude.
- No outstanding lawsuits or claims. A divorce can complicate loan approval.
- Stable employment. Most banks want to see 2 to 3 three years of steady employment.
- Two-paycheck households.
- Your money. They don't want to see borrowed money used for your down payment.

What the Lender Will Do
Evaluate you
In order to do this, they will:
- Perform a credit check, obtaining a credit report. This will show all loans you have opened, paid off and any adverse payment history, such as delinquent and no-payment situations.
- Perform a Loan Analysis, looking at your monthly income and obligations.
- Submit a Verification of Employment to your employer, confirming that you really work there at the job you stated in the application.
- Submit a Request for Verification of Loan to lenders of any loan you may have.
- Submit a Request for Verification of Rent of Mortgage Account to lenders or landlords to assess your previous payment history.

Evaluate the home you plan to build
- Evaluate your blueprints and your construction estimate to ensure that the budget is reasonable.
- Consider your experience and education and ability to complete the project. Some lenders like it when you hire a contractor part-time to help you with the project.

LOAN TYPES

Fixed Rate Mortgage
This is your most desirable loan, if you can afford the payments. The interest rate and the monthly payment stays the same for the life of the loan. If you take out the loan when interest rates are low, you have "inflation insurance." This is especially desirable if you plan to stay in the home for more than 7 years.

Adjustable-Rate Mortgage (ARM)
This mortgage has lower interest rates and payments because the lender can raise the interest rate according to inflation. Since the bank can adjust to changing times, they are more willing to offer a lower rate initially. Fixed rate loans force the lender to add "insurance" points to the loan interest rate. The monthly payment of an ARM is tied to some form of index such as the prime rate, one-year treasury bills (T-bills), or some other standardized cost of funds. Most programs have caps on the maximum amount of interest increase allowed over the life of the loan. Payments may be adjusted quarterly, semi-annual or annually. Take note that ARM's were one of the mortgage time bombs that helped lead to the financial crisis. Try to avoid these graduated loans unless you are absolutely confident in your ability to afford future payment increases.

Graduated Payment Mortgage (GPM)
This mortgage is another sub-prime instrument designed to allow first-time homebuyers to qualify for loans that would be too expensive as a conventional loan. The first few years of loan payments will be lower than the remaining term of the loan or consist of interest only. Payments go up only after the homeowner is settled and more able to pay the larger payment. It sounds attractive at first, but you won't get something for nothing. During the low payment period the principal of your loan actually goes UP not down (negative amortization). You will owe more money after three years than you did when you bought the home. Let's hope that banks begin to take a more conservative approach to these loans in the future.

Mortgage Loan Agencies
FNMA (Fannie Mae)
The FNMA (Federal National Mortgage Association) provides funds for FHA and VA loan guarantee programs and conventional loans. They are the largest single source

of mortgage funds. They underwrite a large portion of the conventional loans that are provided by mortgage bankers. As secondary lenders, they purchase primary loans from other banking institutions. The actual funds come from private sources, mainly the sales of securities. Many of the loan requirements of mortgage companies are dictated by Fannie Mae rules. They publish a set of requirements for the loans they purchase. Most bankers want to make sure their loans meet these requirements so they have the option to sell their mortgages to Fannie Mae.

FHLMC (Freddie Mac)
The FHLMC (Federal Home Loan Mortgage Corporation), like Fannie Mae purchases loans for FHA, VA, and conventional loans.

FHA (Federal Housing Administration)
An FHA loan is not really a loan, but instead a loan guarantee. The FHA insures the loan for the lender that actually provides the funds. The FHA is now a sub-agency of the Department of Housing and Urban Development. An FHA loan is designed to be "homeowner friendly." The terms are very liberal. You can choose from a variety of fixed or adjustable rate.

VA (Veterans Administration)
Like an FHA loan. the VA is a government loan guarantee that provides 100% financing to qualified Veterans, National Guardsmen and Reservists. They guarantee the loans provided by other lending sources. Like the FHA, VA loans are fixed rate and assumable. The assumption policy is a little tricky though. If a non-veteran assumes the loan, the veteran loses his eligibility for another VA loan until the assumed loan has been completely paid off. If another qualified veteran assumes the loan, the original mortgage holder can immediately qualify for a new VA loan. There is no down payment requirement. Obviously, this loan is not available to the non-veteran public.

The Construction Loan
Unless you are one of the few who build with cash, the construction loan is a necessity. It is typical for interest on a construction loan to be a few points higher than a permanent loan. This gives you further incentive to finish the project as quickly as possible. Even though you have jumped through all the hoops for your mortgage, you'll have to do it all over again for the construction loan. (Unless you get a construction-permanent loan. See below.)

WARNING: When you apply for your loan, apply for all the money you will need — and then some. Don't plan on getting extra money later. You are headed for a disaster of mammoth proportions if you run out of money before your project is over. Your lender will likely foreclose on your project and you'll end up with next to nothing. This

also means that once you've done your take-off and are into construction, don't make expensive changes and upgrades unless you can pay for them out-of-pocket.

The Financial Package — Applying for Your Loan
Make sure that your financial package is complete before visiting the lender. As with the mortgage, the construction package should be neat, informative, well organized, and comply with all of the lender's requirements. Your lender will want to be sure that you can keep good records. In addition to your mortgage information, you should include the following:
- Resume or personal biography
- Total cost estimate of the project
- Description of materials
- Survey of the proposed building site showing location of dwelling
- Set of blueprints

The Resume is designed to sell yourself to the lender. Start with a standard job resume but be sure to emphasize any construction interests or experience. List any training courses you have taken (or even this book).

The Cost Estimate will be very important, as your construction loan will be based on it. Make sure to include a 7 to 10% cost overrun factor in your estimates as a cushion for unexpected expenses. Many lenders can provide you with a standard cost sheet for you to fill out. If not, use the one included in this book.

The Description of Materials form will aid the lender's appraiser in estimating the appraised value of your property. Many lenders use the standard FHA form for this purpose, but check to see if your lender uses their own special form.

The Survey should be obtained from the seller of the property or from the county tax office. If the property is very old, you may have to request that a new survey be done. Indicate on a copy of the survey the approximate location of the house.

A Set of Blueprints complete with all intended changes should be submitted. Give your lender the set without the coffee stains.

Start-Up Money
You will need to put up a lot of your own money before you can expect to receive your first draw. Typically your initial outlay will include:
- Lot cost to be paid before the construction begins
- Closing costs (construction loan closing)
- Permits (building, water, sewer, etc.)
- Site preparation (clearing, grading & excavation)
- Foundation

If you can't do this out of your pocket before going to the lender, you aren't ready to begin building.

About Draws

Unlike a car loan where you receive all your cash at once; construction loans are almost always handled on a draw basis. The lender will make you earn the money that you borrowed — bit by bit. Each time the lender sees that you have completed say, 5% of the project, you are allowed a "draw" of 5% of the construction loan. This nuisance is, in fact, a way of making sure that money isn't wasted carelessly up front. Whenever you would like a loan draw, you must contact the lender and request one. The lender has an inspector who will visit the building site to see what work has been completed. The lender will normally have a fixed draw schedule which allows a certain amount of money to be "drawn" for each major phase of construction. Many lenders inspect on Thursdays and distribute draws on Fridays. Builders are usually allowed only five draws free of charge. If you desire more draws than allowed per project, you are usually charged a nominal fee per draw.

Money is drawn after proof is shown that such money is deserved. Hence, you must go into the project with some start-up money of your own. The need for this "pocket money" can't be underestimated. It will be used to buy permits and licenses, dig your foundation and perhaps get your foundation built before money is drawn. It will also be used to cover unexpected expenses, such as price increases, last-minute additions you may want to make and underestimates. Be aware that draw schedules are normally back ended, meaning that you will spend more than you can draw up front, while receiving a 5% draw for painting your mailbox near the end.

The Conventional Construction Loan

These loans allow 80-90% LTV on fixed or adjustable rate loans. Borrower must have a minimum of 10% down payment of the combined land and construction costs. A portion of the loan may be used for land payoff/purchase. The time allowed for construction can range from 6 to 12 months. You may obtain a short term construction only loan or combine the construction and permanent loan together. Construction loans generally have a slightly higher interest rate and fees.

The Construction Permanent Loan

When possible, apply for a Construction Permanent Loan (also known as a convertible loan or rollover loan). This type of loan eliminates one of two closing costs related to separate loans for construction and permanent financing. This is done by putting both the construction loan and the permanent loan in your name. When the permanent loan is ready to close, the construction loan is "converted" to a permanent mortgage. Shop loans carefully. Time here is well spent. Many institutions now offer 15-year payoffs that only cost a little extra each month over a standard 30-year loan. Charging points, a lump-sum cost when you originate a loan, is only a way for lenders to get a little bit more money out of you without raising the interest rate.

Dealing with the Construction Lender

When going to the lender, be prepared:
- Have your financial package ready.
- Know exactly how much money you need.
- Have all requested information typed.
- Dress up (preferably dark suit).
- Show up at the scheduled time.
- Don't mention how much pocket money you have on the side — the lender will want you to put it into the construction loan.
- Don't talk too much. Answer questions only when asked and don't volunteer any information not asked for.
- Have a list of your subs and backups.
- Have all lender forms typed.

The Appraisal

The lending institution will have its capital at risk. As such, they will want to be assured that the house you are planning to build is worth more than the construction loan. They will also test your skill at estimating costs.

The lender will appraise your home in one of two ways; the Market Approach (most likely) or the Cost Approach.

Market Approach — This method of appraisal bases the value of the home on "comparables" in the area. The appraiser will look for comparable houses to yours in the surrounding neighborhood that have similar features, style, square footage and property values. Therefore, it is advantageous to build your house near a neighborhood with homes similar but higher in price than yours. Don't get too far-fetched with your design. The appraiser may not be able to find suitable comparables and may have to guess at the value; usually to your disadvantage.

Cost Approach — This is similar to the cost take-off or a determination of value by square footage. After the value is determined, the lender will then loan you a fixed percentage of the appraised value, usually up to 70% or 80%, to build the house. Very seldom will the lender give you a loan based totally on what you claim your construction costs to be. However, you should easily be able to build a house for less than 75% of its appraised value.

FHA Appraisal

This appraisal is completely different from a bank appraisal. Banks usually hire independent appraisers who follow the above guidelines. When the FHA is guaranteeing the loan, they will send their own inspector/appraiser to the site. Unlike the other appraisers, the FHA inspector will concentrate on structural construction details as well as market value. You should expect a few items on their list that must be corrected before the loan can close.

THE LEGAL ISSUES

Real estate laws have evolved over thousands of years. Needless to say these issues are complex and are best left to the experts. Make sure to consult the appropriate authorities about any legal issue in order to avoid unpleasant surprises. The professionals you will need to deal with are attorneys, insurance agents and financial lenders. The key legal issues affecting your building project are:

- Filing in the public record
- Free and clear title to the property
- Land survey
- Easements
- Covenants
- Zoning
- Variances
- Liens
- Sub-contractor contracts
- Builder's risk insurance
- Building permits
- Worker's compensation

Filing in the Public Record — Because of the importance and complexity of real estate law, all real estate transactions concerning the transfer of property, liens, and loans are normally filed in the public records at your county's local courthouse. This way you are announcing to the public what property you own. Anyone has access to these records and may research the history of your property to see who owned it before you and if there are any mortgages, judgments or liens on the property. This is for your protection, because you can do the same research on any property you intend to buy. Go to your local courthouse and familiarize yourself with the records. Ask one of the title lawyers in the room (there are always a few attorneys there) to help you learn how to search the records. This will give you an idea of the process of real estate transactions.

Free And Clear Title concerns your bundle of rights to a piece of property, whether it is land or improved property. Before closing on a property, your lawyer will do a "title search" of the property in the public record to determine its ownership history. Most real estate records are kept at the county courthouse and are accessible to anyone. If the lawyer finds someone with an "interest" in the property or a confusion over rights, this will create a "cloud on the title" which must be removed before title can transfer. A title search is required by all lenders and is usually done for a standard fee. Never buy a piece of property without doing a title search. Many so called "bargains" are properties with title problems that the present owner is hoping to "pawn off" on some unsuspecting buyer. Don't be that buyer! You should also get title insurance, protecting your claim to the land.

A Land Survey is another required item that must be done before title can be transferred. This survey can usually be arranged by your lawyer or financial lender. The survey will describe the exact location of your land in a written legal description and scale drawing of the property called a "plat map." These descriptions will be attached to the deed and filed in the public record.

Easements on the property will normally be discovered during the survey and title search. Easements give rights of traverse on your property to others such as local governments (sidewalk easement), utility companies (sewer or electric company easements) or individuals. Make sure to check the location of your construction in relation to the easements to avoid any conflicts.

Covenants are building restrictions or requirements placed on construction in certain subdivisions, such as minimum square footage of the building or the type and style of construction. These restrictions are usually placed on the property by the developer of the subdivision in order to protect the value of the homes in the subdivision. Ask your lawyer to check for any building restrictions in the public record before buying a building lot. You may not be able to build the house of your dreams if the design is not allowed in the neighborhood. Watch out for any restriction stating that the home can only be built by a specific set of builders or in a specific time frame.

Zoning Restrictions, like covenants, control use of land in an area and the type of structures built. These restrictions are placed on property by the local government planning board in order to protect the value of land in the area. Check with your local planning office if you are unsure of the zoning of your property. A zoning map of the county can usually be purchased for a nominal fee and can be very helpful in searching for property. Zoning laws also define the area of the lot that the structure can actually occupy. The builder is required to place the structure a certain minimum distance from neighboring homes. This requirement is called the minimum setback. In many planned subdivisions, these boundaries are closely adhered to in order to make the homes in the area appear consistent and planned. You may also have requirements for a backyard of a certain depth and a minimum side yard of 20'. It is important to find out where your baselines are before you break ground. If you build outside a baseline and are caught, you may be in for a hefty fine.

Variances allow you to "vary" from standard state and local zoning ordinances within prescribed, approved limits. For example, if you want to build your home 30' from the front curb and the local zoning ordinance requires a 40' set back, you must apply for a 10' set back variance. Unless your request adversely affects neighborhood appearance or safety, your variance will normally get approval. To obtain a variance, you must normally pay an application fee and go before a panel of county zoning officials at a variance hearing, where you or your representative describe or illustrate your intentions and the reasons for them. Normally a notice of your variance request will be posted on your lot so that the public is aware of your intent to get a variance.

They have a right to appear at the variance hearing to either support or fight against your variance.

Liens can become extremely important to a home builder because of the effect they can have on the construction project and the subsequent closing of the property. A lien places an interest in a property that will create a cloud on the title, preventing the close of the sale until resolved. For example, if the plumbing contractor does not feel that he was fully paid for his work, he may file a lien in the public record on the property involved. This lien can stay attached to the property until the disagreement is resolved. Many states do not require that the owner be notified of the lien, but you will know about it at closing! The lien will prevent sale or mortgaging of the property until the disagreement is resolved.

To avoid the pitfalls of liens to potential homebuyers, many states now require contractors to sign an affidavit guaranteeing to the purchaser that the contractor has paid all his bills due to suppliers and sub-contractors. If any parties then file a lien, the contractor can become criminally or legally liable not only to the lien holder, but also to the purchaser. Builders or subs can be thrown in jail in many states for this offense.

If you are building for yourself or others, you can protect yourself from liens by requiring your suppliers and sub-contractors to sign a sub-contractor's affidavit. This agreement works like the contractor's affidavit. The sub-contractor agrees that all bills have been paid and that the contractor has no legal claim against your property. This will prevent the sub-contractor from filing any liens on the property. Make everyone you deal with sign this agreement for your own protection.

Sub-Contractor Contracts, like affidavits, protect you from surprises. Get everything important in writing no matter how good your sub's memory or integrity. Subs may have contract amnesia if a problem comes up. Make sure to include the specifications for the work done as well as a schedule of how payment will be made. It is wise to use a statement such as "Final payment will be made when work is satisfactorily completed."

Builder's Risk Insurance is another "ounce of prevention" necessary before starting your project. This insurance will protect against liability if someone should be injured on the building site. Theft of building materials is sometimes covered also. Risk insurance may only cover theft after the dwelling can be locked. Be aware of a very high deductible.

Building Permits must be obtained from your County Building Inspector's office before construction can begin. This will notify the city or county of your project so that the building inspector can schedule the required inspection of the property during construction. This permit must be displayed in a prominent location at the site. Check your local county for details. Normally, building permits cost a certain amount per $1,000 of construction value or per square foot. Obviously, this value is somewhat intangible. You may have to accept the authority's estimated value.

Worker's Compensation is required in every state to provide workers with hospitalization insurance for job-related injuries. This accident insurance must be taken out by any person or company with hired employees. Most states also have special statutes that require builders, even when using subcontractors, to have this insurance protection for anyone working on the building site. At the end of each year, the insurance company does an audit to determine the amount of insurance to be paid. Each building trade is assessed an insurance rate depending on the relative risk involved in the trade. For instance, roofing trades will be charged a higher rate than trim carpenters. Since you are providing insurance to cover these workers on your site, make sure to deduct the amount of the insurance from the price paid to the sub-contractor.

Many of the subs who work for you will have their own worker's compensation policies since they may have employees working for them. If a sub has a policy, make sure to obtain the policy number and expiration date. Most insurance companies allow you a credit on your own rate if the sub-contractor has his own policy. This credit will be applied at the insurance audit. Contact your local insurance agent for more information about worker's compensation in your state.

Don't underestimate the importance of this policy. You are liable for any work-related injuries that occur on your property. Without this policy, a negligence lawsuit could take your life savings. Look for any situations that can void worker's compensation coverage. For example, in some states, drywall subs experiencing accidents while using drywall stilts are not covered. Better safe than sorry.

BUILDING CODES

Building codes were developed to provide minimum quality construction guidelines insuring that structures are suitable, and safe, for their purpose. The building codes were originally developed as reactions to major disasters: fires, earthquakes, and hurricanes. Their requirements cover structural and fire safety items. If you are only changing the look or decor of a house, codes are usually unnecessary. But when you plan a new structural living space, you must obtain a building permit and comply with local code requirements.

Prior to 1994, most localities used one of three different building code specifications:
- The Uniform Building Code (ICBO)
- The National Building Code (BOCA)
- The Standard Building Code (SBCCI)

Since then, most localities have moved to the residential code that is provided by the International Code Council.

The International Code Council

This organization was formed in 1994 to oversee the new consolidated residential building code known as the International Residential Code (IRC). This code attempts to consolidate all the strong features from the three other major code organizations. Check with your local municipality to see if they have adopted this code (most have). Their internet site can be found at www.iccsafe.org.

Be aware that local codes may differ from the main code. The adopting state or municipality may make changes to the code to suit their own purposes. In addition, they may have adopted different issuing years of the code. You can order copies of the building code for your area from the parent organization or your local inspector.

Make sure to ask the local inspector about exemptions from the code standards.

The model code is published on a three-year cycle with supplementary information published in the interim. When you apply for a building permit, the building inspector will schedule several visits to the construction site to examine the work for code violations. The holder of the building permit (usually you or the building contractor) is responsible for fixing any shortcomings. All code violations must be corrected before the inspector will issue an occupancy permit.

You should study the building codes for your area during the design process to insure that your plan meets code requirements. This is one area where a licensed architect can save a few headaches. Having your plan reviewed by a qualified architect or draftsman can save you countless headaches later.

Project Management

HOW TO BE A BUILDER

A lot of people incorrectly believe that builders must know everything about building a house and must pass several tests to become licensed. Many states do not require small homebuilders to be licensed, and in states that do, building your own home is usually exempt from licensing requirements.

Few working people can afford the time to build their house themselves by hand. This book is directed at the person wanting to function as the contractor. The contractor can be thought of as manager of the project, making sure that subcontractors and materials show up at the site when needed. Leave the difficult and tedious tasks to the experts: They have spent years learning their trade.

Home-building may challenge your abilities in ways yet untested. It will test your ability to:

- Get people to do what you want and when you want
- Be fair, yet tough
- Save money in creative ways
- Manage an exciting, major project. You are the boss.

Before you embark on this major project, you must become educated. After you have studied this manual, you must decide whether you can handle all the details, people, surprises, insanity and periodic frustration associated with home-building. Building is not for the squeamish. This book will not necessarily make home-building a bed of roses for you. Home-building favors hard work, perseverance, guts, the ability to "horse trade", common sense and anything else you can muster to get the job done. To get yourself on the right path to being a builder, get your mind set for doing three things right from the beginning:

- Act like a builder
- Keep good records
- Keep a builder's attitude

Act Like a Builder

Unless special licensing is required in your area, you are just as much a builder as the next guy. But check with your county Building Inspection Department for a copy of the local building code and information on licensing. If you are building your own house, even areas with licensing requirements will probably not require you to have a license.

Get comfortable using building terms and present yourself in a self-confident and professional manner. The respect of your subs, labor and suppliers is essential to your effectiveness as a builder. The better you relate to those you work with, the easier your job will be. As good practice, get in your jeans, construction boots and your favorite plaid shirt and walk around a few construction sites. Get some mud on your shoes. This is a dress rehearsal. Get used to the smell of lumber, lots of mud, and nails by the box. Talk to subs and imagine they're working for you. Figure out what makes them tick; and keep telling yourself you're a builder.

How you are perceived by others in this industry can make a difference. Much of a contractor's job is human relations, so acting like a builder will help you in getting along with others involved with your project. Order a set of business cards. It is amazing how many doors a simple business card will open. The business card will help convince suppliers that you are indeed a "professional." The only time you should wear a suit while working on this project is when you're at the bank.

Keep Good Records

If this manual accomplishes anything, it should be to emphasize the need for keeping track of costs and expenses. You will be at a distinct advantage over the common builder if you keep good records. Some of the primary records are:

Cost Take-Off — The single most important tool you can use to save money on your project is the cost take-off. The cost take-off will tell you approximately what your house will cost before starting the project. It also gives you that checklist of materials necessary to make sure that all essential materials are ordered at the proper time. The take-off has a further advantage of giving you a cost goal to strive for when purchasing materials. When bargaining with materials suppliers don't hesitate to point out that "I've got to stay within my cost take-off budget."

Purchase Orders — Make sure to use purchase orders when ordering materials. This will eliminate possible future confusion over whether certain items were ordered and from whom. Purchase orders and invoices are essential when figuring the house's cost basis for the IRS.

Invoices — Most of the checks you write will be initiated by invoices. It is your responsibility to make sure they are accurate and that you pay them on time, especially if discounts apply. Match your purchase orders to invoices to control payments.

Receipts and Canceled Checks — Lots of money is going to flow out on invoices. Keep them in a book. You'll need to compare them against purchase orders to avoid paying for things you didn't order and paying for things twice. Compare what was delivered with the invoice. Keep all proofs of payments together in an organized manner. Try to get a receipt — not just a canceled check. Your canceled check may take a month to get back to you as proof of payment.

Photographs — Consider taking photographs of your home at various times during construction. This allows you to see changes take place and record them visually. Use a tripod and mark three spots on the ground. This way, you will always be taking a picture from exactly the same location.

Contracts — Nothing will make or break the project like being able or unable to lay your hands on a good, sound written contract when a dispute arises. A man's memory is only as good as the paper it's written on. Contracts and their related specifications can almost never be too detailed. You can specify anything you want; but if you don't specify it, don't count on it being done.

Worker's Compensation Records — You must keep track of which subs have their own worker's compensation policy. For all those who don't, you should retain approximately 6% of the total charges. For subs with their own worker's compensation, you should make note of their policy number and its expiration date.

A Small Pile of Bank Loan Papers — These papers will vary from one lender to another, but will include all or some of the following:
- Loan documents
- Draw schedules
- Funds drawn
- Original take-off
- Surveys and plats
- Title search and insurance

Keep a Builder's Attitude

Your attitude and that of others working on your project is important. You may not always be able to see it, but a poor attitude can cause:
- Delays
- Liens
- Poor workmanship
- Additional costs due to carelessness
- Lots of frustration and headaches

Undertaking such a bold project takes a special frame of mind. And once you break ground, there is no turning back. Here are some important attitudes you'll need to build successfully:
- Persevere. Don't let anything get you down. When problems arise; tackle them immediately. Keep the project moving forward as close to schedule as possible. Realize that time is money and that you are responsible.
- Be Firm. When you have a dispute with a sub over sloppy work or a high bid, stick to your guns. You are the boss. Don't let anyone talk you up against a wall. Remember the golden rule: "He who has the gold makes the rules."
- Don't Be a Perfectionist. You want the work done on your home to be of high quality, but don't be upset every time you see a bent nail (provided you don't see too many of them). Chances are, the flaws you see in your home will not be noticeable by others when completed.
- Be Frugal. Always look for new and creative ways to save money on anything pertaining to the project. Small savings can really add up in a project of this size.
- Be Thorough. Think your ideas through and be meticulous. Pay attention to detail. Have the desire to keep detailed records organized and up to date.
- Don't Get Mad. Avoid chiding your subs on the job as this will reduce their desire to cooperate with you.
- Don't Worry Too Much. Instead of worrying, do something about the problem.

The Builder's Responsibilities

The laborers and subcontractors on your project are responsible for performing quality work and providing materials as specified and when specified by written contracts. That is their single most important responsibility. As the contractor, your responsibilities include the following:
- Provide clear, detailed specifications for those who work for you.
- Pay laborers and subcontractors promptly.
- Schedule material to arrive when or just before needed.
- Schedule labor and subs to work when needed.
- Minimize costs while maximizing quality.
- Provide a safe site in which to work.
- Maintain control of the project and the subs.
- Inspect work as it proceeds and is completed.
- Maintain accurate, up-to-date records on all project matters such as:
  - Insurance
  - Drawings and plans
  - Inspections
  - Bank papers
  - Purchase orders and invoices
  - Legal documents such as variances, permits and licenses
  - Payments, receipts, bank documents and other financial matters.

How to be a Part-Time Builder

Most people have a full-time job and cannot take time off to construct a house. How can you have your cake and eat it too? There are two good options available.

Option I — Hire a Supervisor

Check the want ads for retired carpenters or builders who would like some light work in the construction industry. They have the time and experience to do things right, and chances are, they've been in the building industry most of their lives. Even if they are too old to do physical work, they can spot good work — and bad. This is also a great source for subs and the occasional use of a pickup truck.

Arrangements can probably be made on an hourly, daily or even monthly basis. The supervisor can be around at three o'clock in the afternoon when the concrete is about to be poured or a load of bricks needs to be delivered when you are at work or out of town. These individuals can play an important part in your project. Without them, you may not be able to build at all. If this route is chosen, involve the individual early in the process, when designs and project are being planned. This will allow you time to determine the individual's credibility, skill and honesty. If you would like more control over your project, keep in touch with the supervisor via cell phone.

Most subs can do work unsupervised when not being paid by the hour. Your "supervisor" only needs to confirm that they are doing the proper work, answer questions and assist in inspecting the work. He's there to keep the job rolling along and answer questions that would otherwise bring work to a halt.

Option II — Pay a Builder for His Assistance
Another option is to pay another builder to help run your project. This allows you:
- Use of his subs. This saves you the time and hassle of finding reputable subs. In return, let him put up his building sign at your site.
- Assistance during your work hours. If you can be at the site before and after work, this can be a helpful bridge during those unattended hours.
- Assistance in securing a construction loan.

In return there may be a number of favors you can do for the builder when you don't have anything to do at your site. Also, your site may attract another prospective buyer for your builder. Let him keep a copy of your plans if he likes them.

THE HOME-BUILDING PROCESS
Regardless of the size and location of the home you are building, there are certain inevitable characteristics shared by all construction projects.

Your Project Will Proceed with Decreasing Degrees of Risk
The first steps, like excavation, footings, and foundation work will have more impact if done improperly than if a door is installed improperly. Hence, your utmost care should be taken in the early stages of the project when every step is critical — particularly through the framing stage.

Your Project Will Proceed Erratically
As the result of unforeseen circumstances, many weeks may pass with no progress. Material shortages, poor weather, or subcontractor scheduling can all contribute to delays. Then for no apparent reason, three different subs and all your materials will show up at the site at one time.

You Will Make Changes and Unplanned, Impromptu Decisions
You will probably change the position of a door, window, wall, or major appliance. You may put in a dormer, a skylight or a screen porch. A prefab fireplace may be substituted for a masonry one at the last minute. There is no limit to the things that can change once a project is started. Try to avoid these changes with good advance planning, but don't let the inevitable change order frighten you. Details not necessarily shown on your plans will unfold before your very eyes. These trivial issues are normal and always occur when an idea on paper is converted to reality.

You Will Spend Lots of Money and Use It as a Tool
Until you have paid a subcontractor, you have his undivided attention. If he doesn't fix those squeaky stairs, that leaky pipe, or that crooked window, you simply won't pay him. Once you have paid too much, too early, you have killed his incentive to return and you have failed to use a builder's most important tool.

You Will Make the Difference
Small jobs will appear with no one assigned to do them — the list can be endless. You become the laborer of last resort. You may sacrifice a lazy morning in bed to pick up some brick, test out paint samples, or get a free load of slate somewhere. If you are determined to get the job done, you will get up at 5AM if that's what it takes. You will crawl in bed some nights aching — but knowing you did what had to be done to make a lasting contribution to a structure that may well last a hundred years.

Common Problems for Builders
Being a builder means handling problems. Every project is different and has its peculiar quirks, but there are several problems that occur to some degree on just about every construction project. Although you may not be able to avoid these problems altogether, it's helpful to know them ahead of time so you can take precautions to minimize their impact. The most common problems are:

Subs are often late.
- Let subs tell you when they'll be there.
- Call the sub the day before he is scheduled to arrive at the site.
- Meet the sub at the site at a certain time.
- Give incentives for early completion. Cash talks.

Subs don't show at all.
- Inform subs up front that you expect prompt attendance and that you will spread the word if they cancel with no warning.
- Have other subs as backups.

Subs proceed slowly.
- Give subs a bonus for finishing work on or before schedule. A cash bonus makes a great incentive.

Subs do work incorrectly or disagree on work to be done.
- Ask them to repeat instructions back to you.
- Have detailed specifications and drawings where needed.
- Don't let too much work be completed without checking on it. Visit the site daily.
- Discuss detailed drawings with subs and make sure they understand them completely.
- Keep a cell phone handy so subs can call you when questions arise.
- Have detailed specifications and drawings where needed.

Disputes arise over the amount to be paid, especially when changes are involved.
- Put payment amounts and schedules in writing and have them signed by all involved parties.
- Use change orders for alterations to the original plan and have them signed by all involved parties.
- Use lien waivers to control your liability.
- Pay with checks to insure proper records.

Materials are not available when needed.
- Order custom pieces early to insure that they are on site when needed.
- Confirm delivery dates with suppliers and have them notify you of delays several days in advance.
- Determine which suppliers deliver after hours and on weekends.
- Get friendly with the delivery men and schedulers at supply houses.

Wrong material or wrong quantity of material is delivered.
- Use detailed descriptions or part numbers on all purchase order items.
- Print legibly.
- Don't order over the phone unless extremely urgent.
- Have materials delivered early so they can be exchanged or corrected before they are actually needed.
- Have your subs help you to determine actual quantities and materials.
- Get to know salespeople by name and work with them every time.

Bad weather.
- Don't begin the project during or just before the rainy season.
- Get the roof up as soon as possible.
- Make the most of good weather.
- Provide a temporary driveway made of "crusher run" to avoid creating muddy areas.
- Cover materials with rolls of poly (plastic sheet).

Theft and vandalism.
- Visit the site each day.
- Call the local police department and have your site put on surveillance.
- Don't always visit the site at a routine time. Make unannounced visits, especially near quitting time.
- Don't keep too much material on site. This is especially true for windows, doors, millwork, good plywood, and expensive tubs and appliances.
- Get builder's risk and theft insurance with a low deductible.
- Install locks as soon as possible and use them.

Rules For Successful Builders
- Put Everything in Writing — not written, not said.
- Be Prepared before breaking ground and don't get behind.
- Keep Project Binders current with all project details.
- Be Fair with money. Pay subs what was agreed, when agreed.
- Always Get References and check out three for each major subcontractor. You must see their work.
- Don't Try to Do Too Much of the actual physical work yourself. Your time and skills are often better used managing the job.
- Shop Carefully for the best material and labor prices. Don't hesitate to insist on builder discounts.
- Use Cash to your maximum advantage.
- Rely on Your Contractors for advice in their field.
- Don't Assume anything. Material is not on site unless you see it. Subs are not on site unless you see them there. Don't expect subs to know what you want unless it's in writing.
- Use as Few Subs as Possible without sacrificing cost or quality.
- Use the Best Materials you can afford.
- Keep Changes to a Minimum and budget funds for them carefully.
- Estimate on the High Side and have a contingency fund.
- Keep Money Under Control and track progress versus your original budget.
- Rely On Your Local Building Code as a primary reference for specifications, inspection criteria, material guidelines and settling of disputes.
- Stay at Least a Week or Two Ahead of schedule. Staying ahead will allow you time to compensate for material delays or time needed to replace unreliable or unsatisfactory subs.
- Use the Best Materials you can afford.
- Let Your Fingers Do the Walking. If it looks like rain, call the roofer. If something can't be delivered, call your subs as soon as possible so they can schedule on another jobsite.

Proverbs for Builders

Below are a few sayings very appropriate to the building industry. As a builder, you are bound to encounter times when one or more of them apply.

Murphy's Law — If anything can go wrong, it will. If there is a possibility of several things going wrong, the one that will cause the most damage will be the one to go wrong. If you perceive that there are four possible ways in which a procedure can go wrong, circumvent it, then a fifth way will promptly develop.

It is impossible to make anything foolproof, because fools are so ingenious.

Dave's Constant — Matter will be damaged in direct proportion to its value.

If you explain so clearly that nobody can misunderstand, somebody will.

Do not believe in miracles — rely on them.

Lester's First Law — Everything put together falls apart sooner or later.

Lester's Second Law — No matter what goes wrong, it will probably look right.

If only one bid can be secured on any project, the price will be unreasonable.

King's Law — No matter how long or how hard you shop for an item, after you've bought it, it will be on sale somewhere, for less.

Everything takes longer than you think.

The Golden Rule — He who has the gold makes the rules.

The Eighty-Twenty Rule of Project Schedules — The first 20% of the task takes 80% of the time and the last 20% of the time takes the other 80%.

Once you have exhausted all possibilities and fail, there will be one solution, simple and obvious, that is highly visible to everyone else.

All delivery promises must be multiplied by a factor of 2.0.

Harrison's Law — The most vital dimension of any plan or drawing stands the greatest chance of being omitted.

Commoner's Law — Nothing ever goes away. When things just can't get any worse, they will. Left to themselves, things tend to go from bad to worse.

Nothing is as easy as it looks.

Why Construction Projects Fail

As a builder, it is up to you to make your project a success, but along with all the glory comes all the responsibility. Listed below are a number of common reasons why construction projects fail and/or encounter severe difficulties:

- Costs are underestimated.
- Lack of budget control. Excessive design changes are made late in the game.
- Project manager is not organized; letting things get out of control.

- Lack of understanding of contracting.
- Acts of vandalism, arson or natural catastrophes not adequately covered by insurance.
- Lack of enthusiasm.
- Workers are paid before careful inspection of work and materials.
- Legal complications due to careless handling of business transactions, such as double invoice payment or payment for work uncompleted or undelivered, damaged or inferior goods.
- Conflicts with spouse or partner.

Building for Yourself vs. Building for Others

One of the first questions that any of your workers are going to ask you is whether you are building this home for yourself or to sell. The difference can be subtle, but more likely the difference will change the house dramatically in terms of workmanship and materials used.

Those who build for a living quickly learn that the only way to make a fair profit is to build into a home only those features and level of quality that is necessary to sell the house and nothing more. Quality materials such as hardwood floors, marble, custom millwork, brickwork, fancy lighting and fancy bath fixtures are used sparingly or not at all.

However, you may defeat at least part of the purpose of building your own home if you don't outspend the neighborhood builder in a few areas. Perhaps you want a skylight, nice brass faucets, fancy wallpaper or better grade appliances. This is your prerogative. You should add a few special touches that will make your house stand out and become easier to sell.

When You Get Nervous

Nervous because you've never done anything so ambitious? Nervous because the carpenter hasn't returned your call yet? Or maybe you just can't put a finger on the source of your uneasiness. Relax.

Remember:

Fear is a natural response to the uncertain and the unknown. A goal of this book is to replace uncertainty with clear, concise knowledge — to remove those cloudy uncertainties. Some individuals maintain a certain level of fear as a natural defense mechanism — to always remain cautious and alert for possible problems.

Individuals with little or no formal education build houses every day — and they do it for a living.

A certain degree of fear is typical when doing anything for the first time. Chances are, after you have built your first home, you'll want to do it again.

Your construction project is staffed by many knowledgeable, professional subcontractors each specializing in one area. It is their job to do the work properly. Don't worry if you don't know everything they do. That is not a requirement. If the job is not done right you can refuse to pay until the items in question are fixed.

The construction approach presented in this book has been time-tested. Using this approach puts you way ahead of many individuals who may either try to brave it alone or otherwise push ahead without adequate knowledge and/or preparation.

Assure yourself that you are doing everything you can to protect your project against the pitfalls described in "Why construction projects fail."

Assure yourself that you are using the Steps, Specifications and Inspection checklists provided.

When You Get Discouraged

You're into the project and things aren't going along perfectly. Your framing crew made some mistakes, you've had four straight days of rain, your roofing sub won't return your call and the price just went up on storm windows. So what else is new? Getting discouraged?

Did you ever hear of any project proceeding from beginning to end without a hitch or two — or twenty? Below are a few things to ponder when you wonder why you didn't just throw this book away and cut your losses.

Look at the money you're saving. You probably have a good idea how much. Just look at that figure for a moment. That's the money you would have paid someone else to do the work you're doing right now. If someone else were performing your job, they'd run into the same kind of problems. When the interest costs are figured in, you will probably be saving one or more year's salary by contracting yourself. Aren't a few months of challenge worth it?

Look at all the problems and solutions you're encountering. One day when your home is completed, you'll look back at them and laugh. You'll tell of some of your experiences the rest of your life.

Look at your accomplishment! Your friends, associates, and family will admire your tenacity. You can honestly say that you built your own home.

Remember — Bad days are always followed by better ones. Hang in there!

CONSTRUCTION & COMPUTERS

Computers

You would have to be living in a cave to miss the tremendous influence that computers and the Internet have had on our lives. The computing power of personal computers that can be purchased for under $1,000 far exceed the multi-million dollar mainframes of a decade ago. This fact is important for you; the potential contractor, because building a home is just a large management project. And computers are perfect for managing the details for you.

Powerful computer software will help you build your house in a variety of ways:
- The Internet can provide a tremendous amount of information about construction suppliers, products, and organizations. You can find detailed information about virtually every major supplier. Most manufacturers now publish detailed specifications, drawings, and installation instructions online. All this information is just a keyboard away.
- Three dimensional design packages such as SketchUp help to visualize the home plan and produce building information models.
- Estimating software can produce accurate and complete takeoffs.
- Project management and accounting software can follow the details of the project and keep it on budget.
- How-to websites can fill in knowledge gaps and provide electronic forms.

Since the Internet moves so swiftly, much of what you read here will be obsolete by the time this book hits bookshelves. That is why we have developed a website so our readers can keep up with the latest information. www.selfmadehomes.com

We invite you to visit www.selfmadehomes.com for updates and electronic documents that will help you to manage your construction project. We welcome your feedback and information on your own construction project.

The Internet

Covering everything you need to know about the Internet is beyond the scope of this book, but we'll discuss a few basic websites that can be particularly helpful. We assume that you have some familiarity with a computer and software; otherwise, most of this information will seem like gibberish to you.

Finding Information (Surfing the Web)

Most often, the easiest way to find the information you need is with a simple search. This approach has become so pervasive that search engines have become verbs. Who hasn't "Googled" some information online? Try to be as specific as possible in your search. Otherwise, you'll be inundated with lists of information that may or may not be helpful. Searches won't find everything. Much of the information provided by manufacturers is targeted toward the professional trades and do not show up high on search lists. Instead, start your search on the following websites:

The National Association of Home Builders: (www.nahb.org)

This is the professional trade group for residential and commercial builders and is filled with valuable construction information. Most material is targeted toward the professional trades and information can be expensive to purchase.

Building America (www.buildingamerica.gov)

This site is sponsored by the U.S. Department of Energy and promotes new building technologies and energy efficiency. Since it is government sponsored, most of the information online is free to the public. You'll find numerous technical articles on the latest building technologies.

PATH (www.pathet.org)

The Partnership for Advancing Technology in Housing is an online resource for homeowners and homebuyers, the homebuilding industry and federal agencies. PATH is a partnership between several government agencies and the National Association of Home Builders. They are dedicated to accelerating the development and use of technologies that radically improve the quality, durability, energy efficiency, environmental performance, and affordability of America's housing. You'll find case studies, technology profiles, links to articles and publications and tips for homeowners.

ToolBase (www.toolbase.org)

ToolBase Services is the housing industry's resource for technical information on building products, materials, new technologies, business management, and housing systems. The NAHB Research Center provides the services, with funding from the Department of Housing and Urban Development (HUD) through The Partnership for Advancing Technology in Housing (PATH) program, and other industry sponsors. You'll find a wealth of technical documents here.

Sweet's Network (www.sweets.com)

Sweet's is the construction industry's primary product information resource and a division of the McGraw-Hill Companies. This is an excellent resource for finding and researching products and manufacturers.

eBuild (www.ebuild.com)

eBuild is an online website of Hanley-Wood publishers, who publish several trade journals for the construction industry. They have an excellent interactive listing of manufacturer of building products, with contact information, and links to technical documents.

Build.com (www.build.com)

Another reference site aimed at the professional trades. Materials can be ordered directly from the internet by registered professionals. They also have consumer sites for lighting, plumbing, rugs and hardware.

Computer Aided Design (CAD) Software

3D CAD

A few years ago, dedicated CAD workstations cost $15,000 and required several days to put in a complete two dimensional house plan. Now, a free CAD package offered by Google called SketchUp has revolutionized the field. With minimal training, you can do something the old CAD programs couldn't; let you view the house from the inside in three dimensions before it is even built. The advances are nothing short of phenomenal.

Even if you don't plan to draw your actual blueprints, the ability to visualize the design in the computer can help you to fix potential design shortcomings. You can place scaled versions of your own furniture in the plan, to see how they fit. You can use this information to make changes to traffic flow, place electrical outlets, even check out the view from the foyer.

If you choose to draw the entire plan, it is advisable to let a licensed architect or draftsman create the initial design to insure there will be no structural or stylistic problems. The software, while powerful, is not aware of structural issues.

To obtain a free copy of SketchUp, visit (sketchup.google.com).

Designers and Architects Working in SketchUp 3D
Contact these designers and architects for more information on home designs and the Google SketchUp Pro design software.

Dennis Fukai, PhD, Architect
16708 SW 132 Lane, Archer, FL 32618
Website: (www.Insitebuilders.com)
Books: *Building SIMPLE: Building An Information Model, Being SUSTAINABLE: Building Systems Performance, Living SMALL: The Life of Small Houses, How A House Is Built: with 3D Construction Models.*

Alan Mascord Design Associates, Inc.
1305 NW 18th Avenue, Portland, OR 97209
Website: (www.mascord.com)
Books: *Mascord Living Spaces*, Mascord Efficient Living, Mascord Home Collection 2007

David E. Wiggins, Architect
2725 Highland Trail, Leander, Texas 78641
Website: (www.wigginsarchitect.com)

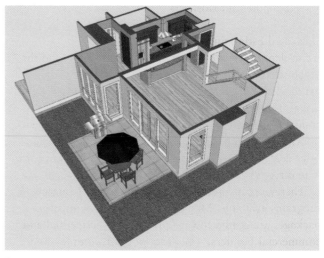

SketchUp can render cutaway three dimensional views of a house to aid in design and evaluation. Courtesy of Alan Mascord Design Associates

2D CAD

These programs fall into the more traditional definition of CAD packages used for construction. They are the ancestors of some older traditional systems that have migrated down from the big CAD workstations. They are the best programs for producing a professional set of blueprints, complete with dimension lines, call outs, and accurate scaled images. They can be output to the large electronic pen-drafting machines, which can draw a full size ¼" - 1' scale blueprint.

Autocad LT
Autodesk Inc.

Baby brother to the bestselling Autocad software, one of the most professional and powerful CAD systems available. Autocad LT is a stripped down version, but still retains the main features necessary to create a full featured drawing. Fairly expensive and best suited for advanced users.

Visio Home
Microsoft Corporation

A very easy to use and productive drafting package. "Smart shapes" makes altering walls and furniture very easy. Drawings are high quality but the program lacks serious drafting features.

Estimating Software

The "Achilles heel" of every construction project has always been the cost estimating. The sheer number of items that go into the house is staggering. A computer can help tremendously with this problem. Don't even think of starting a homebuilding project without a complete and accurate estimate. The bank will be impressed with a well-organized printout of costs.

Computers are perfect for calculating costs and making adjustments when plans change, keeping you within your budget. There are two ways to go:
- Dedicated Estimating Software
- Computer Spreadsheets

Dedicated Estimating Software

An in-depth review of estimating software is beyond the scope of this book. Contact the NAHB for their official list of software suppliers and check the Internet for additional information. Many of the construction software companies have their own web sites and offer detailed information about their products. Several companies even offer demo versions of their software that you can download directly off the Internet.

There are several dedicated construction estimating programs available. Many of the commercial software packages are geared toward professional residential and commercial builders. They can be quite complex and powerful: allowing the user to estimate costs in several different ways, track multiple suppliers, allow for waste and markup percentages, track retainage, estimate groups of materials together, and even enter dimensions using an actual blueprint and digitizing tablet. They range in price from $100 to $1,000. Before you buy one of these packages, make sure you are serious about building and plan to build more than one home. Otherwise, you won't get a good return on investment. Some of the less expensive packages are little more than elaborate spreadsheets, so why not use a generic spreadsheet instead?

Computer Spreadsheets

Another, less intensive method of calculating costs is through the use of a computer spreadsheet such as Microsoft Excel®. This approach will require more effort on your part, but is totally flexible when it comes to making changes. You need to be pretty familiar with your spreadsheet program before beginning, but this is no different from learning the nuances of a dedicated estimating package. You can follow the same format as this book's estimating forms, providing columns for price and quantity, and letting the spreadsheet automatically calculate the formulas as you input the information.

Other Construction Software

The list goes on. There are computer software packages available to track job costing, prepare purchase orders, calculate payroll, manage the scheduling of the entire project, manage bids, calculate excavation requirements, engineer trusses, and satisfy a host of other special needs. Much of this software is designed for professionals and requires a large investment of time and money to use effectively. If you plan to build only one house; your own, then purchasing more elaborate construction software would be a waste of time. Here are a few other less expensive programs that will work well for managing a one-time construction project. They also have additional advantages; they're inexpensive and can be used for other purposes after the construction is over.
- Personal information managers. These programs track names and addresses and can be purchased at any software store. These include such programs as Act®, Microsoft Outlook®, and many others. Set up a special copy of the PIM to track your suppliers' names. You can use the built in appointment scheduler to keep up with the subcontractors and to follow the progress of the job. Computer task lists ensure that you keep up with every detail of the project.
- Word processors such as Microsoft Word®, Openoffice, and WordPerfect® In addition to the obvious function, these programs can create excellent purchase orders. By creating a "mail merge" form, you can automatically generate form letters, bid requests, and purchase orders customized with the supplier's individual names.

- Personal finance packages such as Microsoft Money® and Quicken®. These inexpensive programs will stand in admirably for the more powerful construction accounting software, especially for tracking one job. Make sure to set up a separate checking account just for tracking your construction expenditures. This software can automatically generate checks for payment and categorize construction expenses.

Other Useful Internet Sites

Associations

AHRI
(www.ahrinet.org)
Air Conditioning & Refrigeration Institute

ASHRAE
(www.ashrae.org)
American Society of Heating, Refrigerating and Air-Conditioning Engineers, Inc.

ASID (American Society of Interior Designers)
(www.asid.org)
Interior design industry news and toll free numbers to reach ASID World Wide Referral Service.

Fannie Mae
(www.fanniemae.com)
America's largest source of home mortgage funds

National Wood Flooring Association
(www.woodfloors.org)

WDMA (Window and Door Manufacturers Association
(www.wdma.com)

OSHA - Safety
(www.osha.gov)
The government safety organization.

Resilient Floor Covering Institute
(www.rfci.com)
An industry trade association of resilient floor covering producers.

Southern Pine Council
(www.southernpine.com)
A lumber industry trade council.

The American Institute of Architects
(www.aia.org)
The collective voice of America's architects with over 58,000 members.

The Better Business Bureau
(www.bbb.org)

Energy & Environmental Building Alliance (EEBA)
(www.eeba.org)

The National Society of Consulting Soil Scientists, Inc.
(www.nscss.org)
Directories

Builder Online
(www.builderonline.com)
An excellent directory of product manufacturers created by Builder magazine. Manufacturers list is keyword-searchable.

Clem Labine's Traditional Building
(www.traditional-building.com)
A wealth of data on historical products and services for restoration and traditionally styled new construction

Cost Estimating
(www.rsmeans.com)
R.S. Means Company. Cost estimating information, along with advertising for the company's products.

PlumbingSupply.com
(www.plumbingsupply.com)
Plumbing supplies at discount prices.

WOODnetWORK
(www.woodweb.com)
Web site for woodworking professionals.

BEST PRACTICES

They Don't Make'em Like They Used To

How often have you heard this old adage as applied to house construction? Is it true? The answer is far more complex than you might think. Your ability to specify and recognize good workmanship sums up what this book is all about. Let's explore this adage in more detail.

Are the homes of yesteryear better built than the homes of today? The short answer is absolutely not. The homes of today are far more energy efficient, far more structurally sound, more fire resistant, safer, more convenient, and in many ways more environmentally sound than homes built fifty, a hundred, or two hundred years ago. So why are people so adamant that homes today are a pale comparison to the artistry of the homes of yesteryear? That answer comes down to one word: Craftsmanship.

Up until the housing boom of the fifties, builders were more craftsmen than project managers. They tended to have a broad range of homebuilding skills. Small crews of "craftsmen" would come in and take responsibility for building the house from the foundation to the roof. Their personal pride of workmanship was evident everywhere, from the design to the trimwork. They took personal responsibility for the complete project and usually had a hand (literally) in every phase of construction. Older homes were usually showcases of builder's pride. It was like walking through a museum. They could also be nightmares of structural instability, termite damage, wet, moldy basements, and sagging roofs.

The personal pride and craftsmanship of these early builders simply weren't enough to overcome the structural and safety issues of most homes. Builders back then were laborers and craftsmen, not engineers and scientists. Their homes were works of art, but suffered from high cost (for the day,) slow construction, and non-standard building techniques.

In the fifties, things changed. After World War II the tremendous demand for new housing brought a completely new method of construction to the forefront. Big business brought the efficiency of assembly line construction and the manufacturing sophistication of WWII companies to bear on the insatiable demand for affordable homes. As Henry Ford discovered, it's much cheaper and more efficient to train assembly line workers in one specific skill rather than a range of skills. The result was a spectacular rise in construction productivity, and a drop in housing costs as a percentage of income. The modern housing boom had arrived. The Department of Housing and Urban Development and the National Association of Home Builders became actively involved in engineering homes with better safety, efficiency, and quality standards. Universal building codes began to be adopted by municipalities across the country.

So why are some people not happy with their modern homes? It boils down to one huge side effect of assembly line construction. When a home is built by a team of specialists, no individual specialist takes "ownership" of the project. Their concern is to finish their part of the job as quickly and as cheaply as possible. Subcontractors make their profits on volume, not quality. The quality assessment falls back on the shoulders of the builder, who has morphed into a project manager. The result: Homes today can vary greatly in quality, because the quality of individual subcontractors varies greatly. And the ability of the homebuilder to monitor that quality also varies. Any quality issues buried beneath drywall becomes an issue "out of sight, and out of mind." Buyers of speculative homes have no way of assessing construction that is hidden by trim and paint.

So, are houses better built today than yesterday? Yes, when quality becomes paramount and the best construction practices are followed. Virtually all building material manufacturers today provide detailed brochures that explain the best practices for installing their product. Thanks to the Internet, most of these guidebooks and brochures are available online for anyone to download. The Internet has made effective project managers of us all.

This is where you come in. As the project manager for your self made home, it's your responsibility to make sure that best practices are followed by each subcontractor during their duration on the job. That's the real focus of this book. If you monitor your project and make sure your subs have specifications that demand a high level of quality, then you can be assured of a finely crafted home that will far outlast and outperform the old museum pieces of yesteryear.

Beauty is in the Eye of the Beholder

Don't assume that your subcontractors have the same concept of quality and style that you possess. We all know what happens when you "ass-u-me." Spell out your expectations in the contract so that there's no question.

Richard Karn, the actor who played Al Borland on the TV sitcom Home Improvement, recounts a true story about a tile setter in his book, *House Broken: How I Remodeled My Home for Just Under Three Times the Original Bid*. Richard remodeled his kitchen and had a set of beautiful hand-crafted tiles of grapes inlaid around the kitchen cabinets. The craftsmanship was outstanding, with one exception. The tile setter had laid all the grapevine tiles with the grapes pointing upward. When do grapes ever grow up? The tile setter saw nothing wrong with the incongruency, but Al just couldn't handle the illogical scene of grapes defying gravity. Out went the tiles, and another big portion of his budget.

The moral of the story? Don't assume anything. Spell it out in the specifications of your contract. Make sure subcontractors know your expectations before the project begins. It's far easier for them to tolerate your expectations during the bidding process, rather than verbally later, when they are halfway through their job.

The Seven Deadly Sins of Construction

The Construction section of this book focuses on many best practices and how to identify quality workmanship. The inspection checklists and specifications are good places to start. Also, browse the Internet for installation brochures from the manufacturer on all tricky installations, such as flooring, roofing, floor trusses, rafters, insulation, windows, cabinets, painting, etc. Most of them are readily available and make excellent training tools for the trickiest aspects of your project. Add these brochures to the contract, or include critical excerpts in your specifications. Your knowledge of these techniques will build respect among the subs and prevent many shortcuts.

There will always be some shoddy or substandard work on any project and some is to be expected. Most of it is harmless and will be hidden behind walls. Other "worst practices" are more serious, and the reason why smart builders withhold retainage until after the punch-out. See the Project Management section for more information on these "quality assurance techniques."

What you really want, is to educate yourself against the most egregious and destructive miscues committed by subcontractors. These are the ones that cause the most problems, the ones that are difficult or impossible to repair later. The miscues that will turn your dream home into a nightmare. These are the seven deadly sins. Know them, anticipate them, and eliminate them early. The result will be a happy home and peace of mind.

1. Building on Partial Fill

The foundation of a house literally supports everything else. So, if your foundation has basic structural problems, they will ripple through the rest of the structure, causing cracks in walls, jammed doors, out of plumb rooms, air leakage, and more. One of the most common foundation problems results from improper fill dirt. All fill dirt on a site must be compacted properly with a compaction tool, but it is almost impossible to compress dirt back to its pristine, undisturbed state. Soil to be excavated may have been undisturbed for thousands of years. It will always be more compact than freshly disturbed soil.

Often an excavator will dig out a sloping site for the basement or slab and use that dirt to fill in the low end. This results in a soil base that is half "virgin" soil, and half fill dirt. Any slab poured on such a base will settle

A solid foundation should begin on excavated, virgin soil whenever possible. A crushed rock base for drainage should be applied and well compacted.

more at the fill dirt end and may result in cracking and breaches in the concrete. This will allow moisture penetration and may shift walls and ceilings. Try to avoid this situation whenever possible by digging down far enough to expose virgin soil across the entire footprint of the house. If that is not possible, make sure the fill is compacted perfectly, then expand the size of the footings around the perimeter of the slab to increase resistance to settling.

Improperly compacted soil is less of a problem with poured concrete basements because the poured walls act as girders, supporting the rest of the structure. However, it's still best to insure maximum compaction in any circumstance. A mistake here is almost impossible to repair and will linger for the life of the house. If the lot is a "reclamation," one that has been filled with dirt to reclaim it for construction, you must take extreme care to insure a solid base for the foundation. There must not be any organic materials in the fill, as they will create voids upon decomposition. Make sure the fill has had time to settle. A couple of years is usually required. Have a soil engineer calculate the bearing capacity. It's best to build a house with a full poured concrete basement on such a lot. First, you will be digging down (hopefully) far enough to reach virgin soil, but if not, you can depend on the structural strength of the poured concrete to avoid issues.

Excavators and concrete subs will probably argue about the need for such great care, but it is far better to over-engineer a foundation on tricky soil than to risk problems later.

2. Improper Basement Drainage

This is one of the most common complaints by homeowners, especially in older houses before basement technology improved. It's tempting for a builder to rush through this phase, since any efforts to waterproof will be buried before the house is sold. However, a properly installed basement with good drainage will be completely dry and just as livable as any other part of the home. The most common waterproofing issues with basements are:

- Improper waterproofing on the outside basement wall
- No drainage route for water down the wall. Opt for engineered drainage fabric.
- Improperly installed crushed stone at the foundation with absence of filter fabric to prevent silt clogging.
- A french drain that has been improperly installed with the drainage holes pointing up and with no drainage to daylight
- No vapor barrier underneath the basement floor
- Improper grading of soil to drain away from the house
- Dense clay fill against the foundation, rather than porous sand mixture

Repairing a poorly drained basement later is possible, but difficult and expensive. Get it right the first time.

3. Improper Notching of Floor Joists

Floor joists that are notched or drilled in the wrong places can lose their structural integrity completely. This can lead to floor sagging or in extreme cases, collapse. This is especially important for open web trusses and I-joists where the top and bottom plates (chords) can NEVER be notched. The common culprits in this case are usually plumbing and HVAC subs. They seldom have the engineering expertise to know precisely where and how to notch or drill joists. Often, blueprints may call for a floor joist at the precise location of the toilet drain. What is a hurried plumber to do? Some of them will just saw out the joist so they can finish their job responsibilities and hope no one notices. Their rationalization will often be, "Hey, it's not my fault the framer put this here. Why should I have to suffer? I've got other jobs."

The simplest solution for these problems is to anticipate these framing hot spots ahead of time and make sure joist and wall framing allows for plumbing and HVAC runs. Another is to make sure and have installation guidelines for any engineered truss on site. These guidelines clearly indicate where and how holes and notches may be cut. Open web trusses and the latest open web I-joists help to eliminate these issues by offering plenty of options for running plumbing, HVAC, and electrical.

4. Structural Framing Mistakes

Incorrect structural engineering on load-bearing structures can cause a host of related problems, from sagging roofs, cracking walls, nail pops, jammed doors, vulnerability to weather and earthquake episodes, etc. Some common causes:

- The absence of roof ties. In a stick built roof, roof ties, commonly called collar ties, join opposing rafters together and help to prevent sagging roof lines or racking of walls. Often, framers omit them entirely, or place the roof ties too close to the roof's apex, which renders the tie useless.
- An unsupported central wall. If a central wall extends to the peak of the roof ridge and there is no official "ridge beam," then that wall becomes a load bearing wall and will need to have support all the way to the foundation. Framers used to roof trusses (which do not require central support) may consider the central wall on a stick built home as non-load-bearing.
- Improperly supported rafters or I-joist rafters. Rafters can be tricky to support because of the angles involved. I-joist rafters in particular require special handling to avoid breaching the top and bottom plates. Manufacturers always provide detailed installation instructions for these instances. Make sure the framers are familiar with these instructions.
- Load-bearing walls not properly supported by floor joists. Load bearing walls should be aligned over joists when the wall and joist run parallel. It's easy

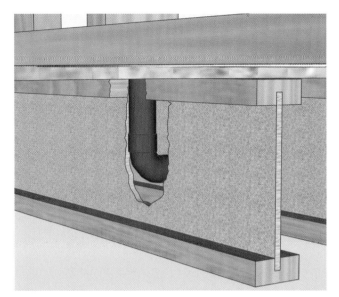

Plumbing breaches made by hurried plumbers can destroy the structural integrity of I-joists.

for a framer to overlook this situation when joist spacing does not align with walls.
- Improperly supported cantilevers. The length and support of cantilevers changes drastically depending on whether the outside wall of the cantilever is load-bearing or not. Make sure framers are aware of the differences.
- Improper support of shed dormers. When shed dormers extend across the back side of a roof, the natural stability of the roof's triangular shape is interrupted. Improper bracing can cause racking and sagging of roof members.

The technical aptitude of framing crews can vary tremendously. Don't just assume that your crew knows all the codes in your area and can handle the engineering challenges of an unusual design. This is one area where your designer or architect will earn their pay. Whenever possible, ask your designer to provide framing details with the blueprints to avoid any confusion.

5. Moisture Damage

Moisture damage can have many causes and the results can be devastating, leading to rot and mold. Only recently, have the devastating effects of toxic mold become evident. Toxic mold can grow anywhere there is a buildup of moisture and an organic food source such as wood or gypsum. Manufacturers have started to sell mold resistant drywall and framing studs, but the best way to prevent the danger of toxic mold is to eliminate the underlying source of moisture. Moisture buildup can come from these causes:

- Improper waterproofing and flashing around windows.
- Absence of moisture resistant house wrap or wrap that overlaps in the wrong direction.

- Absence of a vapor barrier on interior walls, allowing interior moisture to migrate and condense inside exterior walls.
- Insufficient venting of moist air in kitchens and bathrooms.
- Insulation voids at the junctions of floors, walls, and roof.
- Improper flashing around chimneys
- Improper eave flashing and absence of waterproof membrane at the eaves in cold climates, which can lead to ice dams.
- Improper flashing at roof valleys and plumbing vents, resulting in ceiling leaks.

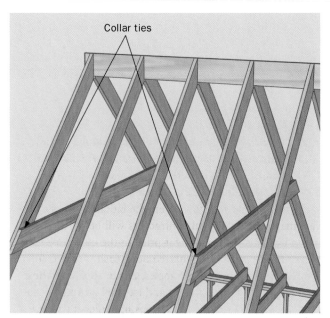

Collar ties help to resist sagging and bowing in roofs under load.

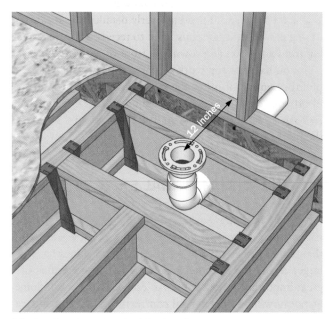

Joist headers to resolve plumbing interference.

I-joists make excellent rafters when installed properly with joist ties and a structural ridge vent.

Poor best practices in all these areas not only increase the danger of toxic mold, but they often result in wasted energy costs and water damage to interior finishes. Leaks and vapor accumulation can be extremely difficult to locate after the home is finished and sometimes goes undetected for years until family members start exhibiting health problems. Nip this insidious problem in the bud by insisting on proper installation and waterproofing techniques during construction. Your family's health may hang in the balance.

6. Termite Damage

This is another insidious problem that can go undetected for years, until irreparable damage has been done. It's not enough to get a good termite insurance policy and treatment. After being detected, termite damage can be extensive and require major structural repair. There are many other ways for termites to invade interior walls. These include:

- Concrete form boards that are left in place after a pour. Concrete subs in a hurry may leave these behind or embedded in the concrete. This wood provides a "bridge" for termites, that migrate through cracks and voids, travel along the form boards and eventually end up in your walls.
- Any place where pipes or cables run through concrete slabs. The termites can migrate along these seams. Termite guards are your best protection against this threat. They consist of special seals that are cast into the concrete during the pour and block all access.
- Improper application of termite poison. Often, soil treatment around the perimeter of the house can be incomplete, or more commonly, landscapers or grad-

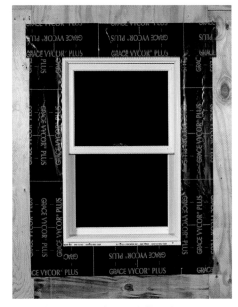

A common source of water infiltration occurs around windows. Modern waterproofing techniques can virtually eliminate this problem.

Frameguard treated lumber uses anti-fungal and borate treatments to guard against moisture damage and toxic mold.

ers will come along after the treatment and disturb the soil, resulting in a breach of coverage.

- Insufficient height on foundation slabs. Termites can build mud tunnels up to a foot or more in length. Monolithic slabs poured too close to the ground increase the likelihood of this problem.

7. Oversized HVAC

As odd as it sounds, having too much of a good thing can be worse than not having enough. Many HVAC contractors size their units for homes based on old formulas from years past, when homes were leaky and poorly insulated. If you've gone to great lengths to make the home air tight and energy efficient, then you may end up with a furnace and air conditioner that is too big for the house. Why is that a problem? For one thing, you've paid more for an expensive appliance than necessary. But there are other problems. Air conditioners are designed to run for a certain length of time to remove moisture from hot, humid air. We feel humidity as heat. An oversized air conditioner will cycle on and off far too often and will not dehumidify the air sufficiently for comfort. The result is a higher energy bill for the life of the home, as family members crank the thermostat down to lower temperatures.

Don't Overdo It

It's tempting to police the construction site like a staff sergeant, checking for the smallest infraction and leaving the crew in a state of nervous apprehension. You've spent a lot of effort building your skill set and maybe you want to show off. But excessive micro-management can be a quick path to ruin. Not only will you become a nervous wreck, but you'll build resentment and distrust among your subcontractors. Resentful contractors are more likely to take their frustrations out by intentionally sabotaging work or by simply abandoning the worksite.

If you've done your job during the bidding process, the vast majority of your subcontractors will be hard working individuals who take great pride in the quality of their work. Their pride can be injured quite easily. Your goal is to look out for the few bad apples, not to spoil the whole barrel. It's best to take a "live and let live" attitude when mistakes or some sloppy work crops up, unless they pose a serious threat to the project. Encourage best practices during the bidding and negotiation stages. It's much easier for a prideful contractor to accept overly detailed specifications during contract negotiations, than to receive a tongue lashing from an overzealous owner in front of his cohorts.

If you have to stop work or correct a major mistake, pull the contractor off to the side, away from others and plead your case in an even and calm manner. Explain the reasons for your concern. You'll sleep better, and so will the workers on your site.

DEALING WITH MATERIAL SUPPLIERS

Dealing with larger companies offers advantages due to the potential savings passed on as a result of volume purchasing. Larger companies may also offer wider selections than their smaller counterparts. Larger companies have a larger reputation to protect. Hence your satisfaction is of great importance. Retail companies like Lowe's and Home Depot now have commercial contracting divisions and are competitive with most dedicated construction suppliers.

Hidden termite damage can destroy the structural integrity of a house.

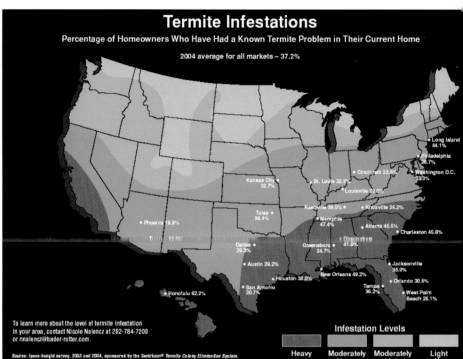

Southern states are particularly vulnerable to termite damage.

Deal direct with manufacturers when possible. Companies that specialize in windows, doors, carpet, bricks or other such items may provide you with merchandise at close to wholesale prices. Be prepared to pick many of these purchases up yourself since delivery may be costly or unavailable.

Limit the number of vendors you deal with in order to reduce confusion, building accounts and contacts. Dealing with a few vendors may mean special discounts due to larger volume purchasing.

Open builder accounts at your suppliers. You are a builder. And builders get discounts. Also, builders don't deal with retail sales personnel. Most large material supply houses have special contract sales personnel in the back of the store. Builder accounts often imply special payment schedules. For example, in one case you may pay the open balance on the first of each month. In this case it is to your advantage to purchase your materials early in the month to gain a full month's use of the money. The time value of money is on your side in this case. Paying accounts within a specified period is a surefire way to earn early payment discounts. Keep track of them and earn them. That's money in your pocket.

Inspect material as it arrives on site. If something is damaged or just doesn't look right, insist on having it returned for exchange or credit. This is normal and expected: You're paying good money and you should expect good quality material in good condition. Indicate returned items on the bill of lading as proof. Generally,

once installed, the materials are your responsibility. This includes such items as doors, tubs, sinks, etc.

Get several bids on all expensive items or whenever you feel you could do better. The old saying "Only one thing has one price: A postage stamp" is alive and well in the building material business. Shop around and chances are, you'll find a better price. Check with the local builders in your area and ask them where they buy their materials.

Determine what is returnable. In the normal course of construction, you will more than likely order too many of certain items accidentally or on purpose to avoid delays due to shortages. Find out what overages can be returned to suppliers. Ask for written return policies. Certain cardboard cartons, seals, wrapping or steel bands may need to be intact for an item to be returnable. There may be minimum returnable quantities for such items as bricks in full skids (1,000 units). Restocking charges may also be incurred when returning materials.

Manage your purchases with purchase orders. Explain to your suppliers up front that you plan to control purchases with "PO's" and that you will not pay any invoice without a purchase order number on it. Purchase orders help to ensure that you receive what you ordered and no more. Make sure to use two-part forms. This makes it impossible for someone to create or alter a form. Pre-numbered purchase orders are available at most local office supply stores.

Minimize inventory and schedule your material deliveries with the project schedule as best you can. This is easier said than done. Minimizing the materials on site minimizes your investment and exposure to theft or damage. Most material suppliers will be glad to store purchased items for you and deliver them when ready for use. Fragile materials such as drywall and windows are good examples of items not to have lying around too long. Try to be on site when material is delivered. Write special delivery instructions on purchase orders. You're the customer; you make the rules.

Have materials dropped close to where they will be used in order to save time. Bricks, loads of sand and gravel are good examples. Place bricks around the house to reduce handling. Have delivery men move plywood, drywall and other heavy items to the second floor if that is where they will be used. Make sure to write your delivery instructions and directions to the site on the purchase order.

Keep materials protected from weather by covering them with plastic or keeping them inside whenever possible. The garage is a great place to keep millwork, doors, windows and other valuable items that shouldn't get wet. Lumber, especially plywood, is susceptible to water damage. Lumber should be covered with heavy-duty plastic and should be ordered shortly before it will be used. Bricks and mortar should be covered carefully with plastic. Wet mortar is ruined and is difficult to throw away. Many suppliers will allow you to purchase materials and keep them in their warehouses until you need them. Rocks, brick or scrap wood should be used to hold the plastic down on windy days.

Protect materials from theft. You should not have material suppliers deliver until shortly before the material is needed. Make sure to lock up all movable items in a storeroom or basement. You will want to install the locks on your house as soon as possible to protect interior items from theft. Instead of giving subs keys, unlock in the morning and lock in the evening.

Get discounts from early payments. Many suppliers give a percentage discount if full payment is made by the end of the month, the beginning of the month or within 30 days of purchase. Make sure you don't lose these discounts. Take advantage of them because they can really add up. Deal with lumber yards and other suppliers who provide payment discounts.

Use float as much as possible. Many suppliers will invoice you at a certain period each month and will only charge you a finance charge thirty days later. If you pick up materials at the beginning of the month and they invoice you for them at the end of the month, you have had thirty days' free use of their money.

Order early for prompt delivery. Orders are normally queued up for next day delivery. The earlier in the day you get your order in, the earlier your shipment should arrive the next day. Ask each supplier for a reliable delivery time.

Give plans to your lumber supplier. In many areas, you can arrange for your supplier salesman to drop by the site every other day or so to keep your framing crew properly stocked.

DEALING WITH LABOR AND SUBS

Labor generally refers to employees who are working for you and have employment tax and withholding taken from their pay. If you use employees, get a good bookkeeping service to figure withholding taxes for you. The subcontractors you use are in business for themselves and withhold their own taxes. Your subs may consist of workers who supply labor only, such as framers and masons, or businesses that provide materials and labor, such as heating and air conditioning contractors. Use contractors whenever possible to avoid the paperwork headaches that go along with payroll accounting. Using contract labor will also save you money because you are not paying their withholding taxes. The IRS may require you to file a "1099" form on your subs. This form shows the amount you paid your subs so that the IRS can be sure that they are paying their share of taxes. Check with a Certified Public Accountant to obtain the proper forms.

Locating Workers

Finding good subs and labor is an art. The following is a surefire list of sources for locating subs and who have the best combination of skill, honesty and price:

- Job Sites — (an excellent source because you can see their work.) As your building site shows progress, it will act as a billboard, attracting all sorts of subs looking for work.
- References from Builders and Other Subcontractors — Some builders may be hesitant to share their valued subs with a stranger, but if you have connections or can convince them to cooperate, you can usually get a premium list of reliable help.
- The Yellow Pages — Look under specific titles. Do this only when you are in a pinch because these folks always seem to charge a bit more.
- Classified Ads — Either in newspapers or online.
- On the Road — Look for phone numbers on trucks. Busy subs are normally good subs.
- Material Supply Houses keep lists of approved subs. Many of them keep a bulletin board where subs leave their business cards.

References are essential in the home-building business. Nobody should earn a good reputation without showing good work and satisfied customers. Insist on at least three references; one of them being their last job. If their last reference is four months old, it may be their last good reference. Some feel that companies which use a person's name as the title, such as "A. J. Smith's Roofing" will tend to be a "safe" choice over "XYZ Roofing" under the premise that a man's name follows his reputation. But this obviously cannot be taken at face value. Contact references and inspect the work. You may even wish to ask the customer what he paid for the job, although this may sometimes be awkward. A good local standing is essential.

Background Information
Acquiring reference information should be a rule. It is not a bad idea to have your "candidate" subcontractors fill out a brief form in order for you to find out a little about them. By asking the right questions, you can get a fairly good idea of what type of character they are.

Accessibility
It is important to be able to reach your subs. Try not to hire Avisit to fix something, it may not be worth their time. Your subs must have a cell phone where they can be reached on the job and a home phone for after-hours contact.

Paying Subs
When you have your checkbook out to pay your subs, remember Lester's Law, "The quality of a man's work is directly proportional to the size of his investment." If you want your sub to do the best job possible, give him or her ample motivation. Do not pay a subcontractor until you are completely satisfied with the completed job and materials. Once paid, a subcontractor loses incentive to perform for you. Even the most honorable and hard working subcontractor still makes choices and sets priorities, and those priorities usually focus on accounts receivable. Subcontractors make a living by completing a lot of jobs, not by taking too much time on any one job. Below is a suggested schedule for payment:
- 45% after rough-in.
- 45% after finish work complete and inspected.
- 10% retainage held for about two weeks after finish work. (This is to protect yourself if a problem is detected later.)

Let them know your pay schedule in writing before you seal the deal.

Paying Cash
Paying with cash gets results. Telling a sub that "you will pay him in cash upon satisfactory completion of the job" should be ample incentive to get the job done. Make sure when obtaining cash for work that you keep documentation of the work done for the IRS. Write a check in the sub's name and have it cosigned. Then cash the check. This way you will have an "audit trail" to prove that you paid the sub for work done. As an alternative audit trail, have the sub sign the purchase order "paid in full" or "paid minus retainage."

Be Fair and Prompt
Be fair with money. Pay what you promise, when you promise. Pay your subs and labor as work is completed. Not everyone can go a week or two without money. However, never pay subs for more than the work that has already been done. Give them a reason to come back and finish the job.

Retainage
Construction jobs are seldom "complete" until the certificate of occupancy is obtained. Most jobs will have a few loose ends that must be finished at the last minute. When a subcontractor's job is finished, you should inspect the job and create a punch list of final details. At that point, you can make the second "finish" payment. However, many items on the punch list can't be completed until finishing is done, trash cleaned up and all utilities turned on. That is why it's customary to "retain" 5-10% of a sub's total payment until the punch list has been completed. This is called retainage in the trade and is the essential guarantee of performance. Avoid the temptation to assume the sub will return to the job site days or weeks after the bulk of the job is complete. Retainage keeps the "pot sweet."

Specifications
Don't count on getting anything you don't ask for in writing. Remember, not written, not said. A list of detailed specifications reduces confusion between you and your sub and will reduce call backs and extra costs. Different subs have different ways of doing things. Make sure they do it your way.

Paperwork
Remember that many subs live from week to week and despise paperwork. You will have difficulty getting many subs to present bids or sign affidavits. Many are stubborn and believe that "their word is their bond." You must use discretion in requiring these items. Generally your more skilled trades such as HVAC, Electrical and Plumbing will be more business-like and will cooperate more fully with your accounting procedures. These trades are the most important ones for which to obtain affidavits because they purchase goods from other suppliers that become part of your house. Make sure that all your subs provide you with an invoice for work done.

Licenses
Depending upon who you have work for you, you may want them to furnish a business license. If required, make sure that plumbers, electricians, HVAC subs and other major subs are licensed to work in your county; if not, have them pay to obtain the license or get another sub.

Arrangements for Materials

Material arrangements must be agreed upon in a signed, written specification. If subs are to pay for all materials or some of the materials, this must all be spelled out in detail. If a sub is supposed to supply materials, make sure he isn't billing the materials to your account, expecting you to pay for them later. (Refer to Lien Waivers in the Legal section.) If you are to supply the materials, make sure they are at the site ahead of time and that there is sufficient material to do the job. This will be appreciated by the subs and will generate additional cooperation in the future.

Ask your sub about any special materials he needs. For instance, many framers use nail guns, which require special nails. Make sure to take this into consideration when ordering materials.

Keep Your Options Open

Don't count on one excavator, framing crew, plumber or anyone for that matter. Have at least one or two back-ups who can fill in if the main subcontractor fails to show. It is best to be up front with your primary subcontractors. Let them know that you expect them to show up on time and complete the job per specifications or you will hire an alternate.

Be Flexible With Time

Because of variables in weather, subs and their work, expect some variation in work schedule. Chances are, they are doing four or five other jobs at the same time. Expect some problems in getting the right guy at exactly the right time.

Equipment

Your work specifications should specify that subs are to provide their own tools including extension cords, ladders, scaffolding, power tools, saw horses, etc. Before hiring a sub, make sure he has all the tools. Don't plan on furnishing them. Defective tools you supply can increase your liability for injuries incurred when used by others. If you don't have sawhorses available for your framing crew and if they didn't bring any, count on them spending their first ten minutes cutting up your best lumber to make a few.

Worker's Compensation

Make sure your sub's worker's compensation policy is up to date before the job is started and get a copy for your records. If the sub doesn't have worker's compensation make it clear to him that you intend to withhold a portion of his payment to cover the expense. This can be a sore spot later if you forget to arrange this in advance.

Worker's compensation should be deducted from each check when paying a sub more than once. Typically, you must set a deduction percentage that matches the rate for the particular sub. Check with your insurance agent to determine the proper percentage.

Poor Workmanship

This can be one of your worst headaches. A sub you've contracted is halfway through a job and you notice poor workmanship. The work will have to be completely redone, so you fire the sub. He may file a lien on the project, halting or hindering further progress. Make sure you inspect often and well; and know your subs.

MAKING CHANGES

Changes are second nature. Most people can't even order lunch without changing their minds. We all see things and perceive how they could be made better. That's why a good building information model and careful thought are so important. The idea here is to make changes on paper, not lumber. The cost of ripping out walls, adding windows or "that breakfast counter" can be staggering. In many cases, the cost of changes after construction begins becomes economically impractical. Be good to yourself. Plan ahead and make modifications and additions while you're still using an eraser.

Let's look at three examples. Suppose you have decided to have a bay window in the kitchen as opposed to the original idea of a flat window. For the sake of simplicity, let's assume the two windows are of the same dimensions.

Change During Planning

You think the change over and decide that the $110 material differential is well worth the money. You take out your eraser and update your drawings, take-off and contract specifications. Not bad.

Change Prior To Installation

The flat window has already been delivered to the site but is uninstalled. Now you will have to negotiate with the subcontractor who already signed the contract to install a flat window. The sub may charge you extra for the bay window installation, but that's part of the game. You hope the material supplier will exchange the flat window for a new bay window. You'll have to pay the $110 differential and any restocking fee which was not in your original budget. This is an out-of-pocket expense that will impact your construction draw. The change will cost you money, but at least you aren't undoing work.

Change After Installation

The flat window has been installed. Now, you decide to change to a bay window. This could be expensive for several reasons:

- You may damage the flat window removing it, so it may not be returnable.

STANDARD BIDDING PROCESS

Most of the steps below will apply to the effort of reaching final written agreements with the subs you will be using.
These steps will be referred to in each of the individual trade sections.

BD1	REVIEW your project scrapbook and locate potential subcontractors.
BD2	FINALIZE all design and material requirements affecting subcontractor.
BD3	PREPARE standard specifications for job.
BD4	PREPARE subcontractor packages (Subcontractor Information Sheet, Standard Contract Terms and Specification Sheet).
BD5	CONTACT subcontractor and discuss plans. Mention any forms to be completed and specify deadline (should be written on forms). Get a good understanding of who will provide what.
BD6	RECEIVE and evaluate completed bids.
BD7	SELECT the best three bids, based on price and your personal assessment of the sub.
BD8	COMPARE bids against budget.
BD9	NEGOTIATE with prospective sub. Ask him what his best "CASH" price is. When you can't get the price any lower, get more service out of the same cash by asking him to do other items in lieu of a cash discount.
BD10	SELECT sub. Make it clear that he has been selected and that he needs to commit a specific period of time for the job. Explain that you are fair with money but expect good, timely work according to a written contract.
BD11	CONTACT subs and tell them when you expect them at the site. Discuss upcoming schedule and keep in touch with them or you will lose them. If the building schedule changes, let the sub know as soon as possible so that he can make other plans. This courtesy will be returned to you when you need the sub at the last minute.

- You have to purchase the bay window, say $700, out-of-pocket. You or the bank didn't count on this.
- You will have to pay to have the flat window removed and the bay window installed. Pray that you didn't apply paint or wallpaper yet.

In addition, you have to prepare the change order form. Note that in this case you are at the mercy of the subcontractor. He knows you want the bay window and may charge you heavily for it. If his price is too high, you have three alternatives:

- Pay the man and try to forget this ever happened.
- Do it yourself. This can be dangerous if you don't know what you're doing. If the sub damages the bay window during installation, it's his problem. If you damage it, that bay window may cost you another $200-$300.
- Live with the flat window. (This is frustrating. So long as you live in that home, that flat window will be a constant reminder of a careless mistake.)

Advice

Consider last minute changes carefully. Is it really important enough to warrant changes in budget and scheduling? Delays will ripple through the rest of the project. If you must make changes, act as early as possible.

Common Pitfalls When Making Changes

- When changing ceiling heights, pay close attention to stairs. Added ceiling height changes the number of steps used in a staircase. Make sure you have room for the extended length of the staircase.

- When walls are moved, play close attention to the effect on structural loads. Load bearing walls cannot be moved easily without providing some other means of supporting the wall. Even small changes in walls can make a big difference. Most walls are located directly above and below each other. If you move a wall one or two feet, the floor between them may sag over time.
- Any changes to walls in the house can affect the roof framing. Check with a home designer or an architect when moving any wall.
- When adding shutters to windows, make sure that the window spacing will allow it without changing the balance or symmetry of the house.
- Examine the door swing of each door in the house. Moving a door may cause the door to block entrances or other doors when opened.
- Adding brick to the outside of a house designed for siding will change the spacing and size of the foundation and roof overhang.

SAFETY AND INSPECTIONS

Inspections and Inspectors

During the course of your project, you will probably run into two types of inspectors:

- County and local officials
- Financial institutions

In general, inspectors are on your side. They are there to help ensure that critical work meets minimum safety and structural standards. Hence, an approved inspection does not assure you that a job has been done to your specifications, only to the specifications of the inspector. If certain work fails inspection, you will be required to correct it, re-schedule the inspection and probably pay a re-inspection fee.

County and Local Officials

A certain number of inspections are normally required by state or county officials during the construction process. When you apply for your building permit, ask for a schedule of required inspections. Some common inspection points are listed below in order of occurrence:

- Before pouring concrete footings
- Before pouring concrete foundation
- After framing rough-in
- After electrical, HVAC and plumbing rough-in
- After laying sewer or septic tank lines and before filling in
- Final completion

Financial Institutions

If you borrowed money from a financial institution for your project, they have an active interest in seeing that the money is being spent properly, as intended, and at a rate such that the money is used up just as the project is being completed. In addition to requiring approval of state and local inspections, bank inspectors will normally want to examine the site each time a loan draw is requested. As your financial institution sees certain phases of work completed, they will release a certain amount of loan money based on the percentage of completion of the project. Hence, they will make sure that you have "earned" the right to use their money as intended. Each lender sets up their own "draw schedule" allowing a certain percentage for each construction task.

Pre-inspection

Nobody enjoys unpleasant surprises. Before you have anyone show up for an inspection, make sure that you are ready to pass the inspection. Check the work yourself. Use checklists and your subs. This will insure that the formal inspection is a mere formality. This eliminates rejected inspections, embarrassment, re-inspection fees and a bad builder reputation. Inspectors will see you in a favorable light if you are always fully prepared for their inspections. Your major subs — plumbers, HVAC and electricians — normally schedule their own inspections, but you should follow up.

Guidelines for Working with Inspectors

- Make sure the work to be inspected is ready for inspection.
- Determine how much advance notice is required and always give proper advance notice.
- Notify the inspector to cancel or reschedule the inspection if work is not ready.

- Be present for the inspection so that you can determine exactly what is incorrect and what must be done to correct it if necessary. If you don't proceed in this manner, you may have to wait for a form letter in the mail and then follow up.
- Stay on good terms with all your inspectors. Be friendly and courteous. Don't argue when they ask for something to be fixed. They can give you good advice. Cooperate.
- Have the site cleaned beforehand. This will put your inspector in a good frame of mind and will let him know that you are organized and efficient.

Project Safety

Few disasters can affect your project more adversely than the death or severe injury of you, one of your subs, a neighbor's child or a building partner on site. As they say, "project safety is no accident." Hence, project safety should play a major part in your overall project planning and routine supervision. The best way to make sure that serious accidents don't happen is to make safety a big issue with your subs. Remain aware of safety and emphasize this to your workers and site visitors. Follow these rules and you should be okay:

1. Use common sense. If something seems a little risky, back off and give yourself more time. Most accidents happen when someone is attempting to perform a short-cut to save time.
2. Wear proper clothing at the site: Sturdy leather shoes with sturdy soles. Preferably steel-tipped.
3. Watch out for kids. Construction sites are magnets for children. "Caution: Construction" signs should be posted at the site.
4. You should have builder's risk insurance covering your building site and accidents that may occur there.

SITE SAFETY CHECKLIST

- ☐ Site safety rules communicated to all subs.
- ☐ First aid kit available? Familiar with basic first aid practices.
- ☐ Phone numbers for police, ambulance and firehouse available.
- ☐ Temporary electrical service grounded.
- ☐ All electrical tools grounded.
- ☐ All electrical cords kept away from water.
- ☐ Warning and danger signs posted in appropriate areas.
- ☐ Hard hats and steel tipped shoes worn where needed.
- ☐ Other protective gear available and used. Goggles, gloves, etc. Always use protective goggles when flying fragments are possible.
- ☐ Set a good example as a safety-minded individual.
- ☐ Adequate slope on edges of all ditches and trenches over 5' deep.
- ☐ Open holes and ditches fenced properly.
- ☐ Open holes in sub-floor properly covered or protected.
- ☐ Scaffolding used and secure where needed.
- ☐ Workers on roof with proper equipment.
- ☐ Excess and/or flammable scrap not left lying around.
- ☐ No nails sticking up out of boards or other materials.
- ☐ Gas cans have spark arrest screens.
- ☐ Welding tanks shut off tightly when not in use. Stored upright.
- ☐ Area where welders working checked for smoldering or slow burning wood. Houses have burned down because of this!
- ☐ Proper clearance from all power lines.

ESTIMATING COSTS

The cost estimate or "take-off" will be one of the most time-consuming and important phases of your construction project. Jumping into your first construction project without a realistic estimate of your costs could spell financial disaster. Spend a lot of time working on this one before continuing with your project.

The first "materials list" you encounter may be included with your construction plans. These lists are usually offered as an extra item at an additional price. These lists are often inaccurate or based on outdated construction techniques. There is no way of telling whether the list is accurate or not. The best use of this type of take-off is as a checklist while running your own figures.

One important source of information will come from the subcontractors you plan to use. These subs are familiar with the type of construction in your area and will be very knowledgeable in calculating the amounts of materials needed. In fact, most of your subs will offer to calculate the amount of materials needed for you. This can be very valuable, especially for framing, because of the sheer number of different materials used. Don't rely solely on your sub's estimate. Make sure to run your own figures to compare against their estimate. Subs have a tendency to overestimate just to be on the safe side. They hate to run out of material and have to sit around while you run to the supplier for reinforcements. Just make sure that you do add in at least 5-10% waste factor to each estimate.

Another good source of estimating information is your local materials supplier. In previous years, suppliers used to do quite a bit of estimating for their customers. If you can find a supplier who still does this, he is worth his weight in gold.

When figuring your take-off, use the enclosed materials sheet as a checklist to make sure that you don't forget any necessary items. As you figure each item, enter it onto your purchase order forms, taking care to separate materials ordered from different suppliers onto different invoices.

Excavation

Excavation is usually charged by the hour of operating time. Larger loaders rent for higher fees. Determine whether clearing of the lot is necessary and whether refuse must be hauled away. Below are approximate times for loader work. Remember, these are only approximations. Every lot is different and may pose unique problems. Your grader may supply chain saw labor to cut up large trees.

Larger loaders can be an advantage and can be cheaper in the long run, especially if the lot is heavily wooded or has steep topography. If the lot is flat and sparsely wooded, use the smaller loader for the best price.

One Half Acre Lot:	Small ldr.	Large ldr.
Clearing trees & shrubs	2-5 hrs.	1-4 hrs.
Digging trash pit	2-4 hrs.	1-3 hrs.
Digging foundation	2-3 hrs.	1-3 hrs.
Grading building site	1-5 hrs.	1-3 hrs.
Total	7-17 hrs.	4-13 hrs.

Concrete

When calculating concrete, it is best to create a formula or conversion factor that will simplify calculations and avoid having to calculate everything in cubic inches and cubic yards. For instance, when pouring a 4" slab, a cubic yard of concrete will cover 81 square feet of area. This is calculated as follows:

1 Cubic Yard = 27 Cubic Feet = 46,656 Cubic Inches
1 Sq. Ft. of Concrete 4" thick = 576 Cubic Inches
  (12"×12"×4")
46,656 Cu. In. ÷ 576 Cu. In. = 81 Sq. Ft. of Coverage

By doing this equation only once, you now have a simple formula for calculating slabs 4" thick that you can use for driveways, basement floors and slab floors. For instance, to determine the concrete needed for a 1,200 sq. ft. slab simply divide by 81.

1200 Sq. Ft. ÷ 81 = 14.8 Cu. Yds. of concrete
NOTE: Always make sure to add a 5 to 10% waste factor to all calculations.

Use this same principle of creating simple formulas for footings, calculating blocks, pouring concrete walls, etc. Just remember to calculate your own set of formulas based on the building codes in your area. For your convenience, refer to the concrete tables to determine the conversion factors for your project.

CU. YDS. CONCRETE NEEDED PER LINEAL FOOT OF FOOTING							
Depth Inches	Width of footing in inches						
	4"	6"	8"	10"	12"	14"	16"
4"	0.0004	0.006	0.008	0.010	0.012	0.014	0.016
6"	0.006	0.009	0.012	0.015	0.019	0.022	0.025
8"	0.008	0.012	0.016	0.021	0.025	0.029	0.033
10"	0.010	0.015	0.021	0.026	0.031	0.036	0.041
12"	0.012	0.019	0.025	0.031	0.037	0.043	0.049
14"	0.014	0.022	0.029	0.036	0.043	0.050	0.058
16"	0.016	0.025	0.033	0.041	0.049	0.058	0.066

Footings

Footing contractors will be hired to dig the footings and to supervise pouring of the footings. Footing subs generally charge for labor only and will charge by the lineal foot of footing poured. Pier holes will be extra. You must provide the concrete. These subs may charge more if they provide the forms. If you use a full service foundation company, they will charge you for a turn key job for footing, wall and concrete. This makes estimating easy.

Footings generally must be twice as wide as the wall they support and the height of the footing will be the same as the thickness of the wall. The footing contractor will know the code requirements for your area. Ask him for the dimensions of the footing and then figure an amount of concrete per lineal foot of footing. (See the section on calculating concrete.) Here is a typical calculation for supporting an 8" block wall:

Footing dimension: 8" high × 16" wide
8" × 16" × 12"(1 ft. of footing) = 1536 Cu. In.
46,656 Cu. In. ÷ 1536 Cu. In. = 30.38 lineal ft. of footing
   per Cu. Yd.

With this size footing, figure one cubic yard of concrete for every 31 lineal feet of footing. Make sure to include 4 extra feet of footing for every pier hole. Add 10% for waste. Refer to the footing table for concrete factors.

Concrete Floors or Slabs

Use the conversion factor of 81 sq. ft. per cubic yard of concrete for 4" slabs or basement floors. If your slabs must be more or less than 4" make sure to calculate a new conversion factor or refer to the table.

SQUARE FOOTAGE OF SLAB THAT 1 CU. YD. OF CONCRETE WILL FILL			
Slab	S.F.	Slab	S.F.
1"	324	7"	46
1½"	216	7½"	43
2"	162	8"	41
2½"	130	8½"	38
3"	108	9"	36
3½"	93	9½"	34
4"	81	10"	32
4½"	72	10½"	31
5"	65	11"	29
5½"	59	11½"	28
4"	54	7"	27

Monolithic Slabs

Break a one piece slab into two components — the slab and the footing sections; then figure the items separately using the tables.

Block Foundations and Crawl Spaces

Concrete blocks come in many shapes and sizes; the most common being 8"×8"×16". Blocks that are 12" thick are used for tall block walls with backfill in order to provide extra stability. Blocks 8"-deep and 12"-deep cover the same wall area: .888 sq. ft.

To calculate the amount of block needed, measure the height of the wall in inches and divide by 8" to find out the height of the wall in numbers of blocks. The height of the wall will always be in even numbers of blocks plus a 4" cap block or 8" half block (for pouring slabs). Cap blocks are solid concrete 4"×8"×16" blocks used to provide a smooth surface to build on. If you are figuring a basement wall, take the perimeter of the foundation and multiply by .75 (3 blocks for every 4'); then multiply this figure by the number of rows of block. To calculate the row of cap block multiply the perimeter of the foundation by .75. Always figure 5-10% waste when ordering block.

If your foundation is a crawl space, your footing is likely to have one or more step downs, or bulkheads, as they are called. Step downs are areas where the footing is dropped or raised the height of one block. This allows the footing to follow the contour of the land. As a result, sections of the block wall will vary in height; requiring more or fewer rows of block. Take each step down section separately and figure 3 blocks for every 4' of wall. Multiply the total number of blocks by the total number of rows and then add all sections together for the total number of blocks needed.

Poured Concrete Walls

Your poured wall subcontractor will charge by the lineal foot for setting forms. Usually, the price quoted for pouring will include the cost of concrete; but if not, you must calculate the amount of concrete needed. Determine the thickness of the wall from your poured wall sub and find the square foot conversion factor for the number of square feet coverage per cubic yard. Divide this factor into the total square footage of the wall. Example for an 8" thick wall:

8" Poured Concrete Wall × 12" × 12" = 1152 cubic inches
   per square foot of wall area
46656 cubic inches ÷ 1152 cubic inches = 40.5 square
   foot per cubic yard of concrete
Refer to the concrete wall table to quickly calculate the
   concrete needed.

Brick

To figure the amount of brick needed, figure the square footage of the walls to be bricked and multiply by 6.75. (There are approximately 675 bricks per 100 square feet of wall.) Add 5-10% for waste.

CU. YDS. CONCRETE NEEDED PER LINEAL FOOT OF WALL						
Height of wall	Width of wall in inches					
	4"	6"	8"	10"	12"	14"	16"
4"	0.049	0.074	0.099	0.123	0.148	0.173	0.198
5"	0.062	0.093	0.123	0.154	0.185	0.216	0.247
6"	0.074	0.111	0.148	0.185	0.222	0.259	0.296
7"	0.086	0.130	0.173	0.216	0.259	0.302	0.346
8"	0.099	0.148	0.198	0.247	0.296	0.346	0.395
9"	0.111	0.167	0.222	0.278	0.333	0.389	0.444
10"	0.123	0.185	0.247	0.309	0.370	0.432	0.494
11"	0.136	0.204	0.272	0.340	0.407	0.475	0.543
12"	0.148	0.222	0.296	0.370	0.444	0.519	0.593
13"	0.160	0.241	0.321	0.401	0.481	0.562	0.642
14"	0.173	0.259	0.346	0.432	0.519	0.605	0.691
15"	0.185	0.278	0.370	0.463	0.556	0.648	0.741
16"	0.198	0.396	0.395	0.494	0.593	0.691	0.790
17"	0.210	0.315	0.420	0.525	0.630	0.7535	0.840
18"	0.222	0.333	0.444	0.556	0.667	0.778	0.889

Mortar

Mortar comes premixed in bags of masonry cement, which consist of roughly one part portland cement and one part lime. Each bag requires about 20 shovels of sand when mixing. To calculate the amount of mortar needed for brick, figure one bag of cement for every 125 bricks. For block, figure one bag of cement for every 28 blocks. Make sure to use a good grade of washed sand for a good bonding mortar. Most foundations will require at least 10 cubic yards of sand.

Framing

Estimating your framing lumber requirements will be the most difficult estimating task and will require studying the layout and construction techniques of your house carefully. Be sure to have a scaled blueprint of your home as well as a scaled ruler for measuring. Consult with a framing contractor before you begin for advice on size and grade of lumber used in your area.

This is the time you will want to pull out your books on construction techniques and study them to familiarize yourself with the components of your particular house. Local building code manuals will include span tables for determining the maximum spans you are allowed without support. Your blueprints will also list the sizes of many framing members and may include construction detail drawings, which can be especially helpful in determining the size and type of lumber needed for framing.

Floor Framing

Determine the size and length of floor joists by noting the position of piers or beams and by consulting with your framing contractor. Floor joists are usually spaced 16" on center. Joists spaced 12" on center are used for extra sturdy floors. If floor trusses are used, figure 24" on center spacing. Calculate the perimeter of the foundation walls to determine the amount of sill and box sill framing needed.

Calculate the square footage of the floor and divide by 32 (square footage in a 4×8' sheet of plywood) to determine the amount of flooring plywood needed. If the APA Sturdifloor design is used, order ¾" or ⅝" Tongue & Groove Exterior Plywood.

If you are using a basement or crawl space, you will probably be using a steel or wood beam to support the floor members. If floor trusses are used, this item may not be needed since floor trusses can span much greater distances. Consult with your local truss manufacturer.

Wall Framing

When calculating wall framing lumber, add together the lineal feet of all interior and exterior walls. Since there is a plate at the bottom of the wall, and a double plate at the tip, multiply wall length by three to get the lineal feet of wall plate needed. Add 10% for waste. For precut wall studs (make sure to get the proper length if precut studs are used, there are many sizes), allow one stud for every lineal foot of wall and 2 studs for every corner. Count all door and window openings as solid wall. This will allow enough for waste and bracing.

Headers are placed over all openings in load bearing walls for structural support and are doubled. Add together the total width of all doors and windows and multiply by two. Check with your framing sub for the proper size header.

The second floor can be calculated the same way as the first floor, taking into consideration the different wall layouts, of course. The second floor of a two story house can be figured the same as a crawl space floor with the exception that the joists will be resting on load bearing walls instead of a beam. This will require drawing a ceiling joist layout.

Roof Framing

If roof trusses are to be used, figure one truss for every 2' of building length, plus one truss. If the roof is a hip roof or has two roof lines that meet at right angles, extra framing for bridging must be added.

The roof truss manufacturer should calculate the actual size and quantity of roof trusses at the site or from blueprints to insure the proper fit. Get him to do this and present a bid and material list during the estimating process.

"Stick building" a roof is much harder to calculate and requires some knowledge of geometry to figure all the lumber needed. A "stick built" roof is one that is built completely from scratch. The simplest way of finding the length of ceiling joists and rafters is to measure them from your scaled blueprints. Always round to the nearest greater even length, since all lumber is sold in even numbered lengths.

If you can't measure the length of the rafter from your plans, you can calculate rafter length and roof area using a conversion factor similar to the one used for concrete. Think of a roof as two identical triangles back-to-back. If you know the length of the triangle base, you can calculate the length of the long side of the triangle (the rafter) by using trigonometry. We have spared you this headache by providing a conversion factor for each roof slope (rise/run). See the accompanying table for this multiplication factor. A 10" slope roof means that the roof rises ten inches for every foot of run (distance). This term is sometimes mistakenly referred to as "10 pitch slope."

To use this table, measure the total width of the house (including roof overhang) from the peak of the roof to the cornice edge. Multiply this by the conversion factor to obtain the rafter length. Don't forget to add in waste for lumber cut from the ends of each rafter (see illustration). This same conversion factor can also be used to calculate roof area for shingles and tar paper. Calculate the total area of the ceiling plus overhang and multiply by the conversion factor to obtain the total roof area. By the way, a 10" slope roof is very steep.

PITCH (SLOPE)	CONVERSION FACTOR	PITCH (SLOPE)	CONVERSION FACTOR
1"	1.01	9"	1.26
2"	1.02	10"	1.31
3"	1.04	11"	1.36
4"	1.06	12"	1.42
5"	1.09	13"	1.48
6"	1.12	14"	1.54
7"	1.16	15"	1.61
8"	1.21	16"	1.67

Conversion factors for roofs

When calculating ceiling joists, draw a joist layout with opposing joists always meeting and overlapping above a load bearing wall. Figure one ceiling joist for every 16" plus 10% extra for waste.

Rafters are also spaced 16" on center. The length of the rafters can be determined by measuring from blueprints, making sure to allow for cornice overhang. Where two roof lines meet, a valley or hip rafter is necessary. Multiply the length of a normal rafter by 1.5 to get the approximate length of this rafter.

Gable studs will be necessary to frame in the gable ends. The length of the stud should be equal to the height of the roof ridge. Figure one stud per foot of gable width plus 10% waste. This will be enough to do two gable ends since scrap pieces can be used in the short areas of the gable.

Decking for the roof usually consists of ½" CDX exterior plywood. Multiply the length of the rafter times the length of the roof to determine the square footage of one side of the roof. Double this figure to obtain total square footage

area of the roof. Divide this figure by the square footage of a sheet of plywood (32). Add 10% waste. This is the number of sheets of plywood needed. Remember the square footage of the roof for figuring shingles.

Roofing Shingles

Roofing shingles are sold in "squares" or the number of shingles necessary to cover 100 sq. ft. of roof. First, find the square footage of the roof, adding 1½ square feet for every lineal foot of eaves, ridge, hip and valley. Divide the total square footage by 100 to find the number of squares. Shingles come packaged in one third square packages, so multiply the number of squares by three to arrive at the total number of packages needed.

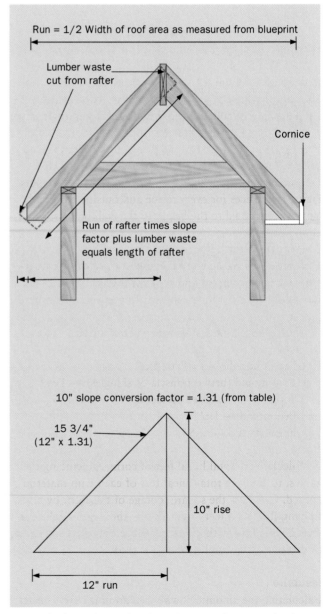

Run = 1/2 Width of roof area as measured from blueprint

Lumber waste cut from rafter

Cornice

Run of rafter times slope factor plus lumber waste equals length of rafter

10" slope conversion factor = 1.31 (from table)

15 3/4" (12" x 1.31)

10" rise

12" run

Length of rafter and area of roof calculated by using conversion factor for a 10" slope roof.

Roofing felt is applied under the roof shingles as an underlayment and comes in 500 sq. ft. rolls. Divide the total square footage of the roof by 500 and add 20% for overlap and waste to determine the number of rolls needed.

Flashing is required around any chimney or area where two roof lines of different height meet and comes in 50 ft. rolls. Measure the length of the ridge if installing roof ridge vents.

Siding and Sheathing

Multiply the perimeter of the outside walls by the height to obtain the total square footage of outside walls. If gables are to be covered with siding, multiply the width of the gable by the height (this figure is sufficient for both gable ends) and add this to the square footage of the outside walls. This figure is the total square footage of area to be sided.

Different types of siding are sold in different manners. Sheathing and plywood siding are sold in 4×8 sheets (32 sq. ft.). Divide this figure into the total square footage to determine the number of sheets needed. Add 10% for waste. Lap siding, on the other hand, is sold by squares or 1,000's. Make sure to ask if siding is sold by actual square footage or by "coverage area" (the amount of area actually covered by the siding when applied). Add 10% for waste or 15% if lap siding is applied diagonally.

Corner trim boards are generally cut from 1×2 lumber. Figure two pieces for every inside and outside corner; the length being equal to the height of the wall being sided.

Cornice Material

To estimate cornice material, you must first determine the type or style of cornice and the trim materials to be used. Consult blueprints for any construction detail drawings of the cornice.

The following materials are generally used in most cornices:

- Fascia boards — 1×6" or 1×8"
- Drip mould (between fascia & shingles) — 1×4"
- Soffit — ⅜" exterior plywood
- Bed mould — 1×2"
- Frieze mould — 1×8"

Calculate the total lineal feet of cornice, including the gables, to find the total lineal feet of each trim material needed. Calculate the square footage of the soffit by multiplying the lineal feet of cornice by the depth of the cornice. Divide this by the square footage of a plywood sheet to determine the number of sheets needed.

Insulation

Calculating the amount of wall insulation is easy — batts of insulation are sold by square footage coverage. Simply multiply the perimeter of the exterior walls to be insulated by the wall height to determine the total square footage to be insulated.

To calculate the amount of blown-in insulation needed, first determine the type of insulation to be used. Each type — mineral wool, fiberglass, foam, and cellulose has its own R-factor per inch. This must be known in order to determine the thickness of fill. Then multiply the depth in feet (or fraction thereof) by the square footage of the ceiling area to arrive at the cubic foot volume. Blown-in insulation is sold by the cubic foot. If batt insulation is used in the ceiling, it can be figured in the same manner as the wall insulation. Most insulation contractors will quote a price including materials and labor when installing insulation. Ask the contractor to itemize the amount of insulation used, for comparison with your figures.

Drywall

Gypsum drywall is sold by the sheet (4×10, 4×12, etc.) but is estimated by square footage. To find the total amount needed for walls, multiply the total lineal feet of inside and outside walls by the wall height. Make sure to count each interior wall twice, since both sides of the wall will be covered. Count all openings as solid wall and add 10% for waste. Some subs will charge by the square foot for material and labor — then add extras for tray ceilings and other special work.

For the ceiling, simply take the finished square footage of the house. Add 10% for waste and then add this to the wall amount for the total amount of Drywall needed. If the house has any vaulted or tray ceilings, extra wallboard and labor must by figured in.

Joint finishing compound comes pre mixed in 5 gallon cans and joint tape in 250 ft. rolls. For every 1,000 sq. ft. of drywall, figure one roll of joint tape and 30 gallons of joint compound.

Trimwork

Base moulding comes in many styles, with clamshell and colonial the most common, and will be run along the bottom of the wall in every room. Therefore, the lineal foot figure of walls used to calculate wallboard is equal to the lineal feet of baseboard trim needed. Add 10% for waste.

Make sure to measure the perimeter of any room that requires shoe moulding or crown moulding. Shoe mould (quarter round) is usually used in any room with vinyl or wood flooring to cover the crack between the floor and wall. Crown moulding around the ceiling may be used in any room but is most common in formal areas such as foyers and living and dining areas. Add 10% for waste.

When ordering interior doors, you may want to purchase prehung doors with pre-assembled jambs. These doors are exceptionally easy to install and already have the jamb and trimwork attached. Study the blueprints carefully to be sure that the doors ordered open in the right direction and don't block light switches. (There are right-hand and left-hand doors.) To determine which door to order, imagine standing in front of the door, walking in. If you must use your right hand to open the door, it is a right-hand door. Most blueprints will have the size and type of each door marked in the opening.

Flooring

Because of the complexity of laying flooring materials — carpet, hardwood floors, vinyl or ceramic tile — it is essential to get a flooring contractor to estimate the flooring quantities. An approximate figure would be equal to the square footage of the area covered plus 10% for waste. Carpet and vinyl are sold by the square yard, so divide by 9 to determine square yardage. Keep in mind, however, that carpet and vinyl are sold in 12' wide rolls. This dimension may affect the amount of waste in the estimate.

Miscellaneous Coverings

This includes ceramic tile for bathrooms or floors, parquet, slate for foyers or fireplaces and fieldstone for steps or fireplaces. Virtually all these materials are sold by the amount of square footage area covered. Subcontractors who install these items also charge by the square foot; sometimes with material included and sometimes without. Grouts and mortars used in installing these items (with the exception of rock) are premixed and indicate the coverage expected on the container.

Paint

When hiring a painting contractor, the cost of paint should be included in the cost of the estimate unless you specify otherwise. If you should decide to do the painting yourself, you will need to figure the amount of paint needed. The coverage of different paints varies considerably, but most paints state the area covered on the container. Simply calculate the square footage of the area to be covered and add at least 15% for waste and touchup. Also figure the amount of primer needed. If colors are custom mixed, make sure that you have sufficient paint to finish the job. It is sometimes difficult to get an exact match of a custom color if mixed again at a later date.

Cabinets

Most blueprints include cabinet layouts with the plans. Use this layout if you are purchasing your own cabinets. If the cabinets are installed by a subcontractor, he will take care of the exact measurements. Be sure that painting, staining and installation are included in the price.

Wallpaper

One roll of wallpaper will safely cover 30 square feet of wall area including waste and matching of patterns. The longer the repeat pattern the more waste. Some European wallpapers vary in coverage. Be sure to check coverage area with the wallpaper supplier. When figuring the square footage of a room to wallpaper, count all openings as solid wall. Also be careful to note if you are buying a single or double roll of wallpaper. Most wallpaper is sold in a roll that actually contains 2 true rolls of wallpaper, or 60 sq. ft. of coverage. Always buy enough paper to finish the job. Buying additional rolls at a later time can cause matching problems if some of the rolls are from a different dye lot. Look for the same lot numbers (runs) on all rolls purchased to assure a perfect match. Use only vinyl wallpaper in baths and kitchen areas for water resistance and cleanup.

Millwork and Miscellaneous

The ordering of windows, lights, hardware and other specialty items will not be covered here as they are fairly straightforward to calculate. Other estimates such as heating and air, plumbing, electrical and electrical fixtures must be obtained from the contractors themselves since they provide the materials and labor.

Subcontractors

The following section outlines the way most subcontractors charge for their services. Some subs may or may not include the cost of materials in their estimates. Make sure you know your particular sub's policy.

BIDS BEAT ESTIMATES EVERY TIME

Now that you know how detailed an estimate can be, look at the other side of the coin. You can spend a day calculating how many cubic yards of concrete you will need and how many approximate hours of labor it will take; or you can call up a full-service foundation company and nail down a bid. The moral here is to use bids whenever possible in completing your take-off. The bottom line is what counts; if your estimate comes out to be $600 to tile a floor and your lowest bid is $750, you've obviously missed the boat — and wasted a little time.

SUBCONTRACTORS

FRAMING	Frame house, apply sheathing, set windows and exterior doors. Charges by the square foot of framed structure (includes any unheated space such as garage). Extras include bay windows, chimney chase, stairs, dormers, anything else unusual.
SIDING	Apply exterior siding. Charges by the square of applied siding. Extras: diagonal siding, decks, porches and very high walls requiring scaffolding.
CORNICE	Usually the siding subcontractor. Applies soffit and fascia board. Charges by the lineal foot of cornice. Extras: Fancy cornice work, dentil mould. Sometimes sets windows and exterior doors.
TRIM	Install all interior trim and closet fixtures and set interior doors. Charges a set fee by the opening or by lineal feet of trim. Openings include doors and windows. Extras: stairs, rails, crown mould, mantels, book cases, chair rail, wainscoting and picture moulding.
FOOTINGS	Dig footings, pour and level concrete, build bulkheads (for step downs). Charges by the lineal foot of footings. Extras: Pier holes.
BLOCK	Lay block. Charges by the block. Extras: Stucco block.
BRICKWORK	Lay brick. Charges by the skid (1,000 bricks).
STONEWORK	Lay stone. Charges by the square foot or by bid.
CONCRETE FINISHING	Pour concrete, set forms, spread gravel and finish concrete. Charges by the square foot of area poured. Extras: Monolithic slab, digging footing.
ROOFING	Install shingles and waterproof around vents. Charges a set fee per square plus slope of roof. Ex.- $1 per square over slope on a 6/12 slope roof = $7/square. Extras: Some flashing, ridge vents, and special cutout for skylights.
WALLPAPER	Hang wallpaper. Builder provides wallpaper. Charges by the roll. Extras: High ceilings, wallpaper on ceilings, grass cloth.
GRADING	Rough grading and clearing. Charges by the hour of bulldozer time. Extras: Chain saw operator, hauling away of refuse, travel time to and from site (drag time).
POURED FOUNDATION	Dig and pour footings, set forms, and pour walls. Charges by the lineal foot of wall. Extras for bulkheads, more than four corners, openings for windows, doors and pipes.
PEST CONTROL	Chemically treat the ground around foundation for termite protection. Charges a flat fee.
PLUMBING	Install all sewer lines, water lines, drains, tubs, fixtures and water appliances. Charges per fixture installed or by bid. (For instance, a toilet, sink and tub would be three fixtures.) Installs medium grade fixtures. Extras: Any special decorator fixtures.
HVAC	Install furnace, air conditioner, all ductwork and gas lines. Charges by the tonnage of A/C or on bid price. Extras: Vent fans in bath, roof fans, attic fans, dryer vents, high efficiency furnaces and compressors.
ELECTRICAL	Install all switches and receptacles; hook up A/C compressor. Will install light fixtures. Charges by the receptacle. Extras: Connecting dishwasher, disposal, flood lights, door bells.
CABINETRY	Build or install pre-fab cabinets and vanities and apply Formica tops. Charges by the lineal foot for base cabinets, wall cabinets and vanities. Standard price includes Formica counter tops. Extras: Tile or marble tops, curved tops, pull-out shelves, lazy Susans.
INSULATION	Install all fiberglass batts in walls, ceilings, floors. Charges by square foot for batts, by the cubic foot for blown-in.
DRYWALL	Hang drywall, tape and finish, stipple. Charges by the square foot of Drywall. Materials are extra. Extras: Smooth ceilings, curved walls, tray and vaulted ceilings and open foyers.
SEPTIC TANK	Install septic tank. Charges fixed fee plus extra for field lines.
LANDSCAPING	Level with tractor, put down seeds, fertilizer and straw. Charges fixed fee. Extras: Trees, transplanted shrubs, pine straw, bark chips.
GUTTERS	Install gutters and downspouts. Charges by lineal foot plus extra for fittings. Extras: Half round gutters, collectors, special water channeling and gutters that cannot be installed from the roof.
GARAGE DOOR	Install garage doors. Fixed fee. Extras: Garage door openers.
FIREPLACE	Supply and install prefab fireplace and flue liner. Extras include gas log lighter, fresh air vent and ash dump.
PAINTING AND STAIN	Paint and stain interior and exterior. Charges by square foot of finished house. Extras: High ceilings, stained ceilings, painted ceilings.
CERAMIC TILE	Install all ceramic tile. Charges by the square foot. Extras include fancy bathtub surrounds and tile counter tops.
HARDWOOD FLOOR	Install and finish real hardwood floors. Charges by square foot. Extras include beveled plank, random plank and herringbone.
FLOORING	Install all carpet, vinyl, linoleum and pre-finished flooring. Charges by the square yard. Extras include contrast borders and thicker underlayment.

STEP	DESCRIPTION	QTY.	MATERIAL		LABOR		SUBCONTR.		TOTAL
			UNIT PRICE	TOTAL MATL	UNIT PRICE	TOTAL MATL	UNIT PRICE	TOTAL MATL	
1	LAYOUT								
2	STAKES								
3	RIBBON								
4	ROUGH GRADE								
5	PERK TEST								
6	WELL AND PUMP								
7	Well								
8	Pump								
9	BATTER BOARDS								
10	2 × 4								
11	1 × 8								
12	Cord								
13	FOOTINGS								
14	Re-Rods								
15	2 × 8 Bulkheads								
16	Concrete								
17	Footings								
18	Piers								
19	FOUNDATION								
20	Block								
21	8"								
22	12"								
23	4"								
24	Caps								
25	Headers								
26	Solid 8's								
27	Half 8's								
28	Mortar Mix								
29	Sand								
30	Foundation Vents								
31	Lintels								
32	Portland Cement								
33									
34	WATERPROOFING								
35	Asphalt Coat								
36	Portland Cement								
37	6 Mil Poly								
38	Drain Pipe								
39	Gravel								
40	Stucco								
41	FOUNDATION								
42	(Poured)								
43	Portland Cement								
44	Wales								
45	TERMITE TREAT								
46	SLABS								
47	Gravel								
48	6 Mil Poly								
49	Re-Wire								

Form 1: Cost Estimate Summary

| STEP | DESCRIPTION | QTY. | MATERIAL | | LABOR | | SUBCONTR. | | TOTAL |
			UNIT PRICE	TOTAL MATL	UNIT PRICE	TOTAL MATL	UNIT PRICE	TOTAL MATL	
50	Re-Rods								
51	Corroform								
52	Form Boards								
53	2 × 8								
54	1 × 4								
55	Concrete								
56	Basement								
57	Garage								
58	Porch								
59	Patio								
60	AIR CONDITIONING								
61	House								
62	FRAMING SUB-FLOOR								
63	I-Beam								
64	Steel Post								
65	Girders								
66	Scab								
67	Floor Joist								
68	Bridging								
69	Glue								
70	Plywood Sub Floor								
71	⅝" T&G								
72	¾" T&G								
73	½" Exterior								
74	Joist Hangers								
75	Lag Bolts								
76	Sub-Floor								
77	FRAMING—WALLS								
78	Treated Plate								
79	Plate								
80	8' Studs								
81	10' Studs								
82	2 × 10 Headers								
83	Interior Beams								
84	Bracing								
85	1 × 4								
86	½" Plywood								
87	Sheathing								
88	4 × 8								
89	4 × 9								
90	Laminated Beam								
91	Flitch Plate								
92	Bolts								
93	Dead Wood								
94	FRAMING—CEILING/ROOF								
95	Ceiling Joist								
96	Rafters								
97	Barge Rafters								
98	Beams								

Form 1: Cost Estimate Summary

STEP	DESCRIPTION	QTY.	MATERIAL		LABOR		SUBCONTR.		TOTAL
			UNIT PRICE	TOTAL MATL	UNIT PRICE	TOTAL MATL	UNIT PRICE	TOTAL MATL	
99	Ridge Beam								
100	Wind Beam								
101	Roof Bracing Matl.								
102	Ceiling Bracing								
103	Gable Studs								
104	Storm Anchors								
105	Decking								
106	⅜" CDX Plywood								
107	½" CDX Plywood								
108	2 × 6 T&G								
109	Ply. clips								
110	Rigid Insulation								
111	Felt								
112	15 #								
113	40 #								
114	60 #								
115	FRAMING—MISC.								
116	Stair Stringers								
117	Firing-in								
118	Chase Material								
119	Purlin								
120	Nails								
121	16d CC								
122	8d CC								
123	8" Glv.Roof.								
124	1½" Glv. Rf.								
125	Concrete								
126	Cut								
127	ROOFING								
128	Shingles								
129	Ridge Vent								
130	End Plugs								
131	Connectors								
132	Flashing								
133	Roof to Wall								
134	Roll								
135	Window								
136	Ventilators								
137	TRIM—EXTERIOR								
138	Doors								
139	Main Entrance								
140	Secondary								
141	Sliding Glass								
142	Doors								
143	Windows								
144	Fixed Window								
145	Glass								
146	Window Frame								
147	Material								

Form 1: Cost Estimate Summary

| STEP | DESCRIPTION | QTY. | MATERIAL | | LABOR | | SUBCONTR. | | TOTAL |
			UNIT PRICE	TOTAL MATL	UNIT PRICE	TOTAL MATL	UNIT PRICE	TOTAL MATL	
148	Corner Trim								
149	Window Trim								
150	Door Trim								
151	Flashing								
152	Roof to Wall								
153	Metal Drip Cap								
154	Roll								
155	Beams								
156	Columns								
157	Rail								
158	Ornamental Iron								
159	NAILS								
160	16d Galv. Casing								
161	8d Galv. Casing								
162	4d Galv. Box								
163	LOUVERS								
164	Triangular								
165	Rectangular								
166	SIDING								
167	CORNICE								
168	Lookout Mat'l								
169	Fascia Mat'l								
170	Rake Mold								
171	Dentil Mold								
172	Cont. Eave Vent								
173	Screen								
174	⅜" Plywood								
175	DECKS & PORCHES								
176	MASONRY—BRICK								
177	Face								
178	Fire								
179	MORTAR								
180	Light								
181	Dark								
182	Sand								
183	Portland Cement								
184	Lintels								
185	Flue								
186	Flue								
187	Ash Dump								
188	Clean out								
189	Log Lighter								
190	Wall Ties								
191	Decorative Brick								
192	Lime Putty								
193	Muriatic Acid								
194	Sealer								
195	Under-hearth								
196	Block								

Form 1: Cost Estimate Summary

STEP	DESCRIPTION	QTY.	MATERIAL		LABOR		SUBCONTR.		TOTAL
			UNIT PRICE	TOTAL MATL	UNIT PRICE	TOTAL MATL	UNIT PRICE	TOTAL MATL	
197	Air Vent								
198									
199	PLUMBING								
200	Fixtures								
201	Misc.								
202	Water Line								
203	Sewer								
204	HVAC								
205	Dryer Vent								
206	Hood Vent								
007	Exhaust vent								
208	ELECTRICAL								
209	Lights								
210	Receptacles								
211	Switches-Single								
212	Switches—3 Way								
213	Switches—4 Way								
214	Range—One Unit								
215	Range—Surface								
216	Oven								
217	Dishwasher								
218	Trash Compactor								
219	Dryer								
220	Washer								
221	Freezer								
222	Furnace—Gas								
223	Furnace—Elec.								
224	A/C								
225	Heat Pump								
226	Attic Fan								
227	Bath Fan								
228	Smoke Detector								
229	Water Htr/Elec.								
230	Flood Lights								
231	Door Bell								
232	Vent-a-Hood								
233	Switched Receptacle								
234	150 Amp Service								
235	200 Amp Service								
236	Disposal								
237	Refrigerator								
238	Permit								
239	Ground Fault Inter.								
240	Fixtures								
241	FIREPLACE								
242	Log Lighter								
243	INSULATION								
244	Walls								
245	Ceiling								

Form 1: Cost Estimate Summary

STEP	DESCRIPTION	QTY.	MATERIAL		LABOR		SUBCONTR.		TOTAL
			UNIT PRICE	TOTAL MATL	UNIT PRICE	TOTAL MATL	UNIT PRICE	TOTAL MATL	
246	Floor								
247	WALLBOARD								
248	Finish								
249	Stipple								
250	STONE								
251	Interior								
252	Exterior								
253	TRIM—INTERIOR								
254	Doors								
255	Bifold								
256	Base								
257	Casing								
258	Shoe								
259	Window Stool								
260	Window Stop								
261	Window Mull								
262	1×5 Jamb Matl.								
263	Crown Mold								
264	Bed Mold								
265	Picture Mold								
266	Paneling								
267	OS Corner Mold								
268	IS Corner Mold								
269	Cap Mold								
270	False Beam Matl.								
271	STAIRS								
272	Cap								
273	Rail								
274	Baluster								
275	Post								
276	Riser								
277	Tread								
278	Skirtboard								
279	Bracing								
280	LOCKS								
281	Exterior								
282	Passage								
283	Closet								
284	Bedroom								
285	Bath								
286	Dummy								
287	Pocket Door								
288	Bolt Single								
289	Bolt Double								
290	Flush Bolts								
291	Sash Locks								
292	TRIM—MISC.								
293	Door Bumpers								
294	Wedge Shingles								

Form 1: Cost Estimate Summary

STEP	DESCRIPTION	QTY.	MATERIAL		LABOR		SUBCONTR.		TOTAL
			UNIT PRICE	TOTAL MATL	UNIT PRICE	TOTAL MATL	UNIT PRICE	TOTAL MATL	
295	Attic Stairs								
296	Scuttle Hole								
297	1×12 Shelving								
298	Shelf & Rod Br.								
299	Rod Socket Sets								
300	FINISH NAILS								
301	3D								
302	4D								
303	6D								
304	8D								
305	Closet Rods								
306	Sash Handles								
307	SEPTIC TANK								
308	Dry Well								
309	BACKFILL								
310	DRIVES & WALKS								
311	Form Boards								
312	Rewire								
313	Rerods								
314	Expansion Joints								
315	Concrete								
316	Asphalt								
317	Gravel								
318	LANDSCAPE								
319	PAINT & STAIN								
320	House								
321	Garage								
322	Cabinets								
323	WALLPAPER								
324	TILE								
325	Wall								
326	Floor								
327	Misc.								
328	GARAGE DOORS								
329	GUTTERS								
330	Gutters								
331	Downspouts								
332	Elbows								
333	Splash Blocks								
334	CABINETS								
335	Wall								
336	Base								
337	Vanities								
338	Laundry								
339	GLASS								
340	SHOWER DOORS								
341	MIRRORS								
342	ACCESSORIES								
343	Tissue Holder								

Form 1: Cost Estimate Summary

STEP	DESCRIPTION	QTY.	MATERIAL		LABOR		SUBCONTR.		TOTAL
			UNIT PRICE	TOTAL MATL	UNIT PRICE	TOTAL MATL	UNIT PRICE	TOTAL MATL	
344	Towel Racks								
345	Soap Dish								
346	Toothbrush Holder								
347	Medicine Cabinet								
348	Shower Rods								
349	FLOOR COVERING—BATH								
350	Vinyl								
351	Carpet								
352	Hardwood								
353	Slate or Stone								
354	MISCELLANEOUS								
355	CLEAN-UP								
356	HANG ACCESSORIES								
357	FLOOR COVERING—OTHER								
358	Vinyl								
359	Carpet								
360	Hardwood								
361	Slate or stone								
362									
363									
364									
365	MISCELLANEOUS								

Form 1: Cost Estimate Summary

4 Construction

PRE-CONSTRUCTION

Introduction

The Construction Section covers each of the trades normally encountered in residential construction. They will be covered in the order in which they normally occur in the construction process. The stages covered are:

- Pre-Construction
- Site Location and Excavation
- Foundation
- Framing
- Roofing and Gutters
- Plumbing
- HVAC (Heating, Ventilation and Air Conditioning)
- Electrical
- Masonry and Stucco
- Siding and Cornice
- Insulation and Soundproofing
- Drywall
- Trim
- Painting and Wallcovering
- Cabinetry and Countertops
- Flooring, Tile and Glazing
- Landscaping
- Decks and Sunrooms

NOTE: *The multitude of combinations and options in a home are as infinite as the ways in which to get them done. The steps and checklists outlined on the following pages attempt to cover most of the possible steps needed to build most homes.*

For this reason, there are bound to be a number of steps in this manual relating to features you may not have planned. For example, you may not have a basement or require stucco work.

Skip steps that do not apply to your project. Likewise, there may be special efforts required in your project that are not covered in this manual. Also, there may be a number of steps you will undertake in special or unconventional circumstances that will not be covered in this book. In many instances, the exact order of steps is not critical; use your common sense.

Preparation

Before you run out and start building, there are a number of preparation steps that you should complete. These are somewhat frustrating because they show no visible signs of progress and are often time consuming. There are a lot of forms to fill out and chores to do here. But you will be better off to get them all out of the way. For most of the steps listed in this section, order of completion is not critical; just get them done.

Give yourself ample time to complete these steps. Rome wasn't built in a day. If you are a first-time builder, it is typical to spend at least as much time planning your project as you spend executing it. That is, if you plan to have your home completed six months after breaking ground, you should spend at least six months in planning and preparation prior to breaking ground.

Once you break ground, there's no turning back and no time to stop, so you'll be relieved that everything has been thought out ahead of time. This section will cover the items that should be completed prior to breaking ground.

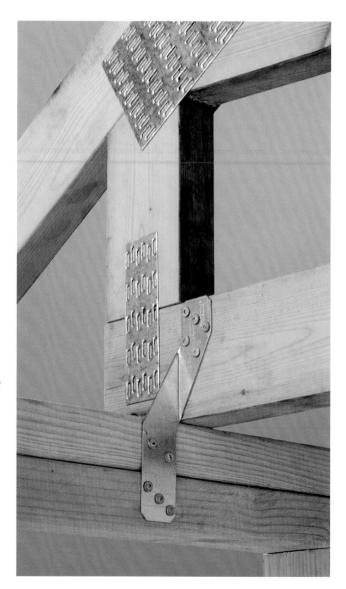

PRE-CONSTRUCTION STEPS

PC1   BEGIN construction project scrapbook. This is where you will begin gathering all sorts of information and ideas you may need later. Start with a large binder or a file cabinet. Break into sections such as:

- Kitchen ideas
- Bath ideas
- Floor plans
- Landscaping ideas
- Subcontractor names and addresses
- Material catalogs and prices

Your "scrapbook" will probably evolve into a small library, a wealth of information used to familiarize yourself with the task ahead, and serve as a constant and valuable source of reference for ideas to use in the future.

PC2   DETERMINE size of home. Calculate the size of home you can afford to build.

PC3   EVALUATE alternate house plans. Consider traffic patterns, your lifestyle and personal tastes.

PC4   SELECT house plan or have house plan drawn. If you use an architect, he will probably charge you a certain fee per square foot. Make sure you really like it before you go out and buy a dozen copies of the blueprints.

PC5   MAKE design changes. Unless you have an architect custom design your home, there will always be a few things you might want to change — kitchen and kitchen cabinets, master bath, a partition, a door, etc. This is the time to make changes.

PC6   PERFORM material and labor take-off. This will be a time-consuming, but important, step. Refer to the Estimating chapter for details on this step. Get bids to home in on true costs.

PC7   CUT costs. After you have determined estimated cost, you may need to pull the cost down. Refer to the Saving Money chapter for details on reducing costs without reducing quality.

PC8   UPDATE material and labor take-off. Based on changes you may have made, update and recalculate your estimated costs based on bids.

PC9   ARRANGE temporary housing. If you own a home and must sell it in order to build, or your current home is too far from your future construction site, consider renting an apartment or home close to your site. If possible, select a place between your site and your office (if you have one). This will greatly facilitate site visits. Since it will take approximately six months to build even the smallest house (unless you are using a pre-fabricated "kit"), you should get at least a six-month lease with a month-to-month option thereafter, or a one-year lease. For first-time, part-time builders, count on ten months to a year before move-in. Delays caused by weather, sub absenteeism, material delays and other problems always seem to add months to the effort. Consider that during the course of the project, you will be paying rent, construction interest and miscellaneous expenses, as well as additional travel costs. Plan extra money to handle this.

PC10   FORM a separate corporation if you want to set your "enterprise" off from your personal affairs. This may give you a better standing among members of the building trade if they feel that they're dealing with a company, instead of an individual. Be advised that your lender may not take kindly to this.

PC11   APPLY for and obtain a business license if required. This may be a legal requirement to conduct business as a professional builder in the county where your site is located. Check with your local authorities for advice.

PC12   ORDER business cards. These will be helpful in opening doors with banks, suppliers and subs. List yourself as president for extra clout.

PC13   OBTAIN free and clear title to land. This must be done before applying for a construction loan. Bankers want to make sure that you have something at stake in this project, too. It is strongly recommended that you only build on land that you own outright.

PC14   APPLY for construction loan. In proper terms, you should be applying for a "construction and permanent loan." This saves you one loan closing since the construction loan turns into a permanent loan when construction is completed.

PC15   ESTABLISH project checking account. This will be used as the single source of project funds to minimize reconciliation problems. Write all checks from this account and keep an accurate balance.

PC16   CLOSE construction loan. This allows you access to funds up to your loan amount. You will pay interest only on money drawn from your reserve. You will not be paying interest on money you have not yet used. Remember, though, that money is paid out as work is completed, based on the judgment of bank officials. Hence, you will need a certain amount of money to begin work (excavation and perhaps some concrete work). Do not begin any physical site work until this is done. Many financial institutions will not even loan you money if you have already started construction prior to contacting them.

PC17   OPEN builder accounts with material suppliers. This will involve filling out forms and waiting for mail replies notifying you of your account number. Read and be familiar with the payment and return policies of each supplier.

PC18   OBTAIN purchase order forms. Look for bound, pre-numbered ones with a carbon copy. You will use these forms for all purchases. Pay only those invoices that reference one of your purchase order numbers. (Make sure to advise suppliers that you are using a purchase order system.)

PC19	OBTAIN building permit. This involves filling out building permit applications with your county planning and zoning department. Build a friendly relationship with ALL county officials. You may need to count on them later.
PC20	OBTAIN builder's risk policy. Your construction site is a liability. You, laborers, subcontractors or neighborhood children could get hurt. Materials may get damaged or stolen. An inexpensive insurance policy is a must. Your financial institution and/or your county officials can point you in the right direction. You should also check with your home and/or auto insurance company for a quote.
PC21	OBTAIN worker's compensation policy. This is normally charged out as a percentage of the cost of your project. There may be a minimum size policy you can purchase.
PC22	ACQUIRE minimum tools. Certain items are a must.

- Assortment of coated, galvanized nails, framing and finish nails
- Builder's level (6" line level and 4' recommended)
- Can of WD-40 (everything is always getting rusty)
- Chisels (a few flat-bladed wood chisels and perhaps a concrete chisel)
- Circuit Tester
- Circular Saw
- Crowbar and/or nail claw (for tearing out framing mistakes)
- Fiberglass hard hat. Preferably yellow or orange. Scratch your name inside.
- Fifty or hundred foot steel tape measure
- First aid kit
- Garden hose
- Good all-purpose 14 oz. framing hammer. Hammers now feature fiberglass construction providing strength and rust-proofing.
- Good ankle height leather construction boots, preferably with steel-tipped toes.
- Heavy duty leather gloves.
- Heavy duty utility markers (for making site markings on framework and leaving notes)
- Large plastic trash can and metal trash can
- Large push broom or shop vacuum
- Orange spray paint (for marking things)
- Pick
- Pliers (rubber-coated handle recommended)
- Plumb Bob (for attaching to string to check for true vertical)
- Pocket knife
- Power Drill and a set of wood drill bits.
- Roll of heavy-duty cotton twine and blue chalk marker
- Set of screwdrivers, including flat-head and Philip styles. (Rubber handles recommended)
- Set of wrenches, crescent wrench or other adjustable wrench (rubber handles recommended)
- Shovels (flat — for clean up and pointed — for digging)
- Small sledge hammer (for pounding stakes and adjusting framework)
- Steel Square
- Strong flashlight
- UL-approved 50-100 foot grounded extension cord and work lamp
- Utility knife and extra blades

If you do not own a pickup truck and can't borrow one, try to buy an inexpensive trailer that can be attached to the back of your car. You can sell it later after the project is over. You can probably borrow a pickup truck from your subs from time to time.

PC23	VISIT numerous residential construction sites in various stages of completion. Observe the progress, storage of materials, drainage after heavy rain fall and other aspects of those sites.
PC24	DETERMINE need for an on-site storage shack. If you do a lot of the work, this could be helpful.
PC25	NOTIFY all subs/labor. Call up all subs and labor to tell them of the date you plan to break ground. This should be done approximately a week in advance. Confirm that they still plan to do the work and at the agreed-upon price. If you must get a backup sub, this is the time to do it — look on your sub list for "number two."
PC26	OBTAIN compliance bond, if necessary. This is primarily for your protection if you have a builder assist you during construction or build for you.

PC27	DETERMINE minimum requirements for Certificate of Occupancy. As soon as you meet these requirements, you can move into your home while you finish it, avoiding unnecessary rent. During final construction stages, concentrate on meeting these requirements. For example, you can do landscaping and certain other tasks later. Don't move in unless you have finished the rooms you plan to live in immediately.
PC28	CONDUCT spot survey of lot. Locate a good, reliable licensed surveyor. Have him furnish a signed survey report. This confirms that the planned project and excavation are within allowable boundaries. Ask your surveyor to help evaluate water drainage if you are uncertain.
PC29	MARK all lot boundaries and corners with 2' stakes and bright red tape.
PC30	POST construction signs. You should have at least one "DANGER" or "WARNING" sign. Have one sign for each street exposure if you have a corner lot. You may also want another sign which has your name, "company" and phone number on it. Your lender may have a sign also.
PC31	NAIL building permit to a tree (that will not be torn down) or a post in the front of the lot visible from the street. You may want to provide a little rain protection for it. You cannot do any work on site until this permit is posted.
PC32	ARRANGE for rental toilet. This is often viewed as a luxury among builders. They rent by the month.
PC33	ARRANGE for a site telephone. This guarantees the ability to contact anyone at the site, regardless of whether or not they have a cell phone. Have the phone company install it with a strong metal case with a lock. You will have to provide a sturdy 4×4 post or tree to hang it on. It should be in a place where you can hear the phone ring if you are inside. Specify a loud bell.

SITE LOCATION AND EXCAVATION

Digging In

Now it's time to really get mud on your new construction shoes. You are now entering that phase of construction where your decisions will be much harder to change. Take great care in locating and excavating your home site. A mistake here will be almost impossible to repair later. Site location consists of several phases:

- Locating the house on the lot.
- Excavating the area where the house is located.
- Marking the foundation lines with batter boards.

Site Location

Before locating the house on the lot, make sure to obtain a site plan from your surveyor. The site plan should indicate the dimensions of the lot, public and private easements and the setback requirements of your county's local building code. With this information in hand, you can insure that you place the house properly with no encroachment on another person's property or on public easements. You can also request that the surveyor place the house on the site plan and install corner stakes when surveying.

If you decide to locate the corner stakes, check the drainage of the property and look for any subsoil rocks or springs. After the house is located and the grading is properly done, the drainage of the lot should drain away from the house, but not onto any other property. If you change the natural drainage of the lot so that water now flows onto someone else's property, you can be held liable for any damage that results. You should know before excavation starts where the top of your foundation will be located. You may want to run a string at this height between the corner stakes to aid in excavating to the proper depth. Use a string level, leveling hoses or a builder's transit to ensure that the string is plumb and level.

Leveling hoses are simple but effective tools to find the absolute level of two distant objects. You can use simple garden hoses for this, but a much better version is available at most do-it-yourself stores for a very reasonable price (much cheaper than renting a builder's transit). Fill the hose with water and mount one end of the hose at the highest corner of the foundation. Raise or lower the other end of the hose (while adding more water) until both ends of the hose are completely full of water. Since water seeks its own level, the two hose ends will then be exactly level. Mark this elevation for string placement.

Once excavation commences, make sure that the depth of excavation will match your intended foundation height. You can measure from the string to ensure proper depth.

To lay out the corner stakes, start from an established reference line, such as the curb of the street. Most houses look best when located parallel to the street. However, you may violate this rule if you are concerned about other factors such as solar orientation. Just make sure to place the house well within the front, side and back setback limits. Establish the front line of the house first by measuring from your reference lines. Then locate the back corners of the house. Make sure that your stakes are reasonably square by measuring the diagonals of the square (see illustration). The length of each diagonal should be the same. To obtain a 90° angle use the old geometric principle of the "3:4:5 triangle." The 3:4:5 triangle has three sides in multiples of 3 ft., 4 ft. and 5 ft. and will form a 90° angle (see lower illustration page 112).

Breaking Ground

This is perhaps one of the most exciting moments during your homebuilding project. This phase actually involves several tasks. You may use your grader four times during the project:

- Clear site and excavate.
- Backfill and smooth out basement floor area.
- Cut the driveway, rough grade and haul away trash.
- Finish grade

The excavation subcontractor is one sub you will pay by the hour. This is usually tracked by the "hour" meter on the bulldozer. Don't let the machine idle while talking with the operator — shut it off. The rates can vary, so shop around. You may be able to get a better rate by hiring excavators to work on "off-hours," such as weekends.

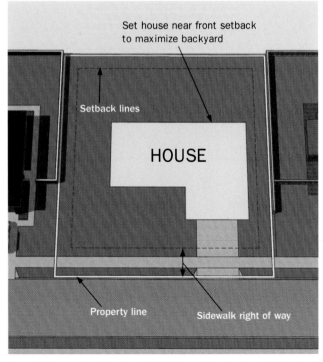

Property lines and setbacks will limit the options for the placement of the house on the lot.

Basement excavation can lead to tight quarters when laying out footings.

Avoid clearing or excavating when the ground is wet or muddy; it will take more time, be less accurate and create a mess. Know your lot. Do you have large, heavy trees which need heavy-duty clearing equipment? Or could you use lighter equipment which rents for less?

Remember: The more "horsepower" the higher the hourly rate, but the more it can accomplish in less time. But never use overpowered equipment for the job or you'll end up paying too much. Negotiate hours with the sub. Does he include travel time to and from the site (drag time)?

If you want to save money cutting the trees down yourself, go ahead. But cut them 4' from the ground; the bulldozer needs a good piece of the tree to pull the roots out of the ground. If you don't have a fireplace, consider selling the wood.

Batter Boards

Once the site is properly excavated, you need to lay out the batter boards. These boards provide a precise reference for the foundation subs to install foundation forms. If you have excavated for a full basement, you may be able to get your poured foundation sub to install forms without the need for batter boards. This is one of the advantages of using a poured concrete foundation.

If you are using a block wall foundation or a slab foundation, batter boards will be necessary for the subs. Since the subs will be laying out the foundation to the dimensions of the batter boards, you must make sure that the layout is completely plumb and square. A mistake here can cause a great deal of frustration and expense to correct later. Crooked slabs are almost impossible to correct without re-pouring. So use a pound of prevention here.

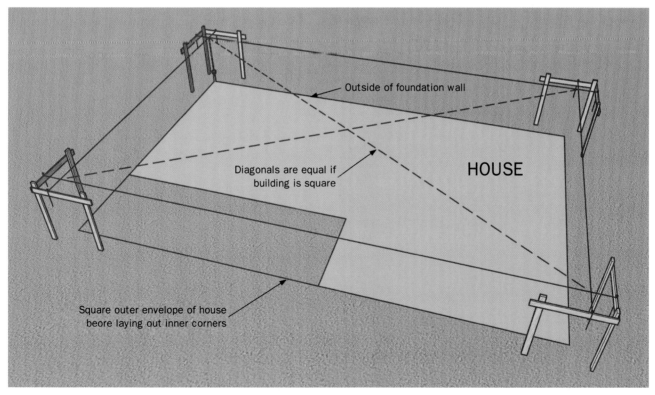

Outside of foundation wall

Diagonals are equal if building is square

HOUSE

Square outer envelope of house beore laying out inner corners

Staking and laying out batter boards

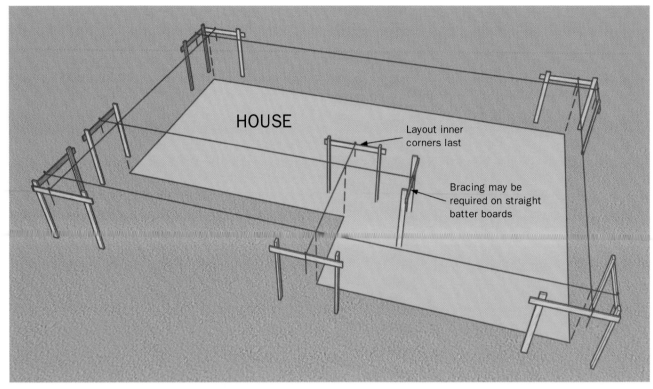

HOUSE

Layout inner corners last

Bracing may be required on straight batter boards

Layout inner boards only after the outer corners are square

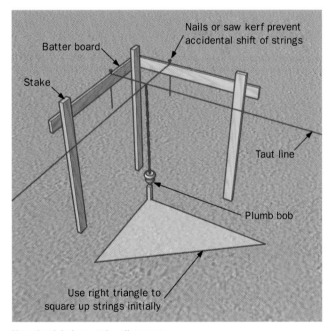

Nails or saw kerf prevent accidental shift of strings

Batter board

Stake

Taut line

Plumb bob

Use right triangle to square up strings initially

Use plumb bob to stake all corners

EXCAVATION — STEPS

EX1 CONDUCT excavation bidding process on all site preparation work

EX2 LOCATE underground utilities (gas main, water line, electrical or telephone cables) if they cross your property. You may want to call a locater service or your local utilities for help in locating utility lines. Mark their location with color-coded flags and mark them on a copy of the plat. Don't place flags down more than two or three days in advance. Kids love to pull them up. Your excavator will need to know where the utility lines run. Damaging the utilities can be time consuming and expensive.

EX3 DETERMINE where the dwelling is to be situated exactly. Your surveyor can help you with this. Make sure to allow both for zoning setbacks and for any utility or highway easements. A landscape architect can help you determine exactly where the home should face from a passive solar and appearance aspect. Mark all corners of the house with stakes so the grader will know where to clear. The grader should dig an extra 3' beyond the corner stakes of the house. For consistency you should have the same setback as the adjacent homes if there are any.

EX4 MARK the area to be cleared. Trees to be saved should be marked with red tape or ribbon. Allow clearance for necessary deliveries and several parked vehicles. Remind your excavator to be careful of knocking the bark off of trees to be saved. This could kill the tree or invite wood-hungry pests. Now is the time to clear an area for the driveway, patio and septic tank (if you plan to have one).

EX5 MARK where curb is to be cut (if it is to be cut). The standard curb cut is 14' wide. This allows for a slight flare on each side of the driveway area. In some cases, if you do not cut the curb, you will have a tough bump to deal with getting into your driveway. If it is to be cut later, you can remedy the problem by piling dirt or gravel up at the curb. You may need a variance if you plan to cut more than one curb area — as in the case of a circular driveway on a corner lot.

EX6 TAKE last picture of site area. This is the last chance you have to see it untouched.

EX7 CLEAR site area. Be present for this step. This is normally not something that is worthwhile for you to do because the grader can clear in an hour what would take you a week or more by hand. Cut down necessary trees and brush. Leave at least 4'-6' of the tree trunk. Tractors need the trunk to act as a lever when pulling the roots out of the ground. Pull stumps out of the ground and remove as necessary. Make sure to go over the plan with the grader. If your plan is reversed, make sure the grader is aware of this fact. At this point you may wish to mark the foundation area with powdered lime or corner stakes. Do not try to save time by cutting down trees yourself; the grader needs the weight of the tree to pull the stump out of the ground.

EX8 EXCAVATE trash pit if allowed. Trash pit location should be a minimum of 20' from the building or driveway. Make sure not to locate the pit near where you plan to add a swimming pool at a later date. A trash pit can save money, but a dumpster is preferred. There will be a dumpster rental fee and a dumping fee. Trash pits attract termites that might eventually be attracted to your home. The area may settle over time as materials decompose. The pit is also a hazard for young children and adults during construction. In most areas it is unlawful to burn or bury trash.

EX9 INSPECT cleared site. Check for proper building dimensions, scarred trees and thoroughness. Have trash, roots, stumps and other debris hauled away.

EX10 STAKE out foundation area. This is the process of locating all corners of the structure. This tells the excavator where to dig. It is customary to excavate two to three feet out in all directions from the house footprint. This will allow room for workers to set up poured foundation walls and to perform waterproofing.

EX11 EXCAVATE foundation/basement area. You must be present when this step is being performed. To execute this step properly, you should have a transit set up. Take depth measurements continuously. This will help you to keep the foundation grade level and at the proper depth. You'll need to shoot the basement depth as you get close to final grade. Cutting too low can cause your sewer line to be lower than the one at the street and can also cause water problems in your basement. Cutting too high can cause you to have a steeper driveway and extra steps at the front porch. Strike a nice balance. But overall, it's better to be 6" too high than 6" too low. This will help prevent water problems later. Have the excavator also clear the topsoil off the driveway and foundation area. This must be done before the crushed stone is put down or the stone will just sink into the loose topsoil. Have the topsoil stacked in its own pile so that you can redistribute it on the lawn area after construction. Your yard will be much healthier with an extra layer of topsoil. Make sure the operator digs down to firm soil and below the frost line, no matter how deep. A poor footing can ruin your house. If you are not sure of the firmness of the soil, use a wider footing to provide more stability.

EX12 INSPECT site and excavated area. Refer to attached inspection sheet and contract specifications.

BATTER BOARDS — STEPS

EX13	LAY OUT outer four corners of house with stakes and square up using a transit or the "3:4:5 triangle." Make sure tops of stakes are level with the proposed top of foundation. Check this with a builder's transit or leveling hoses. Re-measure distances and drive a small finishing nail or mark top of stake at exact corner of the building line. This will act as a reference for the batter boards.
EX14	DRIVE three 2x4 stakes of suitable length in a triangular pattern 4' outside the corner stake with the tops of the stakes 6"-12" higher than the corner stake. Make sure to sharpen the end of the stake before using.
EX15	NAIL 1x4 or 1x6 boards horizontally between the three stakes so that the tops are all level with the tops of the corner stakes. You may want to clamp the horizontal boards initially until final height adjustments are complete.
EX16	PULL a string between opposite batter boards and adjust until the string is exactly over the nails in the corner stakes. Attach string with a finishing nail. Repeat with each batter board until the four outside corners are marked properly.
EX17	RE-CHECK the distance of each string — measuring from the intersections of the strings. Measure the diagonal distance of the corners of the strings — if the distance is the same, the corners are square. Check the height and level of each string. Make minor adjustments. Once the strings are positioned accurately, mark the top of the batter board or saw a kerf in the board where the lines touch the board so that they can be replaced if broken or moved.
EX18	LAY OUT all inside corners and remaining batter boards using the existing batter board strings as a reference. Re-check the measurements, height and diagonals of the strings as you add batter boards.

DRAINAGE — STEPS

EX19	APPLY crusher run to driveway area. This will provide an excellent base for a paved drive and allows for easy site access for your subs and supply trucks when the ground is muddy.
EX20	SET UP a silt fence. Local ordinances will probably not allow mud run-off from your site to drain into the street or adjacent properties. A silt fence prevents mud from flowing into the street and neighboring yards. Two foot stakes will support the fence. Staple the covering to the stakes with a stapler. Rolls of plastic, mesh or burlap are available at material suppliers for this purpose, or you may choose to use bales of hay. These bales can be reused later when landscaping.
EX21 Backfill	PAY excavator. Have him sign a receipt. Pay only for work done up to this point. Give him an idea of when you will need him for the backfill work.
EX22	PLACE supports on interior side of foundation wall to support wall during backfill. Don't backfill until the house is framed in to provide additional support for the foundation walls.
EX23	INSPECT backfill area. Refer to related checklist in this section.
EX24	BACKFILL foundation. Remind excavator not to puncture waterproofing poly with large roots or branches in dirt.
EX25	CUT driveway area.
EX26	COVER septic tank and septic tank line.
EX27	COVER trash pit. Area should be compacted with heavy equipment and covered with at least two feet of dirt.
EX28	PERFORM final grade. Topsoil should be spread over top surface. If you have truckloads of topsoil hauled in, you may want to get a free soil pH test done by the county extension service. This could save you money in buying expensive fertilizers to compensate for inferior soil.
EX29	CONDUCT final inspection. Check all excavation and grading work.
EX30	PAY excavator (final). Have them sign an affidavit.

EXCAVATION — SPECIFICATIONS

- Excavator to perform all necessary excavation and grading as indicated below and on the attached drawing(s) and provide all necessary equipment to complete the job.
- Bid to include all equipment and equipment drag time for two trips.
- Bid to include necessary chain saw work.
- Trees to be saved will be marked with red ribbon. Trees to be saved are to remain unscarred and otherwise undamaged.
- Foundation to be dug to depth indicated at foundation corners. Foundation hole is to be 3' wider than foundation for ease of access. Foundation base not to be dug too deeply in order to pour footings and basement floor on stable base.
- Foundation floor to be dug smooth within 2" of level.
- Excavate trash pit (10' × 12') at specified location as depicted on attached drawing.
- Driveway area to be cleared of topsoil and cut to proper level. All topsoil to be piled where specified.
- All excess dirt to be hauled away by excavator.
- All stumps and trees to be dug up and hauled away. Hardwoods to be cut up to fit fireplace and piled on site where directed.
- Grade level shall be established by the builder, who will also furnish a survey of the lot showing the location of the dwelling and all underground utilities.
- Finish grade to slope away from home where possible.
- Backfill foundation walls after waterproofing, gravel and drain tile have been installed and inspected.
- Builder to be notified prior to clearing and excavation. Builder must be on site at start of operation.

EXCAVATION — INSPECTION

Checklist Prior to Digging and Clearing
- ☐ All underground utilities marked.
- ☐ All trees and natural areas properly marked off.
- ☐ Planned clearing area will allow access to site by cement and other large supply trucks. Room to park several vehicles.
- ☐ All survey stakes in ground.

Clearing
- ☐ Underground utilities left undisturbed.
- ☐ All area to be cleared is thoroughly cleared and other areas left as they were originally.
- ☐ All felled trees removed. All remaining trees are standing and have no scars from excavation. Firewood cut and stacked if requested.

Excavation
- ☐ Excavation includes necessary work space around:
  - Foundation area (extra 3' at perimeter)
  - Porch and stoop areas
  - Fireplace slab
  - Crawl space cut and graded with proper slope to insure dry crawl space.
- ☐ Trash pit dug according to plan if used.
- ☐ All excavations done to proper depth with bottoms relatively smooth. Should be within 1" of level.

Checklist Prior to Backfill
- ☐ Any needed repairs to foundation wall complete.
- ☐ Form ties broken off and tie holes waterproofed (poured wall only).

- ☐ All foundation waterproofing completed and correct. This includes parging, sprayed tar, poly, engineered drain mat, etc.
- ☐ Joint between footing and foundation sealed properly and watertight.
- ☐ Footing drain tile and gravel installed according to plan.
- ☐ All necessary run-off drain tile in place and marked with stakes to prevent burial during backfill.
- ☐ All garbage and scrap wood out of trench and fill area.
- ☐ Backfill supports in place and secure.
- ☐ Backfill dirt contains no sharp edges to puncture waterproofing.

Checklist After Backfill
- ☐ All backfill area completely filled in.
- ☐ Lot smoothed out to a rough grade.
- ☐ Necessary dirt hauled in and spread.

Checklist After Final Grade Completed
- ☐ Two to three percent slope away from dwelling.
- ☐ Berms in proper place, of proper height and form.
- ☐ Water meter elevation correct.
- ☐ Final grade smooth. Topsoil replaced on top if requested.
- ☐ No damage to curb, walkways, driveway, base of home, water meter, gutter downspouts, HVAC, splash blocks and/or trees.

FOUNDATION

The foundation will be one of the most critical stages of construction. You get only one chance to do it right or all the hard work on the rest of the house will be wasted. The concrete structures that you will use will vary somewhat depending upon the type of foundation you choose.

- Basement: footings, foundation wall, foundation floor
- Crawl space: footings, half height wall
- Slab: footings and slab floor, or monolithic slab

Foundation installation will involve individuals from more than one company, which can complicate the job. This phase of construction requires coordination between a small army of subcontractors and all of them must work with you over a relatively short period of time. The following list briefly describes the sequence of events:

1. Footing subs dig footings.
2. Exterminator treats soil in and around footings to prevent termites.
3. County inspector checks footings before pouring.
4. Footings are poured.
5. Foundation walls are laid or poured (basement construction).
6. Plumber and HVAC subs lay water, sewer and gas pipes before slab floor is poured (slab construction only).
7. Plumbing is inspected.
8. Slab is poured by concrete finishers.
9. Basement wall is waterproofed.
10. Basement wall drainage system is installed.

As you can see, foundation work can be an exercise in scheduling and diplomacy. Make sure to contact all involved parties well in advance of the pour and plan around rain delays. If possible, cover the foundation area with poly well in advance to prevent rain from ruining your well laid plans. This poly can be used during the pour as the vapor barrier under the slab, if left undamaged.

Pest Control

Below ground, all slab and footing areas must be permanently poisoned prior to laying any gravel, poly or concrete. Poison must be allowed to soak in after application without rain. Rain will reduce or nullify the effectiveness of the pest control treatment. Your pest control treatment will take one or two visits depending upon how thorough you want to be. There will usually be a second treatment applied around the base of the home after all work is done.

The Concrete Subcontractors

You will likely deal with several different parties when working with concrete for foundations, slabs and driveways:

- A full service foundation company
- A subcontractor to dig and pour footings

- Labor to lay concrete block walls
- Concrete finishers to finish concrete slab, patios, walkways and driveways
- Concrete supplier

Full Service Foundation Company

This subcontractor can be one of the best bargains you will encounter in your project. This sub replaces most of the other foundation subs listed above and may not cost a penny more. A full service foundation company can:

- Layout the entire foundation for you, including installing batter boards. Only the corners of the foundation need to be marked with stakes. The foundation sub will do the rest.
- Layout, dig and pour the footings.
- Pour all basement walls.
- Handle all scheduling of inspectors and concrete deliveries.
- Install waterproofing, gravel, and drain pipe.

Generally, poured foundations are slightly more expensive than other types; but this cost can be offset by savings in time, labor, hassle and interest paid on your construction loan. However, this sub is only cost effective to use if you plan a full basement.

Excavator

The excavating contractor will probably be the one to dig the trench for footings. If your project is on level ground, you may not have an excavator on site. In this case, footing subcontractors will prepare the footing trench.

Foundation forms put in place and braced before pour.

The Footings Sub

The footings sub will normally charge you by the lineal foot of footings required. He will provide either steel or wood forms, but may ask you to supply some of the form material. Form work can cost approximately one-third of the total footing job. In many areas of the country with stable soil, the footings subs can pour directly into the footings trench with no forms required.

Concrete Block Masons

If you do plan to have a basement made of concrete block you'll need this sub. Concrete block masons will likely be a different group than those laying your brick veneer (if you have any). They will charge for labor by the unit (block) laid.

Concrete Finishers

The concrete finisher may be the footing sub, but not necessarily. He will charge by the square footage finished. Costs will also vary in the type of finishes you want. Slick, rough, smooth, trowel finish or pea gravel are a few of the surfaces that this sub can produce. Don't use a crew that has less than three men — this is a tough job. Finishing must be done fast and even faster in hot weather. Concrete can set in half an hour. The finishers will finish the slab and then come back to do drives and walkways later.

Concrete Supplier

The concrete supplier will charge by the cubic yard of concrete used. A standard cement truck can hold 8-10 cubic yards. The supplier may charge you a flat rate just to make the trip, so try to minimize the number of truckloads.

Concrete Footings

Footings provide the base for other concrete structures. Their function is threefold:
• To provide a wider base of support in soft soils
• To provide a flat surface on which to lay block or pour concrete walls
• To provide an edge platform for slab floors

It is a solid, sub-grade rectangle of concrete around the perimeter of the dwelling. Its width and thickness will vary depending on the local building codes, soil and weight of materials supported. Footings are normally twice as wide as the wall they support. Additional footings are provided for masonry fireplaces and column supports. A trench for the footing will be excavated and usually, the footing poured directly into the trench. If a block foundation is used then forms are commonly placed in the trench so that the top of the footing can be leveled easily. This makes the block mason's job easier.

Since most older foundations were built with concrete block, traditional footings became the platform of choice.

However, recent advances in foundations have made the traditional footing questionable in some cases. The upside down "T" shape of a footing acts like a dam that can impede the ability of water to drain away from the bottom of foundations. On all but the softest soils, footings are unnecessary if a poured wall, wood, precast panel, or ICF foundation is used above. In these cases, a well compacted trench of crushed rock will provide just as much stability and allow for natural drainage channel for water.

FOUNDATION TYPES

There has been great advancement in foundation technology in recent years, allowing for several different foundation options.

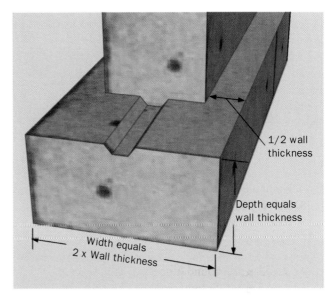

Foundation forms put in place and braced before pour.

New, integrated footing form/drainage channels eliminate the need to remove footing forms after the pour and replace traditional French drain pipe, saving costs and labor.

Concrete Block

This traditional method is normally a little less expensive than a poured wall but requires the services of several different subcontractors, which can slow down the project. Concrete blocks, due to their mortar joints and hollow interiors, are not as strong as other foundation choices and are more prone to moisture issues. It is critical that concrete block walls be waterproofed and coated properly. Concrete blocks are normally cast hollow units measuring 7⅝" wide, 7⅝" high and 15⅝" long. These are often referred to as 8"× 8"× 16" units because when laid with ⅜" mortar joints, they will occupy a space 16" long and 8" high. There are simply better choices for basement foundations. However, concrete block is still a cost effective choice for crawl spaces and raised foundations, where strength and moisture issues are less important.

Permanent Wood Foundation

As a low cost alternative to concrete block, permanent wood foundations can be used for crawl spaces or full basement foundations. They are usually installed by the framing subcontractor and require no footings, instead being placed directly on compacted crushed rock. As a result, they can be installed quickly, inexpensively, and at any time of the year. With only one contractor to worry about, the foundation installation goes quickly and smoothly. Permanent wood foundations are not as popular as they should be, due to the fears of homeowners. Some people can't get over the idea of using wood underground. It's a false fear, however. Years of research have shown that wood foundations are just as durable as any other, when installed properly. The wood is treated against rot and pests and when paired with the latest waterproofing technologies, stays dry and warm. Due to its rarity, it may be hard to find framers familiar with proper installation procedures.

Permanent wood foundations can either be traditional framed structures (although made with specially treated lumber) or as a SIP (structural insulated panel). If you are building the home with SIP panels, your supplier will probably be familiar with basement SIP's and may be able to complete the entire construction project from the ground up. Wood foundations are easy to construct and easy to insulate. Finishing an all-wood basement is no more difficult than working with any wood framed structure, making it a good do-it-yourself project. Foundation construction is simplified since there is no transition from below ground to above ground structures.

Slabs

Slab foundations are the least expensive foundations and a good choice for quality and savings. They require very little grading or excavating and are appropriate on level lots and areas with a high water table. If used in colder

Permanent wood SIP foundations are fast and easy to install in any weather.

climates, care must be taken to insulate the perimeter of the slab with foam sheet to prevent frost heave.

A "floating" slab consists of a traditional footing and a few courses of concrete block. Once the block has set, a concrete slab floor will be poured, using the blocks as the form. This method still requires coordination between excavators, block masons, and concrete suppliers. For this reason, monolithic slabs have become the foundation of choice on flat lots. In this case, the footing and floor is cast as one unified structure, greatly reducing time and labor. With either slab method, care must be taken to position plumbing accurately before pouring concrete. Any plumbing fixtures should be anchored securely to prevent shifting as the concrete is poured. Double-check their precise locations before the concrete has had time to set. The height of a slab should be at least eight inches above ground or higher to prevent the migration of termites up the side of the slab "wall." Pipes or other breeches in the slab should have termite guards installed to prevent termite migration up the side of the pipe. Any areas of the slab that will support load bearing columns should have crushed stone footings installed underneath, before the pour.

Raised Floor Foundations

Also known as a pier and beam foundation, the raised floor foundation is probably the lowest cost foundation available. This foundation was extremely popular in past decades and is still popular in low lying areas where basements and slabs are impractical. These areas include locations prone to flooding and hurricanes. However, a raised floor foundation is appropriate in any climate, if properly insulated. It's also a good strategy if your lot contains some area in the 100 year flood plain. Normally, you are not allowed to build structures in this area. By raising the house

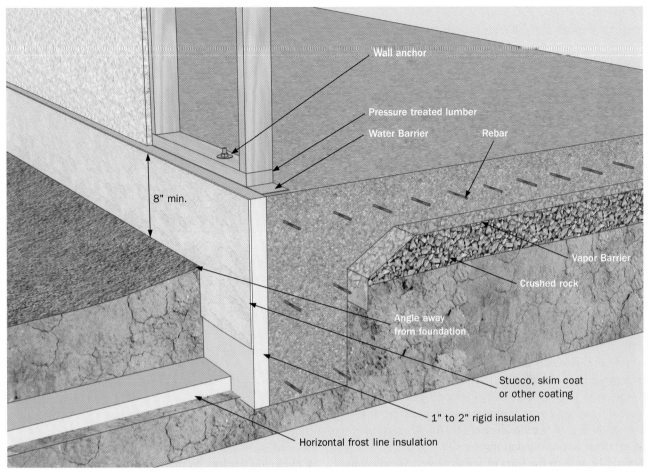

Wall anchor

Pressure treated lumber

Water Barrier

Rebar

8" min.

Vapor Barrier

Crushed rock

Angle away
from foundation

Stucco, skim coat
or other coating

1" to 2" rigid insulation

Horizontal frost line insulation

Combined floor slab and foundation (monolithic slab)

above the flood plain level, it's possible to gain additional construction area on a cramped lot.

The raised floor foundation consists of a set of piers placed on concrete footings. The raised floor area can be left open, or a skirt can be placed around the home for cosmetic purposes. Like a concrete block crawl space, you will likely be dealing with excavators, footing subcontractors, and concrete block masons. The block mason can be eliminated by casting the piers from concrete. Since the piers are spaced out, the amount of concrete used is minimal. In sandy, beach front areas, treated wood posts are often driven deep into the sandy soil in lieu of a concrete footing. Underfloor areas will need to be insulated, and this is a perfect opportunity to use foam-in insulation.

Poured Concrete Foundations

Poured concretes foundations have grown in popularity, due to their simplicity and strength. A single foundation sub will layout the foundation, install the forms and order the concrete. It is more expensive, but the time and money saved by using one subcontractor can recoup much of the expense. If you are planning to have a brick veneer exterior, you must tell the subcontractor to provide for a brick ledge in the forms.

Typical raised floor foundation.

Poured concrete foundations have high strength and resistance to leakage, but they are harder to build out as a finished basement. Furring strips must be installed to provide for drywall installation, wiring, and plumbing.

Precast Concrete Foundations

Basement walls are either cast on-site and "tilted up" or cast in a factory and shipped to the construction site. The precast foundation contractor will take responsibility for the entire foundation install, from laying out batter boards to preparing the crushed rock footing. Because of the natural strength of precast walls, they are usually placed directly on crushed rock footings rather than concrete footings. Most panels are cast using 5,000 psi concrete, rather than the standard 3,000 psi concrete. This denser material is naturally waterproof and will prevent the

Prefab concrete foundations are cast in the controlled climate of a factory and then shipped to the site.

A crane is usually required to place the heavy panels.

Prefab foundation contractors will prepare the site to insure a successful installation.

Precast panels are installed directly on compacted, crushed rock, which provides an excellent base for drainage and moisture control.

A precast panel foundation installed by Superior Walls, complete with ledge for brick façade.

migration of moisture to the interior. The panels come pre-insulated, which eliminates a step later. Precast panels are more expensive than poured foundations, but can be installed quickly, even in inclement weather, and can often pay for their extra cost through labor reduction in other areas of the project.

ICF's – Insulated Concrete Foundations

This new technology uses insulated foam as the form for the poured concrete wall. The form is left in place after pouring and provides excellent insulation and moisture control. The forms themselves are extremely easy to construct, making them a good do-it-yourself option. Framers or other contractors can master this installation technique easily. Forms are usually placed on a concrete footing to insure an even surface, although crushed rock footings are sometimes used. Once the blocks are in place, steel rebar will be anchored to the inside of the forms. The forms must be reinforced during the pour to prevent warping of the lightweight foam blocks.

Steel Reinforcement

Steel rebar is used to strengthen concrete slabs, walls and footings and help them resist the effects of shrinkage and temperature change. Reinforcing comes in two major forms: bars and mesh. Reinforcing bars used for residential construction are normally about ⅜" in diameter. Reinforcing wire, normally with a 6" grid and ¼" wire, is adequate to do the job. For monolithic slabs, mesh is used more often than bar steel, although rebar may be used along the perimeter of the slab at the junction of the thickened footing area. Steel reinforcement should be laid before cement is poured, and secured so that no movement will occur during the pouring process. Normally the steel is located two-thirds of the way up from the bottom of a slab and in the center of poured concrete walls. Your concrete sub and concrete supplier will have primary responsibility for proper reinforcement.

OTHER CONCRETE WORK

Foundation Floors

Foundation floors (with the exception of monolithic slabs) are poured after the footings and walls have properly cured. Coarse, crushed rock is normally used as an underlayment. Poly film is then applied to act as a moisture and vapor barrier. Care must be taken not to tear the film during installation and concrete work. Once the poly is in place, steel reinforcing mesh is installed and raised with rocks or other items so that it is embedded in the center of the slab. Concrete is normally poured to a depth of four inches. Important Note: Backfilling of the basement foundation should not be attempted until the floor slab and ceiling framing is in place. Otherwise, basement walls can shift under the weight of fresh dirt. The slab and framing acts as reinforcement.

ICF foam blocks can be installed quickly and easily by unskilled artisans.

Insulated concrete foundations must be poured in a single continuous process. Note ample bracing to prevent warping.

Driveways

Driveway slabs are generally 4" thick and should be allowed to cure for 2-3 days before any vehicles are allowed to drive on it. Driveways are not usually installed until trim work starts so as to avoid the heavier material supply trucks carrying masonry and drywall.

Permeable Pavement

Every year, millions of gallons of rainwater flow off driveways and sidewalks into city sewers, adding to pollution and putting additional burdens on water treatment plants. Consider using a permeable surface for your walks and driveways that will allow water to drain back into the soil in a natural manner. This will not only alleviate the strain on city utilities but will add moisture to the surrounding yard, helping to feed thirsty plants. Permeable systems consist of a permeable concrete or moulded bricks with voids that can be filled with soil or gravel. Underneath these bricks or tiles is a layer of crushed rock. Sometimes drain tile can be added and routed to underground cisterns for use later in watering the lawn.

Sidewalks and Patios

These are simple structures which should only be poured after the surface ground has had ample time to settle. This is usually done when the driveway is poured.

Avoiding Cracks in Concrete

To avoid cracks in slabs and foundation walls for pennies a yard, use concrete with special fiberglass admixtures. This adds to the tensile strength of the concrete and reduces the potential for cracking. You may still want to use wire mesh for reinforcement. If you don't need to install wire mesh, the concrete contractors may even charge less than for a conventional pour because they have to do less work.

Concrete Joints

Concrete will crack. It's a fact. So make sure the concrete cracks in the places you choose. One of the most common reasons for concrete cracking, aside from unstable soil and tree roots, is cracking due to expansion and contraction. Expansion and contraction joints will absorb the stresses and help to insure that any cracking takes place along the designated joint path.

(1) Isolation (Expansion Control) Joints: These joints are placed between concrete structures in order to keep one structure's expansion from affecting adjacent structures. Common places are between:
- Driveway and garage slab
- Driveway and roadway
- Driveway and sidewalk
- Slabs and foundation wall

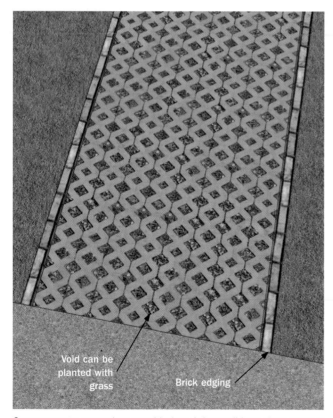

Open concrete pavers have a void where it is possible to plant grass or to fill in with another permeable material like gravel.

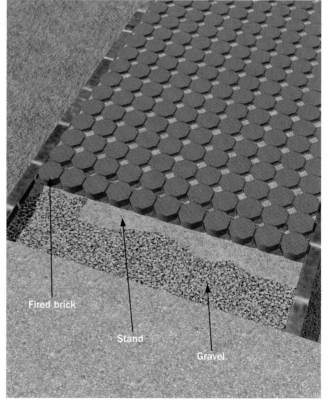

Permeable pavers allow water to drain naturally into the ground, reducing runoff and pollution.

(2) Control (Contraction) Joints: These joints are placed within concrete units, such as driveways (normally every 15'), sidewalks (normally every 5' or closer), and patios (normally every 12').

Construction joints are temporary joints marking the end of a concrete pour due to delays in concrete supply. If you are lucky, you won't need any of these. Your concrete supplier and/or finisher should be responsible for setting all expansion joints, and finishers are responsible for all control joints.

Weather Considerations

Rainy weather can adversely affect almost any concrete work. Heavy rain can affect poured concrete surfaces and slow the drying process. If the soil is saturated with moisture at the time of the pour, it will dry and shrink later, thereby increasing the chances of cracking. Schedule your concrete pouring for periods of forecasted dry weather. Have rolls of heavy plastic available if the weather forecast is wrong.

Cold weather is also a challenge. Concrete work in temperatures below 40° Fahrenheit increases the risk of freezing and frost heave. Work can be partially protected by heating water and aggregate, or by covering the poured concrete with tarpaulins. Frozen ground must be heated to ensure that it will not freeze during the curing process. Mix the heated aggregate and water before adding the concrete. Consider using high-early-strength concrete to reduce the time required to protect the concrete from freezing. Once concrete has been poured, if freezing weather is likely, the concrete should be protected either by special heated enclosures or insulation such as batts or blankets.

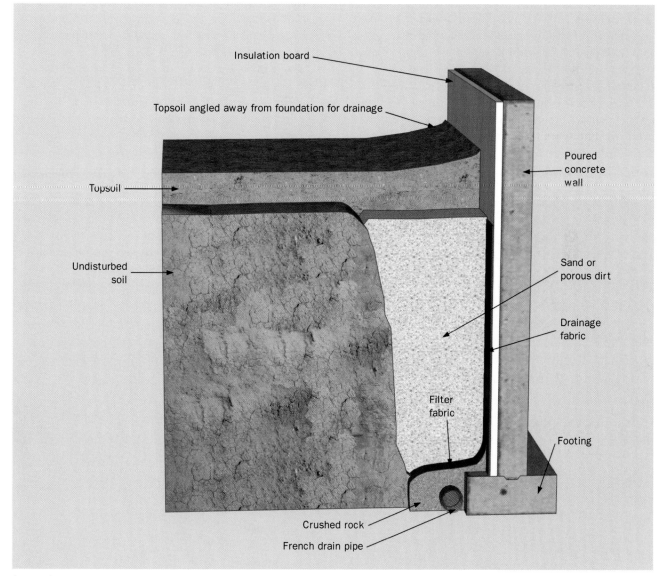

A properly constructed drainage zone will ensure a lifetime of dry basements.

Hot weather is also a challenge. The main problem to prevent here is rapid drying which induces cracking and reduction in concrete strength. Regardless of the external temperature, concrete should not be allowed to exceed 85° Fahrenheit. This can be achieved by using a combination of the following methods:

• Cool the aggregate with water before mixing.
• Mix concrete with cold water. Ice can be used in extreme cases.
• Provide sources of shade for materials.
• Work at night or very early in the morning.
• Protect drying concrete from sun and winds, which accelerate the drying process. Wet straw or tarpaulins can be used.
• Keep the concrete moist.
• Mix in admixtures like fly ash that are designed to slow the curing process.

WATERPROOFING AND DRAINAGE

The vast majority of moisture problems in basements originate from improper waterproofing and drainage methods. Get this right the first time, since it will be difficult to fix later. This is critical to ensure that you will have a dry basement in future years.

Wall Treatment

Wall treatments are applied below the finished grade (below ground level). Normally, this treatment consists of one of the following:

• hot tar
• asphalt or bituminous material
• poly film behind foam insulation board
• filtered drainage mat
• three dimensional drainage membrane

Check with suppliers for the latest membrane technologies. If masonry block walls are used, two ¼" coats of portland cement are applied to the masonry blocks, extending 6" above the unfinished grade level. The first coating is roughened up before drying, allowing the second coat to adhere properly. Masonry blocks and the first coating of portland cement are dampened before applying additional coats. The second coating should be kept damp for at least 48 hours. The joint between the footing and wall should be filled with portland cement before applying the waterproof coating. This is called parging and further protects against underground water and moisture.

Drain Pipe

It's not enough to seal the foundation against moisture. You must also provide a drainage path for water that builds up against the foundation. The hydrostatic pressure from this moisture can apply huge lateral forces against the foundation wall that will encourage cracking and shifting.

The most common solution is a 4" perforated drain pipe (plastic preferred over tile) commonly known as French drain. It is placed around the perimeter of the foundation on top of a bed of gravel that extends down to the bottom of the footing. The holes in the drain must be pointed DOWN to allow the entrance of water. The pipe should be sloped at least 1" for every 20'. The drain pipe will only serve its purpose if it drains off to daylight. You will want to cover the end of the drain with wire mesh to keep out pests.

The drain is then covered with another 8" of gravel and a piece of landscape fabric. This last detail is critical. Quite often filter fabric is omitted and silt from clay will settle and clog the drain, rendering it useless after a few years.

Finally, the foundation should be backfilled with a sandy soil to encourage drainage. If your property is filled with heavy clay, resist the urge to simply push that material back into the hole. Clay can retain a surprising amount of moisture that will apply hydrostatic pressure against the foundation. A sandy mixture is preferable. A final layer of topsoil should be added and angled away from the foundation.

PEST CONTROL — STEPS

PS1	CONDUCT standard bidding process and select pest control sub. Use a company that has been around for at least ten years since you may want to get annual pest control and a warranty. In some cases builders are not charged for the pre-treating service. The pest control sub looks for a contract with the home buyer.
PS2	NOTIFY pest control sub of when to show up for the pre-treat.
PS3	APPLY poison to all footing, slab ground and surrounding area. Since it is hard to test the effectiveness and coverage of treatment, it is advised that someone be present to oversee this critical step. This treatment should only be done if rain is unlikely for 48 hours.

AFTER CONSTRUCTION — STEPS

PS4	APPLY poison to ground-level perimeter of home. Again, it is advised to have someone oversee this step.
PS5	OBTAIN signed pest control warranty and file it away. Many lenders require a pest warranty at closing to insure proper protection.
PS6	PAY pest control sub and have him sign an affidavit.

CONCRETE — STEPS

CN1	CONDUCT standard bidding process (concrete supplier). Bid will be measured in cubic feet. If you use a full service foundation sub, this will probably be unnecessary.
CN2	CONDUCT standard bidding process (formwork and finishing). If you use a full service foundation company, they will lay out the foundation, pour footings and walls and arrange for the concrete. Pricing can get pretty complicated, depending on the height and size of the walls.
CN3	CONDUCT standard bidding process (concrete block masonry).

FOOTINGS — STEPS

CN4	INSPECT batter boards. This is quite critical since your entire project is based on the position of your batter boards and string. This is done with a transit.
CN5	DIG footings for foundation, A/C compressor, fireplaces and all support columns.
CN6	SET footing forms as necessary. Be very careful that the footing sub places footings so that walls will sit in the center of the footing. Subs can either use their own forms or use lumber at the site. If the sub asks you to provide forms material, have three times the foundation perimeter in 16' 2×4s.
CN7	INSPECT footing forms. Refer to related checklist in this section. Make sure that the side forms are sturdy as they will have to support lots of wet concrete. Make sure that the tops of the forms are level as they will be used as a guide in finishing the footing surface. Also make sure that the forms are parallel with each other. Footings are normally coated on the inside with some form of oil so that the concrete doesn't stick to them. Check for this. If you plan to have poured walls on your footings, have the forms sub supply the "key" forms which will be pressed into the footings after they are poured. In many areas, steel rods are inserted vertically in footings prior to pouring.
CN8	SCHEDULE and complete footing inspection. Obtain signed, written approval from local building inspector.
CN9	CALL your concrete supplier to schedule delivery of concrete if you are not using a full service finisher. This should be done at least 24 hours in advance. Make sure good weather is forecast for the next few days. Have rolls of 6 mil poly available in case of rain. Make sure to add calcium carbonate if freezing is a potential problem.
CN10	POUR footings. This should be done as quickly as possible once the footing trenches have been dug so that the soil does not get soft from rain and exposure.
CN11	INSTALL key forms in footings before setting has taken place. Your forms sub should do this. Budget finishers will just cut a groove in the footing with a 2×4.
CN12	FINISH footings as needed. This normally involves simply screeding the top and smoothing out the surface a bit.
CN13	REMOVE footing forms. If subs used your lumber, have them clean and stack it to be used later for bracing. Framers won't use this lumber because it will dull their saws.
CN14	INSPECT footings. Make sure that the footings are level and that there are no visible cracks.
CN15	PAY concrete supplier for footing concrete, unless you are using a full service finisher in which case you will make one payment after the walls have been poured and inspected.

POURED WALL — STEPS

CN16	SET poured wall foundation forms. These are normally special units made out of steel held together with metal ties.
CN17	CALL concrete supplier in advance to schedule concrete. This step will usually be done by a full service foundation contractor.
CN18	INSPECT poured foundation wall. Again, make sure they are level, parallel, sturdy and oiled.
CN19	POUR foundation walls. If more than one truckload is required, make sure to schedule deliveries so that a constant supply of concrete is available for pouring. It is unsatisfactory to have cement setting when wet cement will be poured over it. This can result in a fault crack in the foundation. Make sure that additional loads of concrete are on site or on the way to insure a continuous pour. If the weather is rainy, have a grader available in case you have to push the concrete truck up a muddy slope.
CN20	FINISH poured foundation walls. Again, the tops of the forms serve as screed guides. The surface is smoothed and lag bolts should be set into the surface. Make sure the bolts are at least 4" in the concrete and are standing straight up. Also make sure that the bolts are not in places that will interfere with any studs or door openings. Put wide washers on the sunken end of the lag bolts for greater holding strength.
CN21	REMOVE poured foundation wall forms. This should only be done several days after the concrete has had time to set.
CN22	INSPECT poured foundation walls. Check for square, level and visible surface cracks.
CN23	BREAK off tie ends if metal footing forms are used.
CN24	PAY concrete supplier for poured foundation wall concrete.

BLOCK WALL — STEPS

CN25	ORDER concrete blocks and schedule block masons. Make sure the blocks are placed as near to the work area as possible. Block masons will charge extra if they have to move excessive amounts of block very far.
CN26	LAY concrete blocks. You may want to watch some of this in progress since it will be very hard to correct mistakes when the blocks have set. If you see globs of cement on blocks, wait till it dries to clean it off — it will come off in one piece. Compare against blueprints as the work proceeds.
CN27	FINISH concrete blocks. While mortar is still setting, the masons will tool the joints concave if you like. However, for subgrade work, it is better to have the mortar flush with the blockwork. If any of the block wall is to be exposed, have the block mason stucco the block to provide a smooth, more attractive surface.
CN28	INSPECT concrete block work. Compare work with blueprints. Refer to your specifications and inspection guidelines.
CN29	PAY concrete block masons. Have them sign an affidavit.
CN30	CLEAN up after block masons. Stack or return extra block. Spread leftover sand in drive area. Bags of mortar mix should be protected from moisture.

DRIVEWAYS AND PATIOS — STEPS

CN46	COMPACT driveway, walkway, patio and mailbox pad sub-soil. This must be a solid surface.
CN47	SET driveway, walkway, patio and mailbox forms. You should not pour less than a 3" thick driveway. If you did not lay down crushed stone previously, you may want to do it now. If your electrical or gas service crosses the driveway area, it must be installed before this step. Use scrap lumber for forms.
CN48	CALL your concrete supplier to schedule delivery of concrete. Confirm type and quantity of concrete, delivery time and site location. Schedule finishers.
CN49	INSPECT driveway, walkway, patio and mailbox forms. Pull a line level across forms to determine slope. Forms should be installed so that driveway will slope slightly away from house and toward an area with good drainage.
CN50	POUR driveway, walkway and patio areas. Make sure finishers do not thin concrete for easier working.
CN51	FINISH driveway, walkway, patio and mailbox areas. Make sure that the finishers put isolation joints between the slab, driveway, patio and walk ways. Expansion joints should be put about every 12' in the driveway and every 5' or less in the walkways. Driveway should be formed where it meets the street.
CN52	INSPECT driveway, walkway and patio areas.
CN53	ROPE off drive, walk, patio and mailbox areas so that people and pets don't make footprints. Put some red tape on the strings so that they can be seen.
CN54	PAY concrete supplier for driveway, walkway, patio, and mailbox pad concrete.
CN55	PAY concrete finishers and have them sign an affidavit.
CN56	PAY concrete finishers retainage.

CONCRETE SLAB — STEPS

CN31	SET slab forms for basement, front stoop, fireplace pad, garage, etc. Check this against blueprints as they are being erected. If the slab is the floor of the basement, the walls will act as the forms.
CN32	PACK slab sub-soil. Soil should be moistened slightly, then packed with a power tamper. This has got to be a solid surface.
CN33	INSTALL stub plumbing. If the slab is to be the floor of the house, schedule the plumber to set all water and sewer lines under slab area. Ask the plumber to install a foam collar around all sewer lines that protrude out of the slab. This makes final adjustments to the pipe easier during finish plumbing. If the pipe needs to be moved slightly, the foam can be chipped away instead of the concrete. Make sure any resulting gap is resealed with spray foam insulation. Have the plumber install termite shields over all pipe openings.
CN34	SCHEDULE HVAC sub to run any gas lines that may run under the slab, such as gas fireplace starters.
CN35	INSPECT plumbing and gas lines against blueprint for proper location. Make sure that plumbing is not dislodged, moved or damaged in subsequent steps.
CN36	POUR and spread crushed stone (crusher run) evenly on slab or foundation floor and garage slab area.
CN37	INSTALL 6 mil poly vapor barrier. Can be omitted in garage slab.
CN38	LAY reinforcing wire as required. The rewire should lie either in the middle of the slab and garage slab or at a point two-thirds from the bottom. It can be supported by the gravel or rocks.
CN39	INSTALL rigid insulation around the perimeter of the slab to cut down on heat loss through the slab.
CN40	CALL concrete supplier to schedule delivery of concrete for slab. Confirm concrete type, quantity, time of pour and site location. Finishers should also be contacted.
CN41	INSPECT slab forms. Compare them against the blueprints. Make sure that all necessary plumbing is in place. Check that forms are parallel, perpendicular and plumb as required. Forms should be sturdy and oiled to keep from sticking to concrete.
CN42	POUR slab and garage slab. This should be done as quickly as possible. Call it off if rain looks likely. Have 6 mil poly protection available.
CN43	FINISH slab and garage slab. Finishers should put their smoothest finish on all slab work. This normally requires four steps: • rough screed with a floater • rough trowel • power trowel • finish trowel
CN44	INSPECT slab. Check for rough spots, cracks and high or low spots. High and low spots can be checked with a hose. See where the water collects and drains. Water should never completely cover a nickel laid flat. Make sure any backfill does not cover the brick ledge if one is present.
CN45	PAY concrete supplier for slab and garage slab concrete.

WATERPROOFING — STEPS

WP1	CONDUCT standard bidding process. Invite your full service company to bid.
WP2	SEAL footing and wall intersection with portland cement. Skip if you have no basement.
WP3	APPLY ¼" portland cement to moistened masonry wall. Roughen surface and allow to dry 24 hours. Perform only if you have a masonry foundation.
WP4	APPLY second ¼" portland cement to moistened wall. Keep moist for 48 hours and allow to set. Perform only if you have a masonry foundation.
WP5	APPLY waterproofing compound to all below grade walls.
WP6	APPLY black 6 mil poly (4 mil is a bit too thin) to waterproofing surface while still tacky. Roll it out horizontally and overlap seams. Alternately use three dimensional drainage membrane.
WP7	INSPECT waterproofing (tar/poly). No holes should be visible.
WP8	INSTALL 1" layer of gravel as bed for drain tile. Optional for crawl spaces may use even if you have no basement.
WP9	INSTALL 4" perforated plastic drain tile around entire foundation perimeter. Also install PVC run-off drain pipe.
WP10	INSPECT drain tile for proper drainage. Mark the ends with a stake to prevent them from being buried. Make sure drainage holes are on the bottom.
WP11	APPLY 8" to 12" of top gravel (#57) above drain tile. HINT: Surround water spigot area with any excess gravel to prevent muddy mess during construction. Spigot will be used quite often.
WP12	PAY waterproofing sub and have him sign an affidavit.
WP13	PAY waterproofing sub retainage.

CONCRETE — SAMPLE SPECIFICATIONS

General

- All concrete, form and finish work is to conform to the local building code.
- Concrete will not be poured if precipitation is likely or unless otherwise instructed.
- All form, finishing and concrete work MUST be within ¼" of level.
- All payments to be made five working days after satisfactory completion of each major structure as seen fit by builder.

Concrete Supplier

- Concrete is to be air-entrained ASTM Type I (General Purpose), 3,000 psi after 28 days.
- Concrete is to be delivered to site and poured into forms in accordance with generally accepted standards.
- Washed gravel and concrete silica sand to be used.
- Each concrete pour to be done without interruption. No more than one half hour between loads of concrete to prevent poor bondage and seams.
- Concrete to be poured near to final location to avoid excessive working. Concrete will not be thinned at the site for easier working.

Formwork

- Bid is to perform all formwork per attached drawings including footings for:
  - Exterior walls
  - Monolithic slab
  - Poured walls
  - Bulkheads
  - Garage
  - AC compressor slabs
  - Patio
  - Basement pier footings
- All footing forms to be of 2" or thicker wood or steel.
- All forms to be properly oiled or otherwise lubricated before being placed into service.
- All forms, washed gravel, reinforcing bars to be supplied by form subcontractor unless otherwise specified in writing. All of these materials to be included in bid price.
- All forms will be sufficiently strong so as to resist bowing under weight of poured concrete.
- Form keys to be used at base of footings and for brick ledge.
- Expansion joints and expansion joint placement to be included in bid. Expansion joints: Driveway every 15'. Sidewalks every 5'.
- Pier footings and perimeter footings to be poured to exactly the same level.

- All concrete to be cured at the proper rate and kept moist for at least three days.
- Foundation and garage floor to be troweled smooth with no high or low points.
- Finished basement floor will slope toward drains.
- Garage floor, driveway and patios will slope away from dwelling for proper drainage.
- Concrete to be poured and finished in sections on hot days in order to avoid premature setting.

Concrete Block

- All courses to be running bond within ¼" level.
- Top course to be within ¼" level.
- Standard 12" block to be used on backfilled basement area.
- 4" cap block to finish all walls.
- Horizontal reinforcing to be used on every three courses.
- Trowel joints to be flush with block surface.

Waterproofing — Specifications

- Bid is to include all material and labor required to waterproof dwelling based on specifications described below.
- Two ¼" coats of portland cement to be applied smoothly and evenly. First coat to be applied to moistened masonry blocks and roughened before drying. Second coat to be kept moist for 48 hours to set (block wall only).
- Portland cement to be applied smoothly to intersection of footing and wall.
- Sharp points in portland cement coating to be removed prior to application of tar coating, as they will puncture 6 mil poly.
- Asphalt coat to be applied hot, covering entire subgrade area. Coat to be a minimum of 1⁄64".
- 4" diameter perforated PVC drain pipe to be installed around entire perimeter of foundation. Tile sections to be separated ¼" with tar paper to cover all joints. Drain tile should have a minimum slope of 1" in 20'. Drain holes every 6" in pipe.

Pest Control — Specifications

- Bid is to provide all materials, chemicals and labor to treat all footing and slab area.
- Base of home to be sprayed after all construction is complete.
- Termite treatment warranty to be provided.

Batter Boards

☐ All batter boards installed:
- Exterior walls
- Piers and support columns
- Garage or carport
- Fireplace slabs
- Porches and entryway
- Other required batter boards

☐ All batter boards installed properly:
- Level
- Square
- Proper dimensions according to blueprints
- All strings tight and secure
- All string nails to be marked to prevent movement.

☐ All bulkheads well-formed and according to plan. Step down depth dimension correct.

Checklist Prior to Pouring Any Concrete

☐ All planned concrete and form work has been inspected by state and/or local officials if necessary prior to pouring.

☐ Forms laid out in proper dimensions according to plans. Length, height, depth, etc.

☐ Correct relationship to property lines, set-back lines, easements and site plan (front, back and all sides).

No rain predicted for 6-8 hours minimum.

☐ Forms secure with proper bracing.

☐ Forms on proper elevation.

☐ No presence of ground water or mud.

☐ No soft spots in ground area affected.

☐ Ground adequately packed.

☐ Forms laid out according to specifications: parallel, perpendicular and plumb where indicated.

☐ All corners square.

☐ Necessary portions straight. No bowing. Wood forms should be nominal 2" thick.

☐ Spacers used where necessary to keep parallel forms proper distance apart.

☐ Expansion joints in place and level with top of surface. Approximately 1/2" wide.

☐ All plumbing underneath or in concrete is installed, complete, inspected and approved.

☐ Reinforcing bars bent around corners — no bars just intersecting at corners.

☐ Concrete supplier scheduled.

☐ Adequate finishers available to do finish work.

☐ Holes under forms filled with crushed stone.

☐ Proper concrete ordered.

☐ Forms tested for strength prior to pouring.

☐ Backfill stable and not in the way. Will not roll onto freshly finished cement or get in finisher's way.

☐ Garage, carport, patio, entrance and porch slabs angled to slope away from home.

☐ Any necessary plumbing has been inspected.

☐ Gravel, re-wire and 6 mil poly have been placed properly when and where required.

☐ All water and sewer lines installed and protected from abrasions during pouring of concrete.

☐ Finishers have been scheduled or are present.

☐ Crawl space, if any, has been cleared with a rake.

☐ Foam collars placed over all water and sewer pipes where they protrude out of the concrete slab.

General Checklist After Concrete and Blockwork

☐ Surface checked for level with string level, builder's level or transom.

☐ Corners square using a builder's square.

☐ No cracks, rough spots or other visible irregularities on surface.

☐ Sill bolts vertical and properly spaced (if used). Foundation Wall (Concrete Block)

☐ Masonry joints properly tooled and even.

☐ Blocks have no major cracks or irregularities.

☐ Cap blocks installed level as top course.

☐ Room allowed for brick veneer above, if any is to be used.

☐ Joint between wall and footing properly parged (sloped away from wall) with mortar.

Waterproofing - Inspection

☐ Portland cement is relatively smooth and even with no sharp edges protruding.

☐ Portland cement completely covers intersection of wall and footings.

☐ Asphalt coating completely covers entire sub-grade area, including intersection of wall and footings.

☐ Black 6 mil poly completely covers entire tar coating and adheres firmly to it. Look for any tears or punctures in poly which need to be patched before backfill. Remove large rocks and roots from backfill area so that poly will not be torn during backfill process later.

☐ Tarred area does not go above grade level or where stucco is to be placed.

☐ Poly secured to stay in place during backfill.

☐ No rain in basement.

☐ No wet spots on interior basement walls.

FRAMING

Framing is one of the most visible signs of progress on the building project. Your framing contractor will be involved with the project longer than any other person. The skills of the framing crew will have a large impact on the quality of your home and your ability to stay on schedule.

Take adequate time and choose your framer carefully. A good framer is worth his weight in 2×4s. A good framing crew should consist of at least three people in order to move lumber around efficiently. Check your carpenter's ability to read blueprints. He will be making many interpretations of these plans when framing. Check references and make sure the foreman of your crew has experience as a master carpenter on several jobs and not just as a helper.

Make sure to describe the quality of work you are expecting in the contractor's agreement. Don't just assume that certain procedures will be followed. Get it all in writing. Make sure you can talk the framer's language. Get to know the names and nicknames that are commonly used when referring to lumber pieces. A framer who respects your knowledge will be more likely to listen to suggestions and critiques. More changes occur during the framing of the house than at any other time. You want to insure that your framer is your ally when these changes need to be made in the field.

Lumber

Lumber is used to build a home. Wood is used to build a fire. Lumber comes in lengths starting at about 8 feet and increasing in length by even numbers, usually 2 feet. The most common lengths are 8', 10', 12', 14', 16', 18' and 20'. Lumber is usually priced either by the piece or by the "lineal foot." A lineal foot is the sum total length of all the lumber ordered. For example, 100 2×4s 10' long would be 1,000 lineal feet of lumber. Note that not all lumber is priced the same. Longer lengths of lumber cost more per lineal foot, because the larger trees supplying the lumber are rarer. When you order 1,000 lineal feet of lumber, make sure you know what lengths you will be receiving. You may want to request longer lengths, such as 14' or 16', just to have

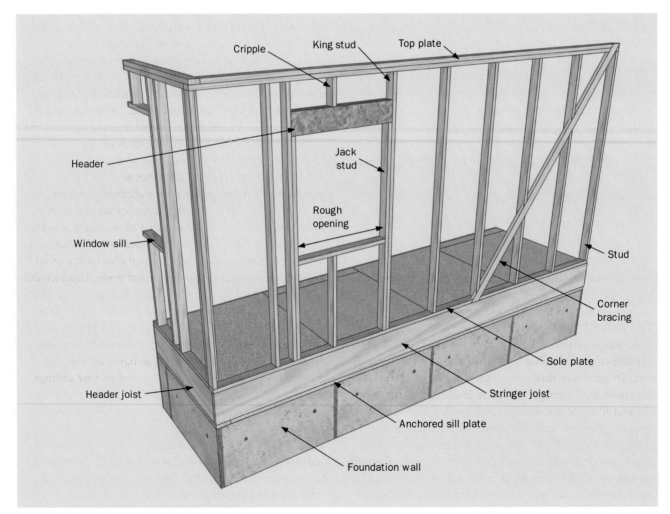

Know your framing technology when talking with framers.

plenty of long pieces when you need them, but note that your costs per lineal foot will rise. It's best to know precisely how many lumber pieces you will need of each length before ordering. This will keep waste and cost lower.

Rough-in lumber used for framing is generally called "nominal" lumber or "dimensional" lumber. It is generally referred to by its width and height. Thus a 2×4 or "two-by-four" is a lumber piece 2"-tall (or thick) and 4"-wide. This can be confusing for many people because lumber's dimensions are measured when the lumber is rough cut, not after it has been planed down to a smooth surface. Therefore, "two-by-four" actually measures 1½" by 3½". To make matters more confusing, these "trimmed down" dimensions are not uniform. You would assume that a 2×8 would be 1½" by 7½", but you'd be wrong. As the dimensional size goes up, the actual size goes down since there is more shrinkage as the wood dries after cutting. A 2×8 actually measures right around 1½" by 7¼". To make matters a bit more confusing, the length of lumber is measured in actual dimensions. So a 2×4, 10' long is really ten feet long. Or not. It's quite common for lumber to vary in length from one piece to another by a ½" in some cases.

Dimensional Lumber Sizes in Inches

Nominal	Actual
2×2	1½ x 1½
2×3	1½ x 2½
2×4	1½ x 3½
2×6	1½ x 5½
2×8	1½ x 7¼
2×10	1½ x 9¼
2×12	1½ x 11¼
4×4	3½ x 3½
4×6	3½ x 5½

Some of this variance results from shrinkage, although lumber shrinks less along the grain vs. perpendicular to the grain. The rest comes from inaccuracies in sawing.

Confused? Welcome to the unique world of the framer.

Shrinkage can also cause warping. Lumber dries unevenly, and that can pull lumber out of its original straight condition. To minimize this, make sure your lumber has been kiln-dried to the proper moisture level (usually between 6% and 8%). Make sure to stack lumber out of the sun or cover it with a tarp. Sunlight will also cause uneven drying. Warping in framing pieces can also be reduced by using a higher grade of lumber and will result in a neater framing job and a happier framing sub. The additional cost is worth it unless you have an unusually careful carpenter willing to spend hours picking out bad lumber. Specify Douglas fir for the best job. Avoid mixing fir and pine lumber on the same job. Pine can be ¼" to ½" different

in size because of different shrinkage rates. Any lumber in contact with the foundation or concrete must be pressure treated to provide moisture and termite protection.

Studs

A 2×4 is different than a stud. A stud is a high-quality 2×4 cut to a specific length, usually 92⅝", although in some states like California, the length can vary slightly. The specific length of a stud is designed to make it easy to frame a wall that will be exactly 8' tall. Therefore the stud length is cut less than the 8' to allow for a single bottom plate and two top plates. Wiggle room is then added back to allow for the thickness of typical flooring. When ordering lumber, make sure to separate the number of studs needed for wall framing from the 2×4s used for other purposes.

Crowning

Make sure the carpenter "crowns" all material during construction. All lumber has a natural curve to it. Observe the concentric arcs when looking at the end of a board. When framing, the carpenter should try to turn all the lumber so that their natural curves point in the same direction. When framing horizontal joists, the top of the curve or "crown" should bow up. This will counteract the natural tendency of wood to bow downward under load. Badly warped or curved lumber should to be set aside and returned to the supplier or used for short framing pieces known as "cripples" or "dead wood." Premium grade straight lumber should be set aside for use in framing the kitchen and bathrooms, where straight walls will be critical.

Payment of the Framing Contractor

The framing contractor does not usually provide any of his own materials for the job except perhaps nails. It is your responsibility to make sure that all necessary materials are on the site when needed. Carpenters usually charge by the square footage of the area framed plus extras or will appraise the job and supply a total bid price. This includes all space covered by a roof, not just heated space. This includes the garage area, enclosed porches, tool sheds, garden houses, etc. The framing charge usually includes setting the exterior doors and windows and installing the sheathing. Special items such as bay windows, stairs, curved stairs, chimney chases, recessed or tray ceilings usually require additional charges.

Types of Framing

The two main types of framing used for residential construction are balloon framing and platform framing. For a more complete description of these two methods, refer to a good carpentry manual. Platform framing is the most widely used type by far and will be the type your carpenter is accustomed to using. In this book, only platform framing will be covered.

Balloon Framing

Balloon framing was more popular at the turn of the century because long spans of lumber were available. These long timbers were used to frame the walls of both the first and second floors. Floor joists were then nailed to these long timbers or rested on a ledger board. This method used less fasteners and created a strong, continuous wall from the foundation to the roof that was an excellent platform for the lap siding and lath finishing used on exterior and interior walls. However, it was difficult to work around the tall timbers. The long vertical channels also became perfect chimneys for fire and a lot of blocking was required to break the free flow of air up the channels. Balloon framing fell out of favor as old growth lumber disappeared.

Platform Framing

In platform framing, floors are stacked on foundations, walls stacked on floor "platforms," ceilings stacked on walls, and so on. This makes it very easy to build a house from the ground up, one layer at a time. The stacking of layers also acts as a fire break. Shorter lengths of lumber are needed for each layer. However, care must be taken to tie all these layers together. The invention of plywood and OSB board helped tremendously in this regard by acting as a "skin" to tie all the framing members together and prevent racking. Since balloon framing is so rare, this is about all you need to know about the history of platform framing. It's all you will see.

Structural Issues

With the exception of the foundation, framing is the most critical phase of your project. Mistakes made here will be difficult to fix later, or may go unnoticed until cracks begin forming in drywall, doors start to sag, or something more serious occurs. It will be the responsibility of your framer

Platform framing makes it easy for small crews to assemble walls on top of prior framing.

and the framing crew to interpret the blueprints and execute the proper structural design of your home (assuming your blueprints are structurally sound to begin with).

If you have chosen a framing subcontractor wisely, he will probably be quite familiar with the building codes in your area. Most of this knowledge is gained through on-the-job experience, rather than from a book or construction course. Unfortunately, customary techniques have a way of leaking into the communal knowledge pool that may be less than "best practice." Most subcontractors are just as concerned about speed and efficiency as they are about quality. While a framing crew may be quite competent when following standard designs, they may become confused by highly engineered framing techniques or tricky structural challenges.

That is why it's so important that you specify in writing, the practices and techniques you expect. That means you must first understand these practices. Study installation brochures and publications. Buy a book dedicated to carpentry and construction. Find the areas that apply to your design and highlight them. Make sure to ask your designer or architect to provide detailed drawings of any complex structural challenge. If you are using a highly engineered framing technique like optimum value engineering or an unusual implementation of floor or roof joists, make sure your framer knows precisely how these techniques are implemented. Download installation instructions from manufacturers of flooring, roofing, and truss products and print them out. Attach copies to your framing contractor's agreement.

If you or your designer has modeled your home in SketchUp or other 3D CAD program, it's a good idea to create and print out wall panel diagrams. These detailed

The long spans of balloon framing are only possible today with the use of laminated strand lumber, such as Boise versa studs.

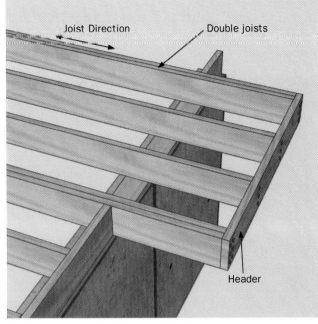

Cantilever as a continuation of floor joists

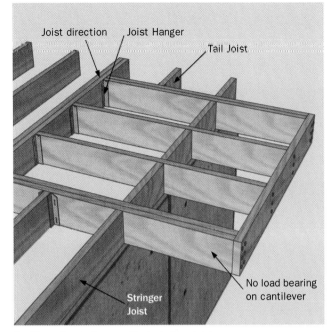

Cantilever framing perpendicular to floor joists

drawings will show the precise measurements of each wall panel that is to be constructed. Some framers may see this as a waste of time or be insulted, but if you have tuned your home design to save lumber, drywall, floor coverings, etc. then you'll need to preserve those design savings with specifics. Otherwise, the framers will interpret the design according to their own rules. On the other hand, if your design follows standard construction practices, it's probably better to trust in your framer, but keep one eye open and your builder's level nearby. As the contractor for your project, it is your responsibility to check the work of framers and identify potential problem areas. To do that efficiently, you need to have a basic understanding of how houses are engineered to support and withstand structural loads.

Loading

Structural framing is designed to support dead loads and live loads. The weight of framing, siding, shingles, carpet, foundation, windows, doors, drywall, and all the other components that are permanent and non-changing make up what is known as dead loads. The weight of dead loading is fairly easy to calculate from the weights of the individual components.

Live loads are more difficult to calculate because they are composed of all the changing and dynamic weights and forces that a structure may encounter, such as the weight of people, furniture, and outside forces like wind and water. Lumber must be able to withstand the unpredictable: such as a crowded deck at a party, winds from an advancing storm, or flooding that presses against foundation walls.

Building codes are designed for probabilistic loads: live loads that are common in most homes. If your home will include large bookcases or a grand piano, you may want to beef up these areas or place these objects to avoid point loads that overwhelm a single joist. For instance, if you have a large bookshelf, make sure it runs perpendicular to the floor joists so that its weight will be carried by several joists.

Load Bearing Walls

The combination of dead loads and live loads will be transferred down the structure by the load bearing walls. These usually consist of exterior and interior walls that are perpendicular to the rafters and joists. All load bearing walls must use headers (also known as lintels) over large openings like windows and doors. Headers are designed to transfer point loads from studs over openings to studs on either side, so the weight can be transferred seamlessly to the foundation. Floor joists function the same way, by transferring live loads from items on the floor, to the load bearing walls that support them. Span tables provide the maximum open areas that joists of different sizes may span without objectionable deflection. While a "spring in your step" may be great on an evening stroll, it's not very pleasant when crossing the living room floor. Creating stiff, stable floors depends not only on using properly sized joists and spacing, but by tying all the individual joists together so that they can share and distribute point loads. One way to do this is to "screw and glue" the subfloor to the joists. The floor then becomes a single, rigid unit that is resistant to squeaks.

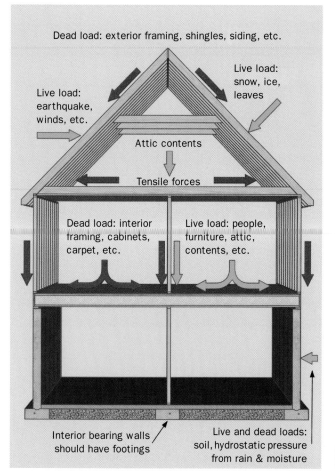

A house structure must be designed to withstand a number of different forces from intermittent live loads like people, furniture, lateral wind, as well as permanent dead loads such as the weight of framing and other building materials.

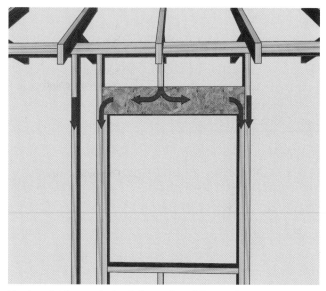

Headers distribute point loads from rafters, joists, and studs to adjoining load bearing structures. They must do this without significant deflection or warping.

Trusses

The trend toward open floor plans demands light, stiff roofs and floor joists that can span long distances. Unfortunately, an opposite trend is occurring in lumber. Old growth forests are disappearing and lumber is getting shorter, smaller, and softer due to the fast growth trees used in sustainable forests. That explains why trusses have become so popular. They span long distances, use a minimum of materials, and can be manufactured in long, continuous lengths. Unlike solid lumber however, their unique properties demand care and a little knowledge of engineering. The top and bottom chords (also known as plates) of trusses do most of the work in supporting the unit as a whole. In fact, the tensile strength of the bottom chord does the lion's share of the

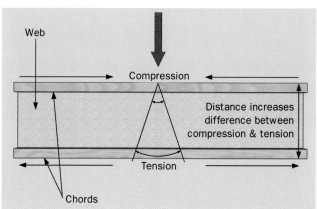

The stiffness of a joist is based on the difference between the compression and tension (stretching) of the upper and lower chords. The web between chords mainly serves as a stiffener and spacer between the two. The web can be solid or triangular.

work. The web between chords is there to provide spacing between the chords and to prevent racking of the truss (see lateral forces). That's why it's possible to have quite large holes in the web of a truss, or even open diagonal members. If you want to span larger distances with trusses, move to taller ones. Since the chords support the majority of the load, it's imperative that they never be notched or breached in any way.

Lateral Forces

A house must support more than its own dead and live weight. It must resist lateral forces as well. This includes wind from hurricanes and storms, and the weight of soil and water against foundation walls. These lateral forces can cause racking, or the sideways collapse of a square object. Diagonal bracing or solid sheathing resists these lateral forces on the house itself. In fact, a roof truss is a huge diagonal brace that is quite strong (which explains why the pyramids are still standing). Bridging, blocking, and rim joists work to resist racking in joist floors.

Anchoring

All this bracing of individual structures will be useless if the house isn't tied together as a solid unit. The weight of a roof helps to hold it in place on the walls of the house and the nails between the rafters and wall plates help to resist lateral sheer forces. However, winds and tornados can easily counteract that weight and pull an unsecured roof right off the house. The shaking of earthquakes can do the same. That is why a house also needs anchoring: between the roof and walls, and between walls and foundation. In earthquake and hurricane prone areas there are strict requirements for anchoring that creates a seamless "strap" around the entire structure. Structural sheathing that extends from a lower wall to an upper wall also acts like strapping. Nine and ten foot sheathing can be used to anchor the first floor to the second floor and reduces labor as well.

Putting It All Together

Think of the framing as the "skeleton" of your home. All the elements must work together to provide a solid structure for the finished "skin." As you inspect the framing, make sure all these elements are tightly joined into a monolithic structure.

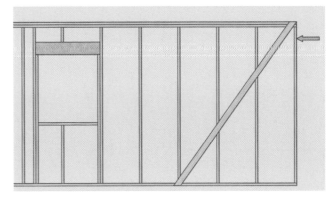

Diagonal bracing resists lateral racking forces from wind.

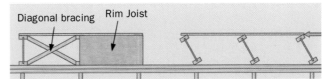

Diagonal bracing resists lateral racking forces from wind.

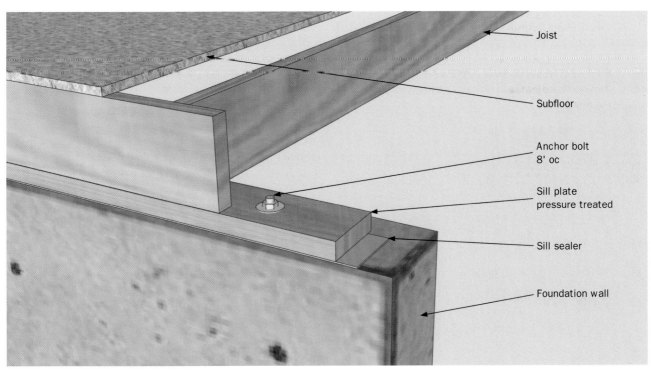

Anchoring and weatherproofing floor system to foundation wall

Joist

Subfloor

Anchor bolt 8' oc

Sill plate pressure treated

Sill sealer

Foundation wall

Hurricane ties anchor the roof rafters to the stud walls and greatly increase resistance to high winds.

Floor Framing

The first "platform" to be framed in a house is usually the first floor, unless you are building on a slab. There are two basic methods of framing with floor joists: solid lumber and floor trusses. Floors framed with solid joists usually consist of 2×8s or 2×10s spaced 12", 16" or 24" on center with bridging in between. Bridging consists of diagonal bracing between the joists to improve stability and to help transfer the load from one joist to the next. Recent advances in framing technology and the advent of truss hangers have reduced the need for diagonal bracing.

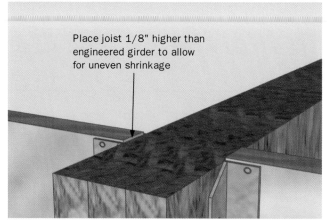

Place joist 1/8" higher than engineered girder to allow for uneven shrinkage

Butted joists detail

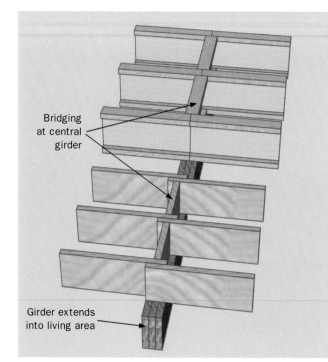

Bridging at central girder

Girder extends into living area

Joist framing above beam

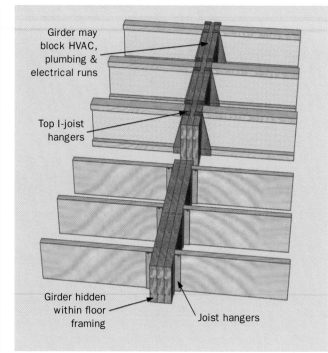

Girder may block HVAC, plumbing & electrical runs

Top I-joist hangers

Girder hidden within floor framing

Joist hangers

Joists butted to side of beam

LENGTH OF MAXIMUM CLEAR SPAN FOR FLOOR JOISTS SPACED AT 24" ON CENTER

	1.0 × 10⁶ PSI	1.1 × 10⁶ PSI	1.1 × 10⁶ PSI	1.3 × 10⁶ PSI	1.4 × 10⁶ PSI	1.5 × 10⁶ PSI	1.6 × 10⁶ PSI	1.7 × 10⁶ PSI	1.8 × 10⁶ PSI	1.9 × 10⁶ PSI	2.0 × 10⁶ PSI
Living areas (40 lb/ft² live load)											
Minimum required bending stress (lb/in²)	1,050	1,120	1,190	1,250	1,310	1,380	1,440	1,500	1,550	1,610	1,670
Joist size											
2×6	7' 3"	7' 6"	7' 9"	7' 11"	8' 2"	8' 4"	8' 6"	8' 8"	8' 10"	9' 0"	9' 2"
2×8	9' 7"	9' 11"	10' 2"	10' 6"	10' 9"	11' 0"	11' 3"	11' 5"	11' 8"	11' 11"	12' 1"
2×10	12' 3"	12' 8"	13' 0"	13' 4"	13' 8"	14' 0"	14' 4"	14' 7"	14' 11"	15' 2"	15' 5"
2×12	14' 11"	15' 4"	15' 10"	16' 3"	16' 8"	17' 0"	17' 5"	17' 9"	18' 1"	18' 5"	18' 9"
Sleeping areas (30 lb/ft² live load)											
Minimum required bending stress (lb/in²)	1,020	1,080	1,150	1,210	1,270	1,330	1,390	1,450	1,510	1,560	1,620
Joist size											
2×6	8' 0"	8' 3"	8' 6"	8' 9"	8' 11"	9' 2"	9' 4"	9' 7"	9' 9"	9' 11"	10' 1"
2×8	10' 7"	10' 11"	11' 3"	11' 6"	11' 10"	12' 1"	12' 4"	12' 7"	12' 10"	13' 1"	13' 4"
2×10	13' 6"	13' 11"	14' 4"	14' 8"	15' 1"	15' 5"	15' 9"	16' 1"	16' 5"	16' 8"	17' 0"
2×12	16' 5"	16' 11"	17' 5"	17' 11"	18' 4"	18' 9"	19' 2"	19' 7"	19' 11"	20' 3"	20' 8"

Allowable spans for simple floor joists spaced 24" on center for wood with modulus of elasticity values of 1.0 to 2.0 x 10⁶ pounds per square inch.

The modulus of elasticity (E) measures stiffness and varies with the species and grade of lumber as shown in the technical note of design. The bending stress (F) measures strength and varies with the species and grade of lumber as shown in the technical note of design.

Source: National Forest Products Association (1977). Span Table for Joints & Rafters.

Note: Use table 8 for joints spaced 16" on center.

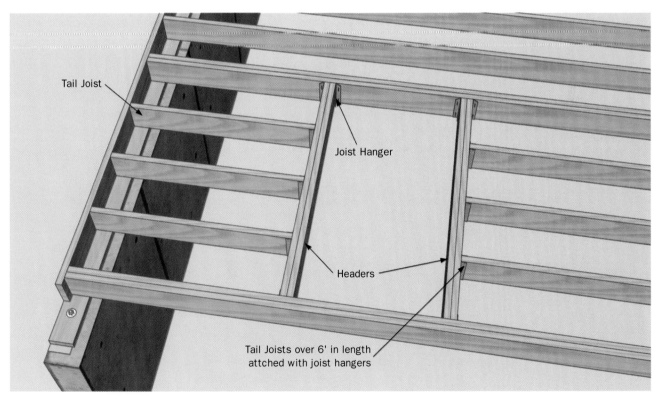

Tail Joist

Joist Hanger

Headers

Tail Joists over 6' in length attched with joist hangers

Floor opening framed with double headers and double trimmer joists

Floor trusses are becoming the choice of the future for many reasons. They have several advantages over solid lumber. A floor truss, like a roof truss, is an engineered frame consisting of two "plates" and a "web" in between. This makes the entire unit very strong for its weight.

Floor trusses have the following advantages over standard joists:

- Many trusses have wider nailing surfaces which allow for glued, squeak-free flooring. The wider surface helps to resist racking.
- Because of their added rigidity, floor trusses can span longer distances without the need for beam support.
- Floor joists can be made from finger jointed lumber and plywood, which is a more sustainable approach to construction. The long spans of large solid lumber joists require old growth trees, which are endangered.
- Wiring and heating ducts may be run within the floor truss, eliminating the need to box in heating ducts below ceiling level. This makes the jobs of the plumber, electrician, HVAC and telephone sub much easier. Make sure to point this out to these contractors when bargaining for rates.

- Open web floor trusses provide better sound insulation between floors because there is less solid wood for sound to travel through.
- Floor trusses warp less, squeak less and are generally stronger than comparable solid wood floor joists.
- Even though floor trusses are a bit more expensive, their advantages can simplify a construction project and can save money in the long run. Your supplier for these units will be the same as for roof trusses.

Because of their engineered construction cutting floor joists can be a problem for some framers. They may not understand the relationship between the joist's integrity and function of plates and webs. The overall strength of a truss is derived from the distance between the top and bottom plates -- the web in between is only there to create that distance. Framers need to understand this relationship because it is never acceptable to notch either plate of a floor truss. On the other hand, cutting holes in the web is acceptable so long as certain rules are followed. Openings around stairs and vents require that joists and headers around the openings be doubled.

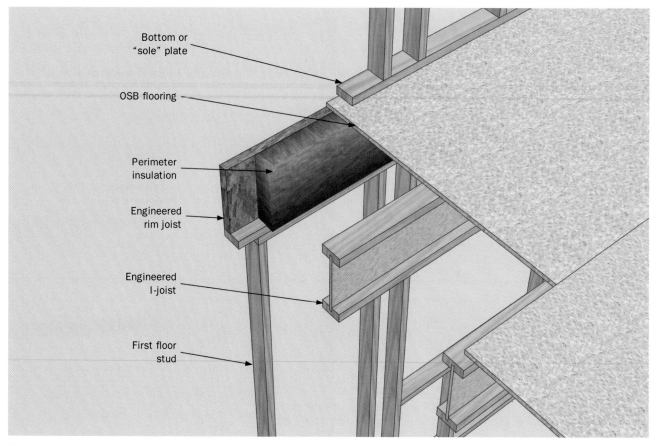

Bottom or "sole" plate

OSB flooring

Perimeter insulation

Engineered rim joist

Engineered I-joist

First floor stud

End-wall framing for platform construction (junction of first floor ceiling with upper-story floor framing)

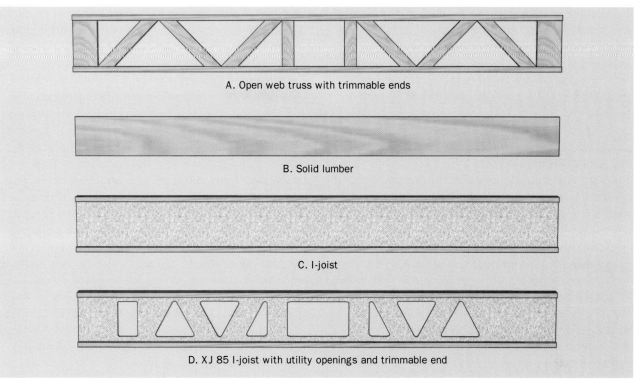

A. Open web truss with trimmable ends

B. Solid lumber

C. I-joist

D. XJ 85 I-joist with utility openings and trimmable end

The latest framing technology provides several choices for cost effective joists.

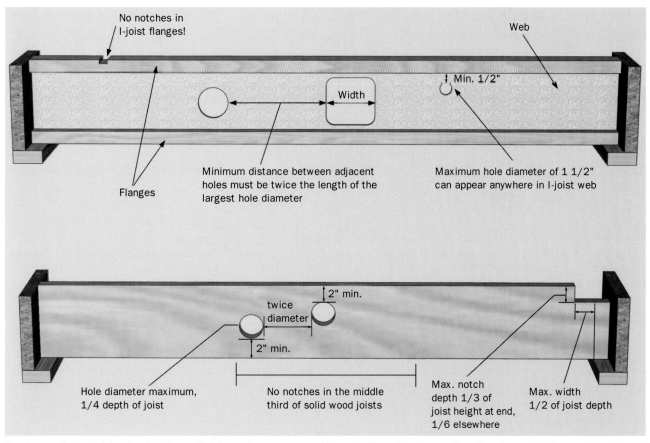

No notches in I-joist flanges!

Web

Min. 1/2"

Width

Flanges

Minimum distance between adjacent holes must be twice the length of the largest hole diameter

Maximum hole diameter of 1 1/2" can appear anywhere in I-joist web

2" min.

twice diameter

2" min.

Hole diameter maximum, 1/4 depth of joist

No notches in the middle third of solid wood joists

Max. notch depth 1/3 of joist height at end, 1/6 elsewhere

Max. width 1/2 of joist depth

Notches and holes in joists for plumbing or HVAC can destroy structural integrity if not placed properly. Follow these guidelines and check local building codes.

Sub-flooring

The sub-flooring that is attached to floor joists creates a solid box that is stronger than the separate parts. The combination of floor trusses and plywood have produced simpler floors that are stronger and quieter than old designs. Most standard flooring consists of a sub-floor, usually of ½" or ⅝" CDX plywood, and a finish floor which can be ⅝" particleboard or another layer of plywood. The second layer is installed after drying in the structure. If this method is used, be sure to install a layer of asphalt felt (the type used in roofing) or heavy craft construction paper between the two layers. This acts as a sound buffer and reduces floor squeaking. Often a two story house will use ½" flooring on the first floor and ⅝" on the second floor.

A potentially superior flooring method has been introduced by the American Plywood Association called the APA Sturdifloor: a single-layer floor that saves labor and materials. This floor is constructed with one layer of Sturdifloor approved CDX plywood (usually tongue and groove) that is nailed and glued to the floor framing members. Tongue and groove (T&G) plywood has edges

Tongue and groove OSB (oriented strand board) has become a popular substitute for plywood sub-floor. Boards will be "screwed and glued" to the underlying I-joists.

that interlock along the two longest edges. This reduces the tendency for boards to flex separately when a load is applied directly to the joint. The act of gluing the plywood to the joists produces an extremely rigid floor and virtually eliminates floor squeaking. This method works extremely well in conjunction with floor trusses or I-joists and produces a sturdy floor with a minimum of time, hassle and materials.

Note: Whatever method you use, make sure you purchase only approved plywood for this purpose. Plywood used for single layer construction will have an approved APA stamp on it. Ask your supplier about this type of plywood. All flooring plywood must be CDX grade or other rated OSB products. The "C" refers to the grade of plywood on one side, the "D" refers to the grade on the other side and the "X" refers to exterior grade. This grade of plywood is made with an exterior type glue that is waterproof and will stand up to prolonged exposure to the elements during construction. If you use a two layer floor with particle board as your second layer, do not leave this material exposed to the weather. It is not waterproof and will swell up like a sponge.

Wall Framing

Wall framing affects the size and orientation of all other finishing materials. You must be sure that everything is in the right place so that surprises do not occur later. If you make any last minute changes to the plans, consult with your framer or architect to insure that you don't create new problems. Even a simple change of framing height from 8' to 9' can cause many design problems. Windows may not appear proportional. The length of the stair's "run" will change because more steps will be needed to reach the second floor. Your plan may not have the room for additional steps.

Stud Spacing

The vast majority of houses in the United States are built with 2×4 studs spaced 16" on center. "Stud grade" lumber is more expensive than standard framing lumber. Studs are more uniform and straight with fewer cracks. You should not frame walls with utility grade 2×4s. They will be more uneven, making drywall application a nightmare. Utility grade lumber also may not be capable of supporting the proper load.

Consider framing with 2×6s to allow more room for insulation. This construction change can increase a wall's insulation factor by 50% or more depending on the type of insulation being used, a significant energy conservation measure. When this method is used, studs are spaced 24" apart. This simplifies construction and reduces the number of studs used by 20-30%. This method is not without its problems however. The biggest problem is that most doors and windows are designed to be set in walls of 2×4 depth. Buying special doors and windows or having custom setting done can cost a considerable sum and can negate many of the cost savings. As 2×6 framing has become more popular, manufacturers have responded with new doors and windows that can adjust to accommodate either wall size. Look for window units that will match the 2' framing width if you want to insert windows between the studs.

Generally, 2×6 stud framing is slightly more expensive than conventional construction, (mainly due to increased insulation) but it does provide a significant energy savings that can recoup your investment quickly. Make sure your

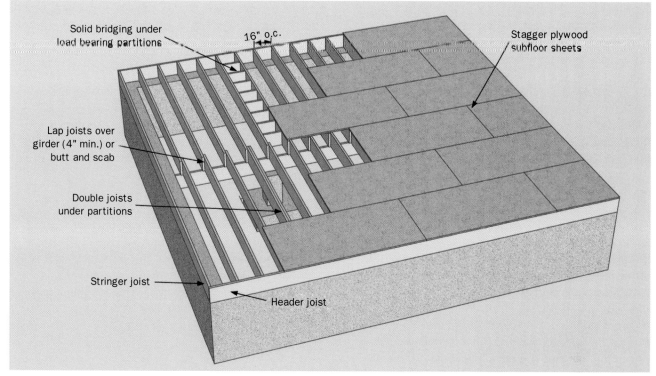

Solid bridging under load bearing partitions

16" o.c.

Stagger plywood subfloor sheets

Lap joists over girder (4" min.) or butt and scab

Double joists under partitions

Stringer joist

Header joist

Typical floor framing with solid lumber.

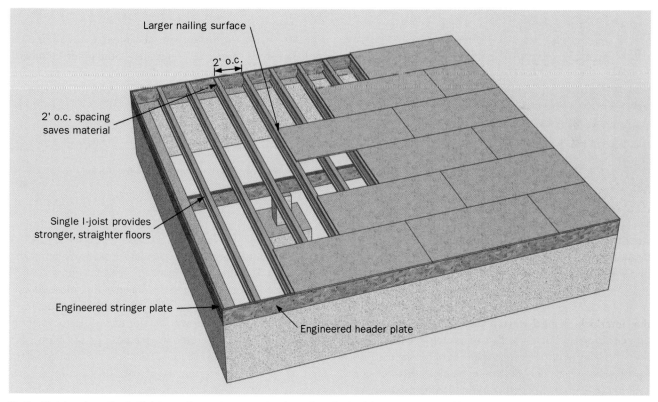

Larger nailing surface

2' o.c.

2' o.c. spacing saves material

Single I-joist provides stronger, straighter floors

Engineered stringer plate

Engineered header plate

"In-line" joist system. Single I-joist over the center support reduces installation labor and quantity of fasteners. It also results in a stronger, straighter floor.

carpenter is familiar with the inherent problems of this type of construction before deciding to use it.

Many of the cost savings of 2×6 framing can be preserved without any of the headaches by going to 2×4s spaced 24" on center and using a denser foam insulation with a higher R-value. This method is approved by all national building codes for single story construction and some local municipalities allow it on two story construction, or at least on the second floor. Check your local building codes to be sure. Spacing studs 24" on center saves time and materials and allows more space for insulation since fewer studs are used. You will find more information on this technique in the Saving Money chapter. If your carpenter is unfamiliar with this method of construction, provide brochures on Optimum Value Engineering available from the NAHB, PATH, or ToolBase. Also refer to Section R602 of the International Residential Code (IRC) for wood wall framing requirements.

Supervision

During construction, make sure that the walls are spaced properly in relation to plumbing that may be cast in a concrete slab. By the time the plumber discovers a problem you might have a major reconstruction project ahead.

Heads Up: If you plan to install one-piece fiberglass shower stalls, you must set them in place before putting up the walls. Some fiberglass stalls will not fit through bathroom door openings.

Make sure to check periodically to be sure that all walls are square and plumb. Purchase a quality builder's level (the longer the better) and check walls periodically. Stairs and fireplaces are other sources of potential problems. Many older carpenters have the habit of installing bridging between studs. Discourage this practice, as it is an unnecessary waste of lumber and can make insulation more difficult to install and less effective.

Your carpenter will be required by building code to brace the corners of the structure for added rigidity against racking forces. This can be accomplished by using a sheet of plywood or OSB in the corners, a diagonal wood brace cut into the studs, or a metal strap brace made for the purpose. Although more tedious, diagonal bracing or metal strapping is advantageous because it is cheaper and does not interfere with the insulating ability of sheathing. Plywood is not a good insulator.

As the house is "dried-in", make sure to cover openings in the walls and ceiling as soon as possible to protect the interior from rain. Place poly over all openings until the windows and doors arrive. Once the windows and doors are installed, lock them with the locksets or nail them shut with finishing nails.

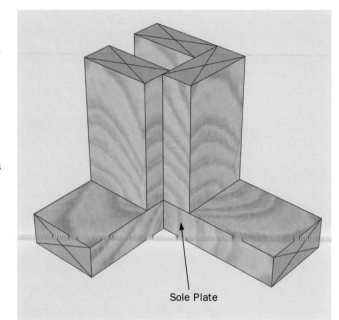

Traditional three-stud corner.

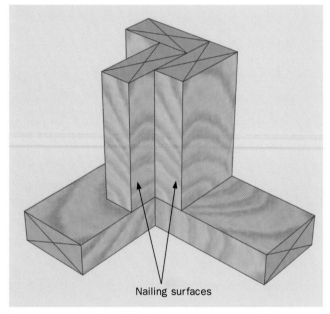

Three-stud corner without blocking allows for more insulation.

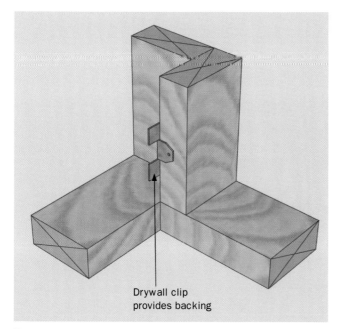

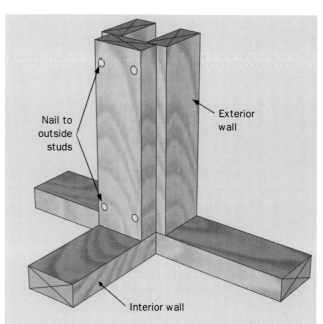

Two stud corner with drywall clips save lumber.

Intersection of interior partition and exterior wall. Double studs in exterior wall

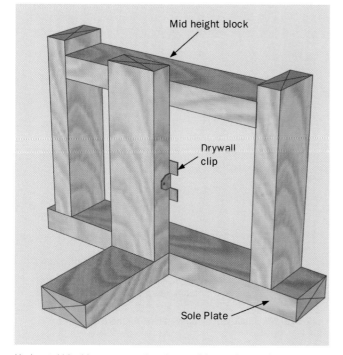

Horizontal blocking supports interior partition and saves lumber without disturbing natural o.c. spacing

Roof Framing

The two basic types of roof framing consist of stick-built roofs and roof trusses. As with floor trusses, roof trusses have the advantage of being very sturdy for their size and weight. In most house designs, roof trusses require no load-bearing walls between the exterior walls. This gives you more flexibility in designing the interior of your home. Roof trusses are also generally cheaper than stick-built roofs because of significant labor savings in construction. Their only major disadvantage is the lack of attic space as a result of the cross member supports. This can be alleviated somewhat by ordering "space saving" trusses that have the inner cross members engineered to provide more attic space. Stick-built roofs are more appropriate for high pitches or oddly shaped roof lines.

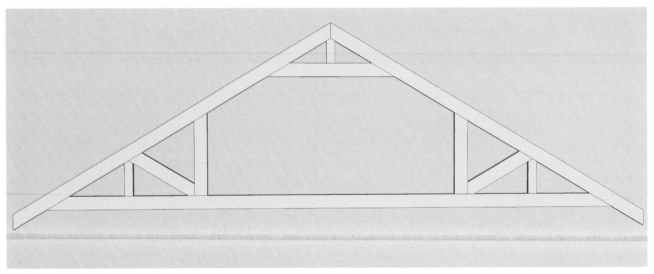

Attic trusses provide more room, but cannot span as great a distance without load bearing support.

Hip roofs (no gable ends) have become very popular in recent years. However, they can add a lot to framing costs and labor. Hip roofs are generally stick built, although some framers have experimented with stick-truss hybrids to frame these complex roof lines. The framing grows in complexity with every valley and corner, because each rafter must be cut to a different length and multi-angled cuts are required where rafters meet ridges. Framers must be experts in this type of roof to be successful and they will charge a premium for that expertise. Hip roofs greatly reduce attic space as well.

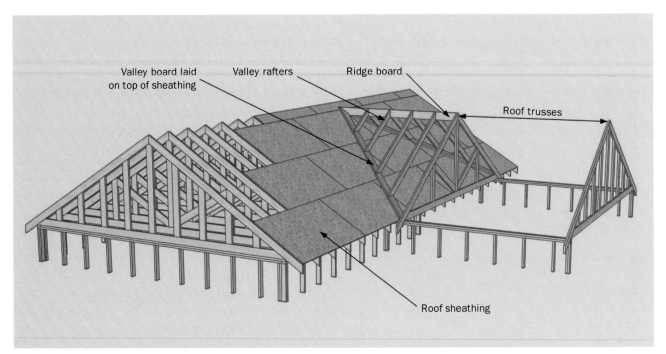

"Closed" rafter framing joining perpendicular truss roof segments. Uses a combination of trusses and stick framing.

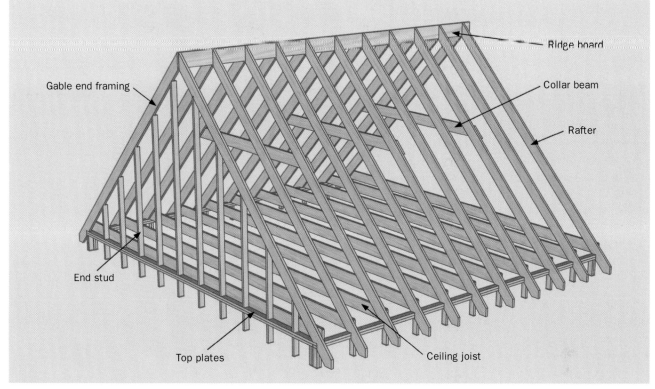

Typical rafter framing for pitched roof.

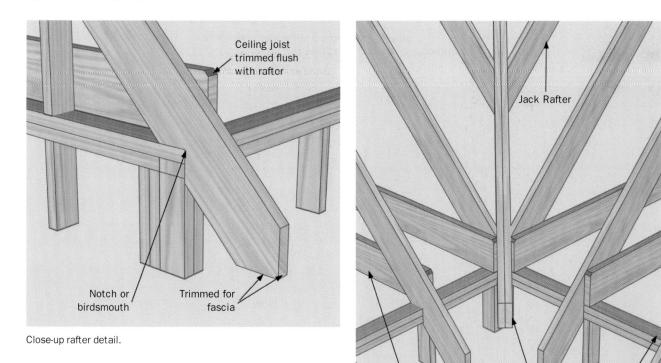

Close-up rafter detail.

Framing at valley in rafter roof.

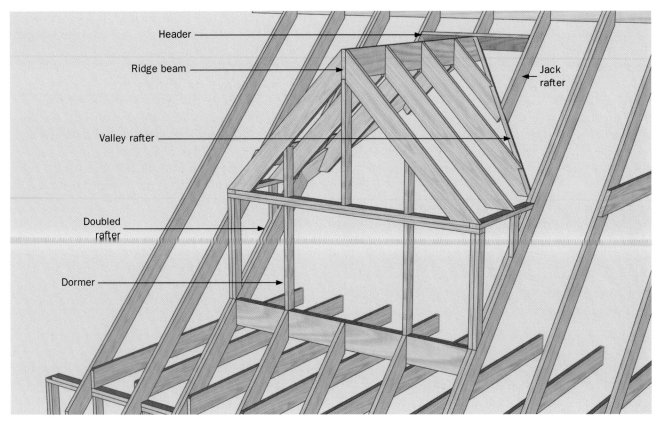

Header

Ridge beam

Jack rafter

Valley rafter

Doubled rafter

Dormer

Gable dormer framing

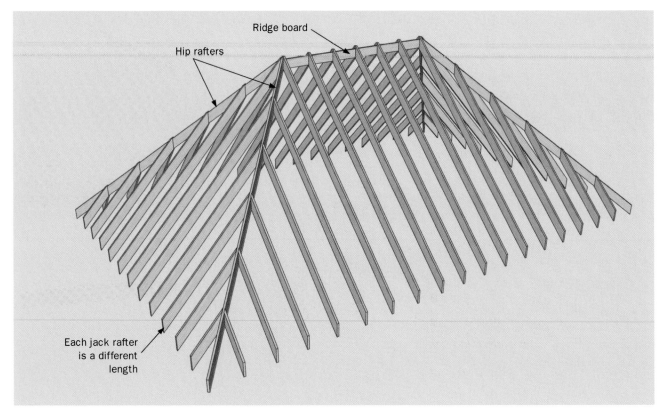

Ridge board

Hip rafters

Each jack rafter is a different length

Hip roof framing requires complex multiple angles, resulting in extra labor and expense.

SHEATHING

Roof sheathing

Your carpenter will install the roof sheathing, which usually consists of ½" CDX plywood or OSB (oriented strand board) made specifically for this purpose. Radiant barrier sheathing is another option that can greatly reduce heat gain in the attic. Make sure the radiant barrier is pointed down toward the attic. This sheathing functions best when there is a layer of air between the radiant barrier and any insulation.

Radiant barrier sheathing can cut heat gain in summer months.

Request that plywood clips be used in between each sheet of roofing material. These clips are small "H"-shaped aluminum clips designed to hold the ends and sides of the plywood sheets together for added rigidity. This will help prevent warping and "wavy" roof surfaces.

Normal roof decking may not be used with cedar shake, slate or tile roofs. So, if you are planning to have such a roof, consult with your framing and roofing subs. When installing the roof sheathing, the carpenter should be aware of the type of roof ventilation to be used. The most convenient and efficient type to use is continuous eave and ridge vents.

Wall Sheathing

Wall sheathing comes in several different types. Depending on the material, sheathing is classified as structural or non-structural. Asphalt-soaked fiber sheathing has been replaced in recent years by OSB board or insulating foam sheathing. Foam sheathing is a good compromise of value and insulation ability. If you plan to stucco, exterior gypsum board is most often used. Talk to your local supplier to see what types are available in your area and their insulation ability. When installing foam sheathing, your carpenter should take care not to rip or puncture the sheathing. If he does, ask him to replace the defective piece. Special nails

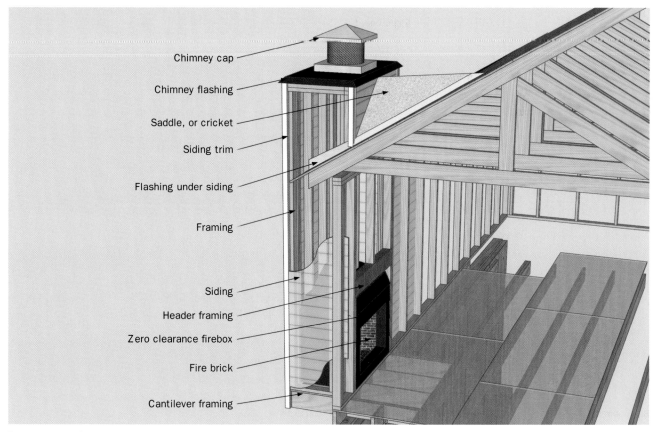

Zero clearance fireplaces allow for easy, low cost installation without the need for foundation support.

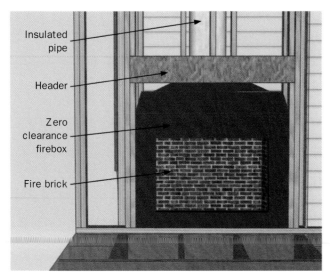

Conventional framing simplifies hearth construction

that have a plate on the shaft should be used to hold the foam sheathing to the stud without puncturing the sheathing's skin.

Structural sheathing, such as OSB board and plywood is required at corners to add structural rigidity and racking strength to the house, but adds little to insulation values. For simplicity, many framers opt to use OSB to sheath the entire structure, but this is unnecessary from a structural point of view and a waste of expensive material. Structural bracing is only needed in the corners. If you want the added insulation of foam sheathing, consider using wood or metal cross bracing at the corners for structural rigidity so that foam sheathing can be used for the entire exterior.

Doors and Windows

After the roof and sheathing have been installed, the framing contractor will be ready to install the doors and windows and "dry in" the structure. The proper installation of doors and windows is probably the source of more irritation and hot tempers than anything. The advent of pre-hung doors has made the job easier, but still not foolproof. With your level, check to make sure each door is completely square and that there is an even gap between door and frame from top to bottom. Make sure the door has plenty of room to swell in damp weather (if it is wood) and still open freely. Measure under the door to make sure there is plenty of room for subflooring, carpet pad and carpet without the door rubbing over the carpet when it is installed. The same amount of care should be taken with the windows to guarantee that they will slide freely within their frames.

If you are using house wrap, make sure the wrap is folded inside the rough-in framing before the window or door is set. Coordinate with the siding subcontractor to insure that the proper waterproofing methods are followed around openings and that flashing is properly installed and lapped.

Waste Disposal

Few projects generate more scrap than frame construction. By using Optimum Value Engineering, you can cut waste significantly. Most waste will have to be hauled off to a landfill where much of it will lie undisturbed for centuries. This can cost a considerable amount of money. An alternative is to turn the waste into mulch. Some modern chipping machines are capable of grinding up wood, drywall, and concrete. These materials can then be used on site as mulch or as aggregate for new concrete structures.

Recycle as much waste as possible. Many small pieces of lumber can be used as blocking and other items. Cornice subs use 2×4s 8" and longer to run cornice framing. Electricians sometimes use 2×4s for blocking around outlets and switch boxes. Plumbers use 2×4s 6" and up to brace pipes, tubs, and other fixtures. Stack all usable scrap in a neat, accessible pile protected from the elements. Framers can use these pieces in lieu of fresh lumber.

Other Trades

Your framing sub must be familiar with the workings of the plumbing, HVAC, and prefab fireplace subs so that he can anticipate where ductwork, pipes and heating equipment must be placed. Make sure that the blueprints indicate the location of furnaces and ducting.

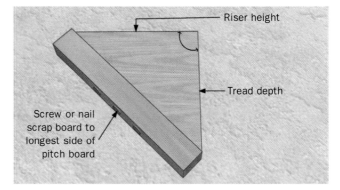

To build a pitch board, cut ¾" plywood to exact dimensions of the tread and riser.

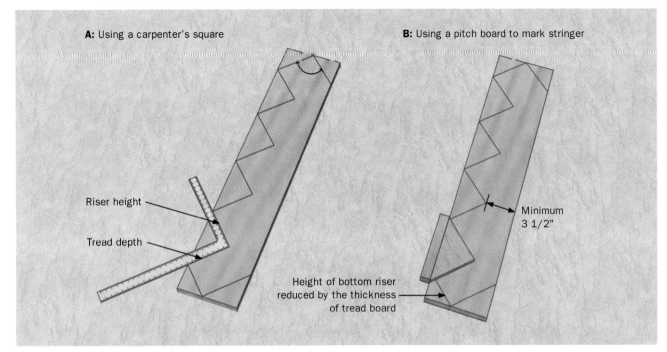

A: Using a carpenter's square   **B:** Using a pitch board to mark stringer

Riser height

Tread depth

Minimum
3 1/2"

Height of bottom riser
reduced by the thickness
of tread board

Two methods for laying out a stair stringer.

FRAMING — STEPS

FR1	CONDUCT standard material bidding process. Find the best package deal on good quality studs, sheathing, press-board, plywood, interior doors (pre-hung or otherwise), exterior doors and windows. Find out their return policy on damaged and surplus material and whether their price includes delivery.
FR2	CONDUCT framing labor bidding process. Ask to see their work and get recommendations. A good crew should have at least three people. Any less and the framing probably won't go as quickly as you would like it. Consider a speed/quality incentive tied to payment.
FR3	DISCUSS all aspects of framing with the crew. This includes all special angles, openings, clearances and other particulars. Also make sure that the electrical service has been ordered. Discuss special order materials such as flitch plates or glue-lam beams and order them as soon as possible.
FR4	ORDER special materials ahead of time to prevent delays later. This includes custom doors, windows, beams and skylights. This is also a good time to make sure your temporary electric pole works — you will need it for the power saws.
FR5	ORDER and receive first load of framing lumber. Have a flat platform ready to lay it on near the foundation. "Line out" framing walls with chalk lines on the foundation to act as a framing guide. Check chalk lines for accurate dimensions and squareness.
FR6	INSTALL sill felt, sill caulk, or both.
FR7	ATTACH pre-treated sill plate to lag bolts embedded in the foundation.
FR8	INSTALL steel or wood support columns in basement.
FR9	SUPERVISE framing process. Check framing dimensions to be sure flooring and wall measurements are in the right place. This is an ongoing process. Check with framer in the morning each day to discuss progress and problems. Check for needed materials. Don't leave excess materials on site — they have a tendency to disappear. Ask framers to cover lumber with plastic if rain is likely. Ask them to put nails and other valuable supplies in their truck to prevent theft. Check wall plates with a level. Also run a span board between two parallel walls and check with a level to insure that walls are all at the same exact height.
FR10	FRAME first floor joists and install subfloor. Place these as close together as you can afford. If you can afford 2×10s instead of 2×8s, this will help give your floor a more solid feel. Consider I-joists to increase stiffness. Tongue and groove plywood is a popular sub-floor. Make sure to use exterior grade plywood since it will be exposed to the weather during framing. At bay windows, joists will extend out beyond the wall.
FR11	FRAME stairs to basement (if any). In many respects a home is built around staircases. This is a critical step.
FR12	POSITION all large items that must be inserted before framing walls prior to setting doorways and inner partitions. These include bathtubs, modular shower units, HVAC units and oversized appliances. Beware of installing oversized appliances that will not fit through the door when you move or when they have to be replaced.
FR13	FRAME exterior walls/partitions (first floor). Keep track of all window and door opening measurements.

FR14	PLUMB and line first floor. This is critical to a quality job in drywall and trim. This involves setting the top plates for all first floor walls with lapped joints while someone is checking to make sure that all the angles are plumb and perpendicular. This involves straightening the walls exactly — fine tuning. Walls are braced one by one. Start from one exterior corner and go in a circular direction around the exterior perimeter of the house, adjusting and finishing each wall.
FR15	FRAME second floor joists and subfloor.
FR16	POSITION all large second floor units prior to setting doorways and partitions. Again, these include bathtubs, modular shower units and other large items.
FR17	FRAME 2nd floor exterior walls/partitions.
FR18	PLUMB and line second floor. Refer to plumbing/lining first floor.
FR19	INSTALL second floor ceiling joists if roof is stick built. Have your window and door supplier visit the site to check actual dimensions of all openings.
FR20	FRAME roof. If you want a space between your window trim and cornice, add a second plate on the top of the second floor wall. Depending upon the size and complexity of your roof, you may either use pre-fab trusses, or stick build your roof.
FR21	INSTALL roof deck. Plywood sheathing will be applied in 4x8 sheets in a horizontal fashion. Each course of plywood should be staggered one-half width from the one below it. Your framers should use ply clips to hold the sheets together between rafters.
FR22	PAY first framing payment (about 45% of total cost). Check for missed items such as framing for attic stairs, fir downs for kitchen cabinets, framing for house fan, hearth framing, stud supports around tubs, etc.
FR23	INSTALL lapped tar paper. This protects your roofing deck and entire structure from rain. At this point, the structure is "dried in" and a major race against the elements is over.
FR24	FRAME chimney chases, if you have them. Ask framer to cover top of chase with plywood or rain caps to protect against rain.
FR25	INSTALL prefab fireplaces. Install a fireplace with at least a 42" opening and 24" depth. Anything smaller will have problems taking full-sized fire wood. Prefab fireplaces are usually installed by the supplier or a separate sub. Since plans seldom indicate the specific type of fireplace, make sure framer is aware of the brand and requirements of your particular fireplace. Store your gas log lighter key in a safe place to avoid losing it.
FR26	FRAME dormers and skylights.
FR27	FRAME tray ceilings, skylight shafts and bay windows if you have them. Skylight shafts should flare out as they come into the room. This allows more light to enter.
FR28	INSTALL sheathing on all exterior walls.
FR29	INSPECT sheathing. Watch for punctures and gaps in the insulation that can be fixed with duct tape.
FR30	REMOVE temporary bracing supports. Bracing supports installed prior to backfill and those used during the framing process are removed and used as scrap.
FR31	INSTALL exterior windows and doors. Make sure house wrap is folded into rough opening before adding windows. Add other waterproofing membranes as windows and doors are installed. The siding and cornice sub may do this operation. If you are using pre-hung doors, this becomes an easier process. A pre-hung door comes with its own door frame, but without door knobs and locks. After installing door knobs and locks, the house can be locked up to prevent theft and vandalism. Just make sure to give your subs temporary keys. If your siding and cornice sub installs your windows, this is covered under another step.
FR32	APPLY dead wood. "Dead wood" consists of short pieces of lumber installed in areas that need backing on which to nail drywall. These include: Tops and sides of windows where curtain hardware will be attached, small areas around stairs, locations of electrical boxes and bracing for ceiling lights or fans.
FR33	INSTALL roof ventilators. This should only be done on the back of the roof.
FR34	FRAME DECKS using pressure-treated lumber to prevent rotting and termite infestation.
FR35	INSPECT framing. You should have been checking on the framing process all along, so there should be no big surprises at this point. This is the LAST chance to get things fixed before you make your last payment to the framing crew.
FR36	SCHEDULE and have your loan officer visit the site to approve rough framing draw.
FR37	CORRECT any problems with framing job. This is an ongoing process that should occur throughout the framing process. The earlier a problem is detected, the easier and cheaper it is to fix. Look for bowed wood that will making drywall attachment difficult. Correct these by shimming or by cutting a slice partway through the stud. At this point, you have one last bargaining tool before making the next payment to the framing crew.
FR38	PAY framing labor. Retain a percentage of pay until all work is completed and final inspection is approved. Remember to deduct any necessary worker's compensation.
FR39	PAY framing labor retainage. Get a signed affidavit when final payment is made.

FRAMING — SPECIFICATIONS

- All materials to be crowned (circular grain goes in the same direction). Where possible, all wall studs have crown pointing the same direction. Floor joists to have crowns aligned pointing up.
- All walls and joists to be framed as specified in the blueprints.
- All joists to be framed 16" (24") OC.
- All vertical wall studs to be framed 16" (24") OC.
- All measurements to be within ¼" and plumb, perpendicular and level.
- Cross bridging or strongback to be used at all joist midspans to increase strength and stability of floor.
- Flooring underlayment should be screwed and glued with approved construction adhesive.
- All lumber in contact with concrete must be pressure treated.
- Double joists to be used under load-bearing partitions and bathtubs unless truss floors are used.
- Floor joists not to interfere with tub, sink and shower drains.
- Four-inch overlap on lap joists supported over girders.
- Blocking used at joist overlaps to resist racking.
- All cuts necessary for HVAC and plumbing work to be braced.
- Hip roof to be 10/12 pitch and gable roof to be 12/12 pitch.
- No cuts or notches to be made in laminated beams or truss plates.
- Notching and drilling of I-joists to follow manufacturer's recommendations.
- All chimney chases framed with a cricket (also known as a saddle) installed to allow drainage of excess water.
- Worker's compensation insurance to be provided by builder.

Framing extras include:
- Bay windows
- Stacked bay windows
- Staircases
- Prefab fireplace opening and chase
- Skylights
- Basement stud walls

Sheathing
- Sheathing to be installed vertically along studs.
- Sheathing nailed every foot along studs.
- All gaps over ¼" to be taped.
- Skin on foam sheathing not to be broken by nail heads, hammer or other means.
- Sheathing nails to be used exclusively.

FRAMING — INSPECTION

Framing inspection will be an ongoing process.
- [ ] Errors must be detected as soon as they are made
- [ ] for easy correction.
All work done according to master blueprints.
- [ ] No framing deviations exceed standard ¼" leeway
- [ ] for error.
- [ ] Vertical walls are plumb.
Opposite walls of rooms are parallel.
- [ ] Horizontal members such as joists, headers and sub-floors are level.
- [ ] Window and door openings are proper dimensions and are square. Rough openings are 2" wider than opening dimension.
- [ ] Adequate room at all door openings for trim to be attached. Opening is not flush against a corner.
- [ ] No loose or non-reinforced boards or structures.
- [ ] Corner bracing used in all wall corners.
- [ ] No room framed before installing large fixtures or appliances that are larger than door openings.
- [ ] All framing is being done to within 1" of on center (O.C.) spacing requested.
- [ ] Wall corner angles are square. This is especially critical in kitchens (where cabinets and counter tops are expressly designed for 90° angles).
- [ ] Check that vertical studs in stud walls are even and not warped. This can be done by pulling a string along each stud wall. The string should just touch all studs at the same time.
- [ ] All joists, vertical studs and rafters are crowned. This means that the natural bow of the wood is running in the same direction for all adjacent members.
- [ ] No excessive scrap.
- [ ] Adequate lumber is available for framers. Have them notify you if shortages are about to occur. Site measurements match the blueprints. If something is to change, it should be recorded on the blueprints in red.

Sub-floor
- [ ] Specified plywood or other sub-flooring used. Proper spacing and nailing of sheets. Tongue and groove plywood where specified.
- [ ] No weak or crushed plywood in floor. Bottom two sheets in a plywood load often are broken when sliding load off truck. Make sure it is not used in the floor.
- [ ] All sill studs are exterior grade and pressure treated.
- [ ] All sub-flooring glued to studs with construction adhesive and nailed adequately. Check from underneath.
- [ ] Bracing between floor joists as specified.

Stairs

- ☐ No squeaks or other noise from treads or risers.
- ☐ Treads deep enough for high-heeled shoes to fit on comfortably.
- ☐ Proper bracing in walls to support rails later.
- ☐ All risers same height.
- ☐ All treads same width or as specified.
- ☐ All framing complete for trim work to begin.
- ☐ Attic staircase works properly.

Fireplace

- ☐ Fireplace area headers in proper position for prefab or masonry work.
- ☐ Framing for raised mantel installed.
- ☐ Flue framed with clearances according to building code.
- ☐ Proper height for raised hearth if used.

Roofing Deck

- ☐ Plywood sheets staggered from row to row.
- ☐ Plywood sheets nailed every 8" along rafters or trusses.
- ☐ Ply clips used on all plywood between supports.
- ☐ Proper exterior grade of plywood used.
- ☐ Plywood sheets and roofing felt cut properly for ridge and roof vents.

Deck or Porch

- ☐ Deck has galvanized joist hangers on joist ends.
- ☐ Pressure-treated lumber used on outdoor decks.
- ☐ Galvanized decking nails or screws used.

Miscellaneous Framing

- ☐ Tray ceiling areas framed as specified.
- ☐ Tray ceiling angles even.
- ☐ Skylight openings and skylights according to specifications.
- ☐ Skylights sealed and flashed.
- ☐ Access door to crawl space and wall storage areas installed according to specifications.
- ☐ Pull-down attic staircase installed and operable.

Sheathing

- ☐ Proper sheathing installed.
- ☐ Sheathing applied vertically.
- ☐ Sheathing nails used.
- ☐ No punctures or other damage to sheathing.
- ☐ Perimeter of sheets nailed adequately to prevent sheathing from blowing off in wind.
- ☐ Sheathing joints tight and taped where necessary.

ROOFING AND GUTTERS

Types of Roofing Materials

Roofing is usually measured, estimated and bid based on squares of shingles (units of 100 square foot roof coverage). Common roofing compositions include asphalt, fiberglass and cedar shakes, although roofs are also composed of tile, slate, metal sheet, fiber cement and other more exotic materials. Fiberglass and asphalt are popular choices due to their favorable combination of appearance, price and durability. Roofing warranties are most often a minimum of 20 years.

Asphalt and Fiberglass

Asphalt and fiberglass shingles are normally 12" tall and 36" wide with two or three tabs to make the units look like several individual shingles when installed. Weights vary from 180 lb. (per square) on the cheap side up to 340 lb. on the high end. The heavier they are, the more durable the roof. Lighter shingles also have a tendency to get blown around in heavy wind. Look for shingles with a wind rating higher than normal. These shingles will have self-sealing asphalt strips along the underside of the shingle that will melt and fuse to the shingle below, creating a wind resistant surface. If shingles are applied in the winter, care must be taken to avoid cracking or splitting them. They become quite brittle in cold weather. Try to avoid installing during periods of high wind. Self sealing strips need the heat of summer to adhere, so shingles are particularly vulnerable in the first year, if installed in cold weather.

Strip shingles normally come in packages of 80 strips. About three packages cover one square of roof. When installed, about 5" of the shingle's height remains exposed, hence a 7" top lap. Architectural shingles, also known as laminated or dimensional shingles, are heavier and more expensive brands. The "architectural" term means that the shingle usually has uneven tabs or multiple layers to give it a look of texture, weight, and depth. They tend to break up the "monotony" of a standard shingle roof. They often have 40 year warranties. The highest rated architectural shingles can often withstand winds up to 120 mph. With modern manufacturing techniques, asphalt and fiberglass shingles have been produced that approximate the appearance of a wide variety of roofing products, at reasonable prices.

Metal Roofs

The popularity of metal roofing has increased tremendously in recent years as new products have hit the market. They have several advantages over standard roofing options, such as fire-resistance, long life, and durability. They are also lightweight and good at reflecting sunlight when painted in lighter colors. Interlocking or "standing seam" metal roofs are also highly wind resistant. The latest designs are manufactured to resemble other roofing products like asphalt, Spanish tile, wood shakes, or slate. From a distance,

The uneven tabs of architectural shingles add texture and character to the roof.

Metal roofs, long popular in commercial settings, are making their way to residential homes.

they can often be indistinguishable from the real thing, at a more reasonable price. Metal roofing can be tricky to install properly, so it's best to hire an experienced roofing contractor for this job.

Tile

As with fiber cement siding, tile shingles have taken the market by storm. They include cement and clay roof tiles that have properties of durability and fire resistance. Cement tiles can be formed to resemble almost any type or style of roof shingle. Houses in fire prone areas can have the look of cedar without the risk. The difficulty of installation varies with the type and style of tile shingle. Clay tiles can be one of the most challenging types to install properly and will require a special underlayment. Cement tiles can be brittle and hard to cut. Because of these demands, tile roofs require installers fully familiar with the technology. They can be expensive but have an extremely long lifespan. If your family plans to stay in your home for generations, this is the shingle for you.

Fiber cement shingles made to resemble cedar shakes.

Traditional clay tile.

Cedar Shake

Once one of the most popular shingles in America, cedar shakes have fallen out of favor because of their high maintenance and risk of fire. If you love the look, don't despair. Convincing substitutes, like metal and tile will allow you to maintain the flavor and texture of an authentic cedar shake roof without the hassle.

Recycled and Green Alternatives

Some manufacturers are now recycling old tires and soft drink bottles into durable tile products that have excellent durability and lifespan. The costs can be high, but for a homeowner who wants to "go green" these are excellent choices. Plastic tiles made to resemble slate can have warranties of 50 years or more.

An even more exciting technology is on the horizon. Solar manufacturers like SRS Energy are integrating amorphous solar cells into tile and slate shingle substitutes. Part of the resistance to solar panels in suburbs has been the large, obtrusive panels jutting out from roof lines. Solar integrated shingles will allow solar technology to become part of the architecture. As prices on these materials come down, expect to see many more solar roofs in the future.

Flashing and Waterproofing

Water leakage at roof shingle junctions is a common and difficult problem to solve. This usually results from the improper installation of flashing materials and can be

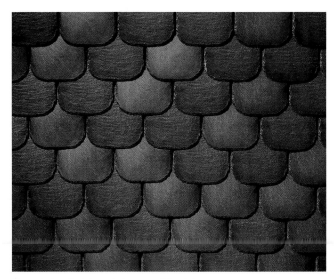
Recycled material made to resemble slate

Integrated solar tile

Solé Power Tile by SRS Energy integrates solar electric into architectural design.

difficult or impossible to solve properly without applying a new roof. What makes this problem even more maddening is the tendency for leaks to migrate along rafters and exit the ceiling several feet or yards away from the original source. It can make leaks difficult to track down.

Make sure flashing is properly installed along the junctions of chimneys and along valleys. Many roofing subs like to "bend" shingles across valleys because it reduces the labor necessary to cut each shingle to the proper length and angle. However, these shingles are prone to cracking and leaks, oftentimes years later when the roofing contractor has moved on. Insist that your roofer install a continuous roll of flashing and/or waterproof membrane down valleys. Shingles should then be cut at angle to overlap the flashing by several inches to allow an unobstructed drainage path for water. Most people don't realize the sheer volume of water captured by a large roof. All that water will migrate to the roof valleys and come rushing down in a torrent. This vulnerable area needs to be installed properly.

Ice Dams

This is a problem unique to northern climates which can cause extensive damage to roofs and cornices. In winter — as snow and ice accumulates — heat from the home will migrate through the roof and cause some of the ice to melt. This moisture will drain down to the cornice area, which usually extends past the heated portion of the roof structure. There, the water will refreeze. This ice "dams up" any subsequent water that travels down the roof. All shingles and flashing depend on gravity to drain moisture away. The ice dam will prevent this process and water will eventually find its way behind shingles and into pockets of the cornice. A few years of "ice dams" can quickly cause extensive rotting and mold.

If you live in a northern climate zone, have your roofer apply an extra layer of waterproofing membrane underneath the last few courses of shingles. Another good way to prevent ice dams is to insulate your roof thoroughly. The less heat that escapes, the less melting that will occur.

Roofing Subcontractors

Your roofing installer should quote you a price with and without material. The steeper your roof pitch, the steeper his price per square installed. Prices start to get steep after about 10/12 pitch, which represents 10" of rise (vertical) for every 1' of run (horizontal). Hence, 12/12 pitch is a 45 degree angle. Appeal is growing for steeper and steeper pitches. Make sure the roofing sub has plenty of worker's compensation insurance or deduct heavily from his fee if you have to add him to your policy. He's one fellow with a very high risk of injury. If you plan to supply the shingles, make sure they are at the site a day or so early.

Caution: Regardless of whether your roofing sub has workman's compensation, you should demand that all installers wear safety harnesses when working on the roof. Falls from the roof are some of the most serious and common injuries on a work site and are totally preventable with proper safety gear. Your potential for liability in the event of a serious injury can outstrip most insurance policies.

NOTE: Since asphalt and fiberglass shingles are by far the most common used, the STEPS on the following pages relate specifically to them. If you use roofing materials other than these two, refer to a handbook and detailed construction methods.

Roof membranes along valleys provide an extra layer of insurance against leaks.

In northern climates, apply waterproofing membranes along cornices and intersections to help guard against ice dams.

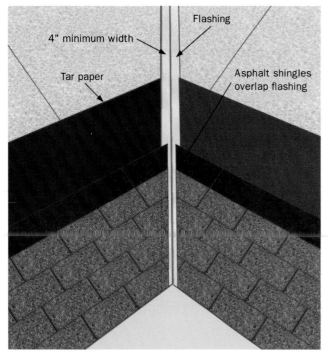

4" minimum width

Flashing

Tar paper

Asphalt shingles overlap flashing

Flashing in roof valleys.

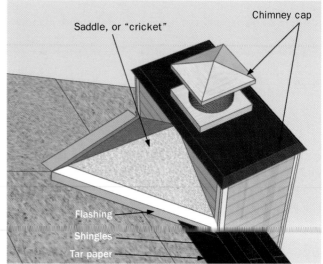

Saddle, or "cricket"

Chimney cap

Flashing

Shingles

Tar paper

Saddle, or "cricket" diverts water away from chimney.

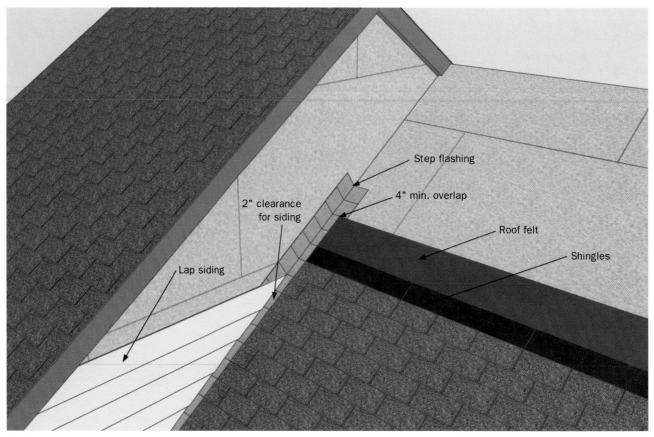

Step flashing

2" clearance for siding

4" min. overlap

Roof felt

Lap siding

Shingles

Flashing at roof and wall intersections.

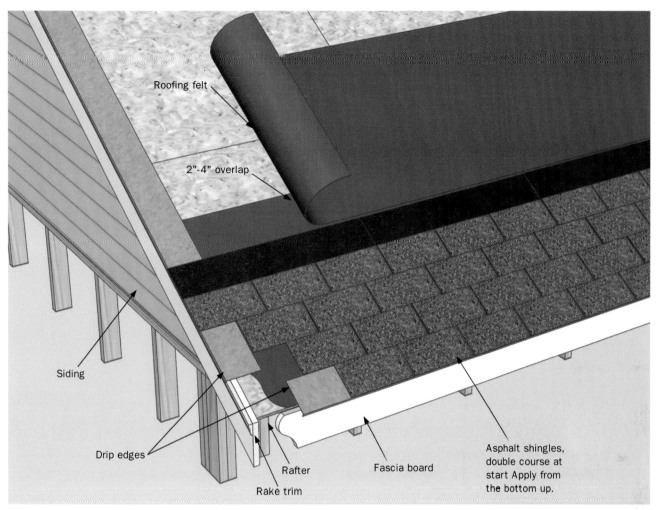

Roofing felt

2"-4" overlap

Siding

Drip edges

Rafter

Rake trim

Fascia board

Asphalt shingles, double course at start Apply from the bottom up.

Roofing underlayment details. In northern climates, apply waterproof membrane underneath the first few courses of shingles to prevent ice dams.

Gutters

Gutters are used to channel rain water to downspouts and then to planned drainage paths. Typically, gutters are made of galvanized iron, vinyl, or aluminum. The number of downspouts will vary. Your gutter subcontractor will help you in determining the number and placement of your downspouts and gutters. If you can, use underground drain pipes instead of simple splash blocks. Care must be taken that they are not completely buried during the backfill stage. A heavy gauge PVC black tubing should be used.

Gutter subs normally charge by lineal feet of gutters and downspouts installed. They may charge you a certain amount for any more than four corners. They will also charge you extra for installing gutter screens and drain pipe. Get your gutters up as soon as possible to start good drainage and to keep mud from splashing on the side of the house. Gutters shouldn't go on until the cornice is primed and painted and the exterior walls finished (because of downspouts).

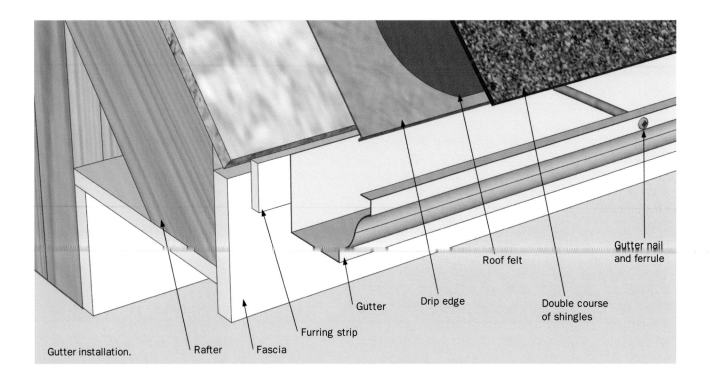

Gutter installation.

Rafter

Fascia

Furring strip

Gutter

Drip edge

Roof felt

Double course
of shingles

Gutter nail
and ferrule

ROOFING — STEPS

RF1	SELECT shingle style, material and color.
RF2	PERFORM standard bidding process for material and labor.
RF3	ORDER shingles and roofing felt (tar paper). This is done only if you have decided to use your roofer for labor only. Check with him to make sure he knows of the brand and how to install them.
RF4	INSTALL 3" metal drip edge on eave nailed 10" on center. Roofing paper will go directly on top of this drip edge (optional).
RF5	INSTALL roofing felt. This should be done immediately after the roofing deck has been installed and inspected. Your framer will install the tar paper as the last thing before finishing the job. This step protects the entire structure from immediate water damage. Until this step is performed, the project should proceed as quickly as possible.
RF6	INSTALL 3" metal drip edge on rake. This drip edge goes over the roofing paper. Also install aluminum flashing at walls and valleys (optional).
RF7	INSTALL roofing shingles. Roofs over 30" wide should be shingled starting in the middle. Shorter roofs can be shingled starting at either rake. Starter strips, normally 9" wide, are either continuous rolls of shingle material that start at the eave or doubled shingle strips. Shingles are laid from the eave, overlapping courses up to the highest ridge. Don't install shingles on very cold days, if you have a choice.
RF8	INSPECT roofing. Refer to contract specifications and inspection checklist for guidelines. You will have to get up on the roof to check most of this.
RF9	PAY roofing subcontractor. Get him to sign an affidavit. Deduct for worker's compensation if applicable.
RF10	PAY roofing sub retainage.

GUTTERS — STEPS

GU1	SELECT gutter type and color.
GU2	PERFORM standard bidding process.
GU3	INSTALL underground drain pipe. Flexible plastic pipe is the type used most often. Make sure it is covered properly.
GU4	INSTALL gutters and downspouts.
GU5	INSTALL splashblocks and/or other water channeling devices.
GU6	INSPECT gutters. Refer to inspection guidelines, related checklist, contract specifications and your building code.
GU7	INSTALL copper awnings and dormer tops.
GU8	PAY gutter subcontractor. Have him sign an affidavit.
GU9	PAY gutter sub retainage.

ROOFING — SPECIFICATIONS

- Bid is to perform complete roofing job per attached drawings and modifications.
- Bid to include all materials and labor and ten year warranty against leakage.
- Apply 240 lb. 30-year fiberglass shingles (Make/Style/Color) to roof using galvanized roofing nails.
- Install two thermostatically controlled roof vents on rear of roof as indicated on attached drawings.
- Install continuous ridge vent per drawing specifications.
- Roofing felt to be applied and stapled to deck as indicated below: No.15 asphalt saturated felt over entire plywood deck.
- All roofing felt to have a vertical overlap of 6".
- Install 3" galvanized eave and rake drip edges nailed 10" O.C.
- All work and materials are to conform to or exceed requirements of the local building code and be satisfactory to contractor.
- Roofer to notify builder immediately of any condition which is or may be a violation of the building code.
- Nails are to be threaded and corrosion resistant. Either aluminum or hot-dipped galvanized roofing nails to be used depending upon shingle manufacturer recommendations.
- Nails to be driven flush with shingles and installed per roofing specs.
- Soil stacks to cover all vents. All vents to be sealed to shingle surface with plastic asphalt cement.
- Flashing to be installed at chimney, all roof valleys, soil and vent stacks and skylights.

GUTTERS — SPECIFICATIONS

- Bid is to provide and install seamless, paint-grip aluminum gutters around entire lower perimeter of dwelling as specified in attached drawings.
- 5" gutter troughs and 3" corrugated downspouts to be used exclusively.
- Gutter troughs are to be supported by galvanized steel spikes and ferrules spaced not more than 5' apart.
- Downspouts to be fastened to wall every 6 vertical feet.
- Gutter troughs are to be sloped toward downspouts 1" (one) for every 12 lineal feet. Gutters should be installed for minimum visibility.

ROOFING — INSPECTION

- ☐ Shingle lines inspected for straightness with string drawn taught. Tips of shingles line up with string.
- ☐ Even shingle pattern and uniform shingle color from both close up and afar (at least 80' away).
- ☐ Shingles extend over edge of roofing deck by at least 3".
- ☐ All roofing nails are galvanized and nailed flush with all shingles (random inspection).
- ☐ No shingle cracks visible through random inspection.
- ☐ Hips and valleys are smooth and uniform.
- ☐ Roofing conforms to contract specifications and local building codes in all respects.
- ☐ Water cannot collect anywhere on the roof.
- ☐ Shingles fit tightly around all stack vents and skylights. Areas are well sealed with an asphalt roofing compound that blends with the shingles.
- ☐ Drip edges have been installed on eave and rakes.
- ☐ No nail heads are visible while standing up on the roof.
- ☐ All shingles lie flat (no buckling).
- ☐ No visible lumps in roofing due to poor decking or truss work.
- ☐ Edges of roof trimmed smooth and evenly.
- ☐ All vents and roof flashing painted proper color with exterior grade paint.
- ☐ All garbage on roof removed.
- ☐ Shingle tabs have been glued down if this was specified. (Used only in very windy areas or where very lightweight shingles are used.)

GUTTERS — INSPECTION

- ☐ Proper materials used as specified. Aluminum, vinyl, etc.
- ☐ Proper trough and downspout sizes used.
- ☐ Downspouts secured to exterior walls.
- ☐ Adequate number of gutter nails used to support gutters securely.
- ☐ Water drains well to and through troughs when water from a garden hose or rain water is applied.
- ☐ No leaks in miters and elbows while water is running.
- ☐ Water does not collect anywhere in troughs and is completely drained within one minute.

PLUMBING

If you are building on a slab foundation, plumbing work will start before the slab is poured. The locations of all waste and plumbing lines must be precisely laid out — it will be very difficult to change these plumbing decisions later. For full basement and crawl space plans, plumbing layouts may be more flexible, but only a bit. Your plumber must still run the water service and waste lines before the foundation is constructed. Bathrooms in basements pose the same position issues as ones built on a slab. Make sure your plumbing plan has been thoroughly analyzed before construction begins. The next phase of plumbing work can start as soon as the house has been "dried-in."

Plumbing is composed of three interrelated systems:
- The water supply
- The sewer system
- The (wet) vent system

The Water Supply

Unless you have opted for an individual water supply (such as a well), you will tap into the public water system. The cost of drilling a well is very expensive, so only consider this option if you have no access to city water or plan to

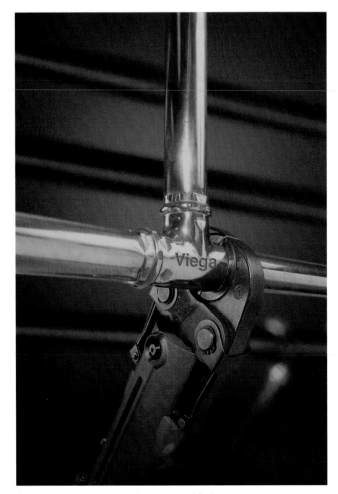

Solderless copper joints using a crimped fitting.

live in the home for eight years or more. Remember, your water bill includes sewer service, too. You'll still pay for that, unless you also opt for a septic tank. If your local water supply is 80 psi or greater, you will need to install a pressure reduction valve at the main water service pipe. High water pressure can and will take its toll on most pipes and fixtures in a short period of time.

Copper Pipe

Copper has been the material of choice for plumbers for many years, but new materials available today have numerous advantages over copper. That said, your plumber will probably be more comfortable working with his old standby than some of the newer technologies. Copper is durable and has antibacterial qualities. However, it is expensive and tricky to install. More than one house has burned down from smoldering embers left from "sweating" copper joints with solder. The industry is trying to address this issue with new solderless connections, but the vast majority of plumbers still do things the "old way."

It's best to use ¾" copper pipe for most water services. Two or more fixtures turned on at the same time can cause a drop in pressure for ½" pipe. When using ¾" pipe try to place the hot water source as close to the fixtures as possible. The larger diameter pipe holds more water volume and will take longer for hot water to reach a shower head or spigot. Copper's excellent thermal conductivity makes this matter worse by allowing great heat loss from the hot water line. Several gallons of water can be wasted each time one waits for hot water to reach a fixture. Tankless water heaters located near the bathroom or kitchen will help to combat this problem.

Copper pipes also have a high coefficient of expansion, which means that they will expand and contract significantly as they heat up or cool down. Make sure your plumber avoids anchoring copper pipes to a wall with brackets. These can cause popping and squeaking noises as the pipes move. Copper is also quite rigid. Then water is turned off at a fixture, the sudden change in pressure can cause a thumping sound called "water hammer." Ask your plumber to install several water hammer arrestors to help alleviate this problem. Make sure to specify certified arrestors. Some plumbers like to build their own arrestors by soldering in a vertical copper pipe that is capped at the end. Air will be trapped in the vertical pipe when the water is turned on, and will act as a piston or shock absorber. Problems arise a few years later because that trapped air will eventually dissolve in the water and the homemade arrestor will cease to function. Certified arrestors contain sealed mechanical pistons that will last for decades.

CPVC Pipe

CPVC, or chlorinated polyvinyl chloride is a white, hard plastic used for both hot and cold supplies. CPVC has been around for decades and is a less expensive choice than copper. It's easy to install, making it a good do-it-yourself project. Joints are glued together with a two part solvent. Some people complain about an aftertaste in the water for the first year, so consider filtering your drinking water. Its cousin PVC, is only rated for cold water and sewer installations.

Polybutylene

The saga of polybutylene is a sad one. Once hailed as the latest innovation in plumbing, this flexible plastic pipe was promoted in the seventies and eighties as a cheaper and easier replacement for copper. The reality was much different. The pipes suffered from flawed connectors that were prone to leaks, sometimes within weeks of installation. Plumbers spent days on callbacks repairing the leaking pipes and replacing soaked drywall. The nightmare continued 15 years later when pipes began to fail. The institute that had certified polybutylene's longevity had used distilled water in their tests. They had failed to test the long range effect of chlorine on the pipe's lifespan. Needless to say, polybutylene is no longer marketed as a viable residential plumbing product.

PEX Pipe

PEX, or cross-linked polyethylene is the latest plumbing alternative to copper and CPVC. Although new in America, it has been used in Europe for 50 years with an excellent history of durability and economy. It is totally inert and imparts no taste to the water. The pipe is flexible, allowing for quick and easy installation. Its flexible nature allows it to be installed with a minimum of joints, which are the most expensive and labor intensive component in plumbing systems. Its flexibility makes it impervious to water hammer and highly resistant to freezing. The joints are installed using several methods of crimping or friction fitting.

You may have trouble finding plumbers willing to work with PEX. Plumbers are slow to accept new innovations and have long memories. The nightmares of polybutylene are still circulated like nighttime stories of the boogie man. It is unfortunate, because PEX is a time tested and reliable plumbing technology that opens up many options to innovative homebuilders. Gun-shy certification agencies have run PEX through the gauntlet and it has passed with flying colors. Check with your local plumbing inspector for plumbers familiar with this new material.

The "home-run" plumbing layout is one innovation made possible by PEX. In home-run plumbing, a separate hot and cold line is run from a central manifold to each individual fixture in the house. Because no two fixtures share the same line, much smaller pipe can be used and water pressure remains stable. This drastically reduces the amount of water that must pass through the line before hot water exits at the fixture. For maximum water flow, locate the manifold near the hot water heater and supply it through large diameter copper pipe.

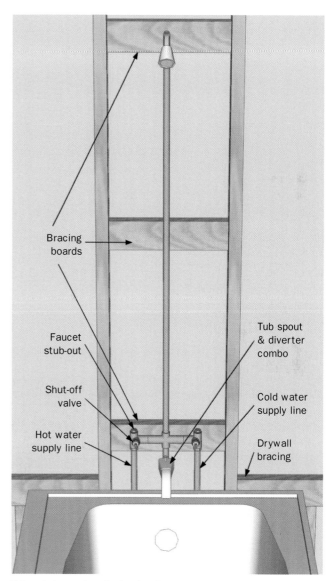

Tub and shower installation details.

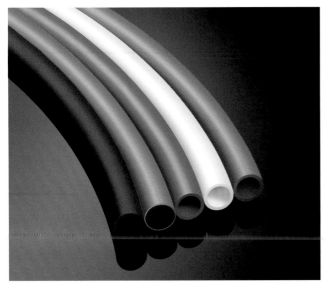

PEX comes in several colors to simplify installation of hot and cold water lines.

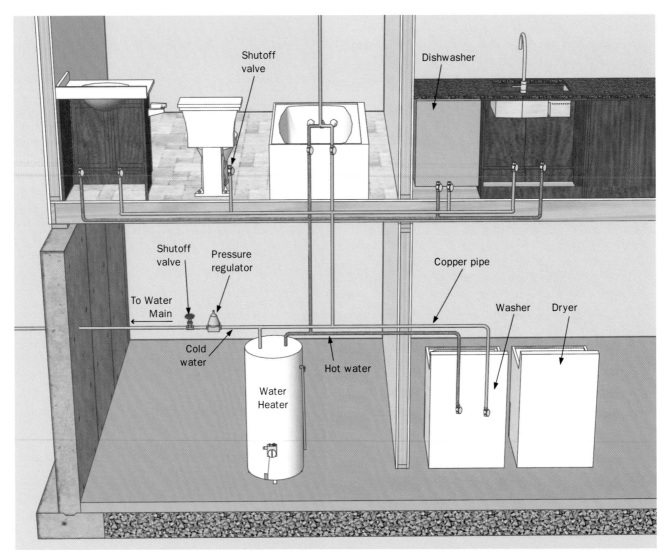

Typical plumbing distribution schematic of a residential copper plumbing system.

Battery powered portable crimping device for PEX.

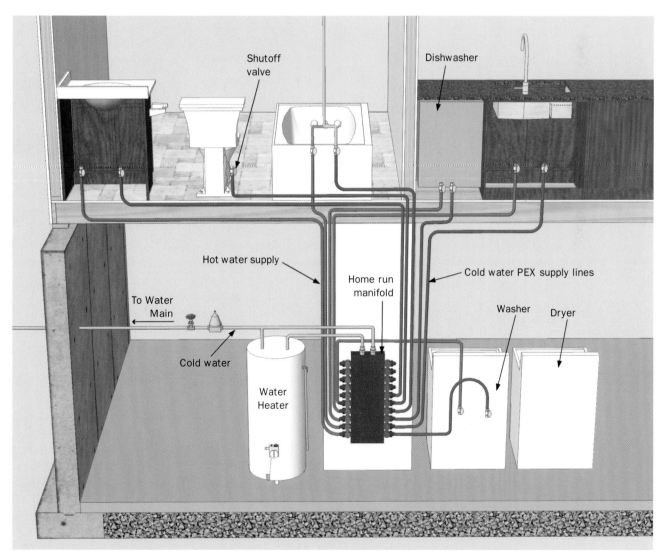

PEX flexible tubing and "home-run" layout offers many advantages: 1) Each supply line can be turned off independently 2) Smaller pipes bring hot water to faucets more quickly with less waste 3) Easier installation 4) Resistance to freezing and water hammer.

WATER HEATERS

Tankless Water Heaters

Tankless water heaters can save on energy costs while providing an endless supply of hot water. They come in both electric and natural gas powered units. The gas units have a slightly lower operating cost and faster "ramp up" speed. Place the unit as close to the bathroom or kitchen as possible. The units come in several sizes and can serve several fixtures at a time.

When paired with a home-run PEX plumbing system, a tankless water heater can give the best cost to performance ratio of any plumbing design. Run a single large diameter copper pipe from the main supply line to the tankless unit and split the line right before entering the heater. From there, the cold water and resulting hot water line (from the tankless heater) can feed directly into a

Foam insulation around plumbing and drain pipes can greatly reduce the sounds of running water.

home-run PEX manifold. PEX will then be routed to fixtures with the shortest possible runs. This saves material and labor and shortens the "time to hot water" to seconds instead of minutes.

Condensing Storage Water Heaters

This represents a newer, more efficient version of the traditional gas water heater that is an excellent alternative to the tankless water heater. An advanced heat exchanging combustion chamber inside the heater keeps combustions gasses in contact with water long enough for the moisture in the flue gas to condense, which releases latent heat. As a result, these heaters can reach efficiencies of 96%. Combustion gasses are cooled to the point that flues can be constructed from ABS or PVC pipe and run long distances. When combined with home-run PEX plumbing, condensing storage water heaters can provide quick and efficient hot water. Some units are even equipped with connections for solar hot water and space heating. Although more expensive than either traditional water heaters or tankless heaters, condensing storage water heaters have great potential when combined with solar water heating and in-floor radiant heating systems. These can save tremendous amounts of energy that can result in a very short payback period.

Heat Pump Water Heaters

This technology is very new and builds off the existing whole house heat pump. In this design, the heat pump system is sized to provide heat not only for the house but for the water heater as well. Since heat pumps only "move" heat rather than create it, heat pump water heaters can be 50% more efficient than their electrical counterparts.

Tankless water heater.

THE WASTE SYSTEM

The waste system collects all used water and waste and disposes of it properly. Unless you opted for a private sewer (septic tank), you will connect to the public sewer system. In most cases, building codes will not allow you to build a private sewer if a public one is available. When designing the house, keep in mind that sewer systems are powered by gravity. Hence, your lowest sink, toilet or drain must be higher than the sewer line. The sewer line must have at least ¼" drop per foot in order to operate properly. This may mean raising your foundation if necessary to accommodate proper drainage.

With sewer pipe materials, technology has created a new problem. PVC pipe is now the most common material used because of its ease of installation; however, plastic pipes transmit the noise of flushing and water draining much more than the old cast iron pipes. Take care to position the pipes in walls that are isolated from bedrooms or you will constantly hear water noises. One way to combat this noise is to ask your insulation subcontractor to place insulation around drain pipes. This can be particularly effective with foam-in insulation. Make sure to discuss the location and run of drain pipes to and through the basement. The living space of many finished basements has been ruined by framing that must be built around poorly placed drain pipes.

The latest water saving toilets use far less water per flush than older units. With less water to carry the waste, drain pipes need to be sized a bit smaller and have a greater drop per foot to aid in proper drainage. Make sure to discuss this issue with your plumber to make sure that old standards are not being used.

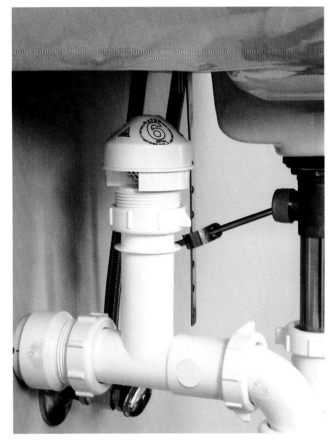

Air admittance valve under sink.

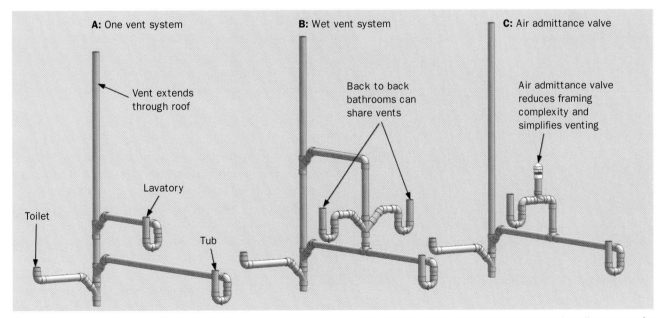

Three different approaches to venting that reduce the number of vent stacks. A: The one-vent system, or "stack venting" uses the soil waste vent for all drains. B: The wet vent system uses another drain's vent as its waste drain. C: The air admittance valve replaces the need for all but one vent, saving material and reducing roof breaches.

Blocking for support
and drywall backiing

Joist interference
with plumbing

Double joist
for support

Placement of joist framing near plumbing is critical. Plan ahead to avoid interference.

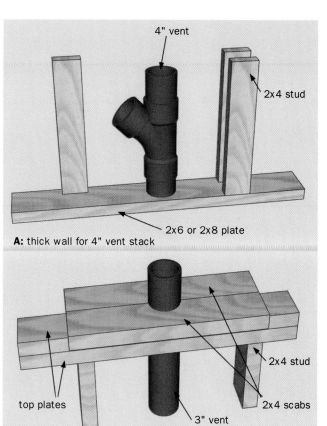

4" vent

2x4 stud

2x6 or 2x8 plate

A: thick wall for 4" vent stack

top plates

2x4 stud

2x4 scabs

3" vent

B: reinforcing for 3" vent stack in 2x4 wall

Placement of joist framing near plumbing is critical. Plan ahead to avoid interference.

PL1   DETERMINE type and quantity of plumbing fixtures (styles and colors). This includes:

- Sinks (kitchen, baths, utility, wet bar, etc.)
- Bathtubs (Consider one-piece fiberglass units with built-in walls requiring no tile. If you go this route get a heavy gauge unit.)
- Shower fixtures
- Toilets and toilet seats
- Water spigots (exterior)
- Water heater
- Garbage disposal
- Septic tank
- Sauna or steam room
- Water softener
- Refrigerator ice maker
- Any other plumbing-related appliance

PL2   CONDUCT standard bidding process. Shop prices carefully; subs who bid on your job may vary greatly on their bid price, depending on what materials they use and their method of calculating their fee. This is where you will decide what kind of pipe to use.

PL3   WALK THROUGH site with plumber to discuss placement of plumbing and any special fixtures needed. For instance, if you have a brick exterior, the plumber must extend faucets to allow for the thickness of brick veneer.

PL4   ORDER special plumbing fixtures. This needs to be done well in advance because supply houses seldom stock large quantities of special fixtures. These will take some time to get in.

PL5   APPLY for water connection and sewer tap. You will probably have to pay a sewer tap fee. Ask the plumber for an adjustment piece that will allow you to connect a garden hose to the water main. Record the number stamped on the water meter.

PL6   INSTALL stub plumbing. This applies to instances where plumbing is cast in concrete foundations. This must be done after batter boards and strings are set and before the concrete slab is poured. The plumber will want to know where the sewer line is located.

PL7   PLACE all large plumbing fixtures such as large tubs, fiberglass shower stalls and hot tubs before wall framing begins. These fixtures will not fit through normal stud or door openings. For security, you may want to chain expensive fixtures to the studs or pipes.

PL8   MARK location of all plumbing fixtures including sinks, tubs, showers, toilets, outside spigots, wet bars, icemakers, utility tubs, washers and water heater. Make sure the plumber knows if you have a regular vanity or pedestal sink. Mark the end of tubs where drain will be located. Mark areas in wall and ceiling where pipes must not be located, such as locations for recessed lights or medicine cabinets.

PL9   INSTALL rough-in plumbing. This involves the laying of hot (left) and cold (right) water lines, sewer and vent pipe. Pipe running along studs should run within holes drilled (not notched) in the studs. All pipe supports should be in place. Your plumber should use FHA straps to protect pipe from being pierced by drywall subs where ever cut-outs are made. FHA straps are metal plates designed to protect the pipe from being punctured by nails. Exterior spigots should not be placed near the point where water mains or other pipes go through the exterior wall, as this will invite water into the home if the spigot freezes or leaks. Make sure plumber extends the spigot if brick veneer is used. The plumber will conduct a water pipe test using air pressure to insure against leaks.

PL10   INSTALL water meter and spigot. This has to be done early because masons will need a good source of water.

PL11   INSTALL sewer line. Mark locations of other pipes (water or gas) so diggers will avoid puncturing them.

PL12   SCHEDULE plumbing inspector.

PL13   INSTALL septic tank and line, if applicable.

PL14   CONDUCT rough-in plumbing inspection. Note: This is a VERY important step. No plumbing should be covered until your county inspector has issued an inspection certificate. Plan to go through the inspection with him so you will understand any problems and get a good interpretation of your sub's workmanship.

PL15   CORRECT any problems found during the inspection. Remember, you have the county's force behind you. You also have the specifications. Since your plumber has not been paid, you have plenty of leverage.

PL16   PAY plumbing sub for rough-in. Get him to sign a receipt or equivalent.

FINISH

PL17   INSTALL finish plumbing. This involves installation of all fixtures selected earlier. Sinks, faucets, toilets and shower heads are installed.

PL18   TAP into water supply. This is where your plumbing gets a real test. You will have to open all the faucets to allow air to bleed out of the system. The water will probably look dirty for a few minutes but don't be alarmed. The system needs to be flushed of excess debris and solvents.

PL19   CONDUCT finish plumbing inspection. Call your inspector several days prior to the inspection. Make every effort to perform this step with your local inspector to see what he is looking for so you will know exactly what, to have your sub correct if anything is in error.

PL20   CORRECT any problems found during the final inspection.

PL21   PAY plumbing sub for finish. Have him sign an affidavit.

PL22   PAY plumber retainage.

PLUMBING — SPECIFICATIONS

- Bid is to perform complete plumbing job for dwelling as described on attached drawings.
- Bid is to include all material and labor including all fixtures listed on the attached drawings.
- All materials and workmanship shall meet or exceed all requirements of the local plumbing code.
- No plumbing, draining and venting to be covered, concealed or put into use until tested, inspected and approved by local inspectors.
- All necessary licenses and permits to be obtained by plumber.
- Plumbing inspection to be scheduled by plumber.
- All plumbing lines shall be supported so as to insure proper alignment and prevent sagging.
- Worker's compensation to be provided by plumber.
- Floor drain pan to be installed under washer if located upstairs or for water heater in attic.
- Plan for gas dryer.
- Install approved pressure reduction valve at water service pipe.
- Stub in basement, plumbing for one sink, tub and toilet.
- Drain pans to be installed in all tile shower bases.
- Install construction water spigot on water meter with locked valve.
- Install refrigerator ice maker and drain.
- Plumb water supply so as to completely eliminate water hammer.
- Water pipes to be of adequate dimension to supply all necessary fixtures simultaneously.
- Cut-off valves to be installed at all sinks and toilets, and at water heater.

Sewer System
- Hook into public water system.
- Install approved, listed and adequately sized back-water valve.
- Horizontal drainage to be uniformly sloped not less than 1" in 4' toward point of disposal.
- Drainage pipes to be adequate diameter to remove all water and waste in proper manner.

Wet Vent System
- Vent piping shall extend through the roof flashing and terminate vertically not less than 6" above the roof surface and not less than 1' from any vertical surface, 10' from and 3' above a window, door or air intake or less than 10' from a hot line.
- All vent piping to be on rear side of roof without visibility from front.
- All vent piping to conform to the local plumbing and building code.

PLUMBING — INSPECTION

Rough-In
- ☐ All supplies and drains specified are present and of proper material (PVC, CPVC, copper, cast iron, PEX).
- ☐ Sewer tap done.
- ☐ No pipes pierced by nails.
- ☐ Hot water fixture on left of spigot.
- ☐ No evidence of any leaks. This is particularly critical at all joints, elbows and FHA straps.
- ☐ All plumbing lines inside of stud walls. Must allow for a flush wall.
- ☐ FHA straps used to protect pipes from nails where necessary.
- ☐ All cut-out framing done by plumber repaired so as to meet local building code.
- ☐ Tub centered in bath properly and secured in place.
- ☐ Tub levelers are all in contact with tub. Tub does not move or rock.
- ☐ Toilet drain at least 12" to center from all adjacent walls to accommodate toilet fixture.
- ☐ Toilet drain at least 15" to center from tub to accommodate toilet fixture.
- ☐ Exterior spigot stub-outs high enough so that they will not be covered over by backfill. Minimum of 6".
- ☐ Ice maker line for refrigerator in place (¼" copper pipe).
- ☐ Roof stacks have been properly flashed with galvanized sheet metal.
- ☐ All wall-hung sinks have metal bridge support located with center at proper level.
- ☐ Water heater firmly set and connected.
- ☐ Attic water heater has a floor drain pan.
- ☐ Water main shut-off valve works properly.
- ☐ Water lines to washer installed.
- ☐ Rough-in inspection approved and signed by local building inspector.

Finish
- ☐ Tub faucets (hot and cold) operate properly. No drip. Drain operates well.
- ☐ Sink faucets (hot and cold) operate properly. No drip. Drain operates well.
- ☐ Toilets flush properly. Fill to proper line and action
- ☐ stops completely with no seepage.
- ☐ Kitchen sink faucets (hot and cold) operate properly. No drips. Drain operates well.
- ☐ Garbage disposal operates properly. Test as needed.
- ☐ Dishwasher operates properly (hot and cold). No leaks. Drain operates well. Run through one entire cycle.
- ☐ No scratches, chips, dents or other signs of damage on any appliances.

- [ ] No evidence of water hammer in entire system. Turn each faucet on and off very quickly and listen for a knock.
- [ ] All water supplies: hot on the left and cold on the right.
- [ ] Turn on all sinks and flush all toilets at the same time and check for significant reduction in water flow. Some is to be expected.
- [ ] Cut-off valves on all sinks, toilets and water heaters.
- [ ] All exterior water spigots freeze-proof and operating properly.
- [ ] All roof and exterior wall penetrations tested waterproof.
- [ ] Pipe holes in poured walls sealed with hydraulic cement.

Note: It is HIGHLY recommended that you obtain the assistance of your local inspector when performing your plumbing inspections. Your plumbing inspector is well trained for such tests.

HVAC

Heating, Ventilation and Air Conditioning is one of the big three trades in terms of cost and impact. Heating and air conditioning systems usually share the same ventilation and ductwork although that's not always the case. A wide variety of passive and active heating systems can be integrated into the heating and cooling system. For homeowners striving for super energy-efficient homes, it may be possible to eliminate active heating and cooling systems altogether.

The efficiency of heating and cooling systems varies widely. Select your systems carefully with this in mind. Contact your local power and gas companies for information regarding energy efficiency ratings used to compare systems on the market. When shopping for systems, check the efficiency rating, and be prepared to pay more for more efficient systems. Your power and gas company may also conduct, free of charge, an energy audit of the home you plan to build, and help suggest heating and cooling requirements.

Heating Systems

HVAC contractors should more appropriately be called "climate control specialists" since it is their responsibility to maintain the interior climate of the home or workplace within the typical comfort zone of 65-78 degrees Fahrenheit. This will usually require a heating and a cooling system. While the methods of cooling a home are limited, the options for heating a home are diverse.

Gas Furnaces

Probably the most popular systems today, gas furnaces are forced air systems that run on natural gas or propane. Propane fueled systems are usually installed in remote areas where a local tank will be refueled each season. Natural gas lines provide fuel to the home through a piping network and are available constantly. The efficiency of gas furnaces is only moderate, which will become a bigger issue as oil prices continue to rise. Much more efficient pulsed gas furnaces were released in the eighties but have suffered from reliability issues and noise concerns. Recent advances in technology may reverse this trend and make the pulsed gas furnace a new player in the energy efficient furnace field. Gas furnaces are typically paired with traditional air conditioning units and both share the same ducting system.

Oil Furnaces

Once extremely popular, oil furnaces have fallen out of favor because of the high cost of oil and their relatively high pollution levels. However, they can still be an option for homes in isolated areas that are not on the "natural gas grid."

Radiant Heat Systems

This is one of the oldest and most reliable heating systems and best known through the old "radiators" used in tenement homes. Radiant heat options have expanded greatly in recent years to include underfloor and ceiling systems. The basic premise is to heat a liquid, usually water or antifreeze, and run it through a radiant source, although less efficient electric systems are also available. Underfloor systems typically use PEX flexible plumbing either adhered to the bottom of sub-flooring or cast into a concrete or gypcrete base. The heat is even, dust free (because there is no moving air involved) and gives a very pleasant "foot warming" heat that is welcome on cold mornings. The choice of floor covering is limited however. Carpet tends to insulate the floor and keep the heat from radiating properly. Underfloor heating works best under tile, vinyl or wood floors.

It's an excellent option for bathrooms, where people are often barefoot and the floor is commonly tile. Radiant systems can be expensive to install, due to the plumbing requirements and the necessity of a separate forced air cooling system. If you want the underfloor option in the bathroom only, consider an electric underfloor system. It's less energy efficient but cheaper to install. The inefficiency really isn't an issue since it will be used intermittently in the bathroom.

If you still want carpet on the floor, consider the newer style "radiators." These are long, low profile baseboard heaters that don't take up wall space like the radiators of old. Take care when exercising this option as many baseboard heaters are notorious for noisy "pops and clicks" when heating up or cooling down. Make sure your HVAC sub is familiar with this system's properties and can guarantee a noise-free environment. Another way to use radiant heat in a carpeted room is to opt for a radiant ceiling system. You lose the comfort of warm floors, but preserve the plush feel of carpet. Radiant ceiling systems also open up the possibility of a radiant cooling system, which is rare, but gaining in popularity.

Radiant systems are a perfect option for energy efficient and passive solar designs. They tend to heat up and cool down gradually, so they work best in highly insulated homes that do not require sudden bursts of conditioned air. When paired with solar water heating units, radiant systems can become virtually energy free. Solar heated liquid is used to recharge an oversized water heater, usually via a heat exchanger (the liquid in the solar unit will usually contain antifreeze to guard against low temperature nights). The oversized water heater can then be used both for typical hot water and as a heat source for radiant heat. Some companies are even experimenting with adding radiant cooling to this system through the use of geothermal wells, where temperatures hover around 55 degrees Fahrenheit.

All radiant systems are expensive and complex to install, so make sure your HVAC sub-contractor has considerable experience and expertise with these systems. Great care needs to be taken during installation to insure against leaks. They can be difficult to repair later. However, if you want the ultimate in comfortable, reliable, and energy-efficient heating, radiant systems may be the choice for you.

Wood Stoves

Another "old world" heating source, wood stoves are having a renaissance of sorts. They use the ultimate "renewable" energy source: wood. When fueled with properly harvested wood, they do not contribute to the carbon footprint, because the carbon in wood is "recycled" rather than originating from oil stores underground. The biggest problem with wood stoves in the past came from their high output of particulates, or soot as it's commonly called. Soot is extremely unhealthy for the environment and can cause lung problems and acid rain. Modern regulations now require woodstoves to include catalytic convertors that greatly reduce particulates. In fact, the latest designs are extremely clean and efficient to operate. Whole house heating could require nothing more than a well placed wood stove or closed cycle fireplace. In well insulated homes with open plans, (few walls) this may be all the heat necessary. For larger demands, consider wood pellet stoves that have an auto-feed mechanism, making them virtually maintenance free. Many pellet stoves have propane gas backup systems in case the pellet bin runs low.

Heat Pumps

Heat pumps are one of the most versatile and energy efficient climate control systems available. The reasons for its efficiency are not well understood by most people. Rather than "creating" heat like a furnace, heat pumps "move" heat from one location to another. The moving of heat requires far less energy than creating it in the first place and is the principle behind most air conditioning units. A heat pump is both heating and cooling system in one. In the summer, it moves heat from the inside of the house to the outside. In winter, heat pumps simply reverse direction and pump heat from outside air into the interior.

The temperature difference in the conditioned air is small, so heat pumps tend to move air at slower speeds for longer periods than gas furnaces. This results in more even temperature swings. You don't get that blast of hot air that is so typical for gas furnaces. The constant movement of low speed air also helps to keep the warmth or coolness evenly distributed throughout the house. Heat pumps are excellent systems when paired with heat recovery ventilators (see page 173). HRV's need smaller duct systems than those required by gas furnaces in order to move air properly, so often require a separate ducting system to be installed. Heat pump ducts can be sized so that they can function for HRV's as well.

The heat pump's one Achilles heel occurs during extremely cold temperatures. Heat pump "heating" efficiency drops considerably when outside temperatures are below 30 degrees Fahrenheit. Most systems provide back-up electric heating elements for those times. If you live in a climate with sub-freezing temperatures for more than 30 days per year, you may want to consider other options, or a move up to geothermal heat pumps.

Geothermal heat pumps are designed to keep heat pumps working at their maximum efficiency. In the typical geothermal installation, several wells are dug in the yard or pipes run at the bottom of ponds or water sources. One hundred feet underground, the temperature stays at a constant 55 degrees Fahrenheit. Geothermal heat pumps tap that constant temperature source and are able to operate at maximum efficiency. Geothermal heat pumps can also be tied into solar hot water heaters and underground cisterns to achieve even greater efficiencies. The initial installation cost is high, but the investment can usually be returned in a few years through energy savings.

Air Conditioning

Air conditioning units, like heat pumps, operate by transferring heat from the air inside your home to the outside via a refrigerant fluid. Freon was the most common refrigerant until the last decade when scientists discovered that it destroyed the ozone layer. Nowadays, other refrigerants are used. Make sure your system has been fully charged before operating to avoid damage to the compressor.

Regardless of the heat source, most homes in the US have air conditioners installed, with the exception of far northern states. If your heat source is not based on moving air, you will need to install a ducting system for the air conditioner.

Air conditioners consist of an evaporator unit and condenser. The evaporator is commonly attached to the fan unit of gas furnaces or to its own dedicated fan unit, if no furnace is available. The condenser resides outside the home, under the floor, or in the attic and is cooled by a large fan. Pipes carry refrigerant back and forth from the condenser to the evaporator. Make sure these pipes have been carefully insulated to improve energy efficiency. If your unit is a dedicated air conditioner and not a heat pump, it's best to locate the condenser in the coolest possible location. This would commonly be in a crawl space or shady area outdoors. Condensers lose efficiency when placed in hot attics, unless a conditioned attic is used. (see Insulation chapter)

Other Climate Control Devices

At your option, your HVAC sub will also install other climate control devices. These include heat recovery ventilators, air filters, and humidifiers. Consult with your

HVAC sub-contractor. He may be able to supply a unit that combines some or all of these functions in one unit, which will save money.

Air Filters

All forced air units have rudimentary air filters installed, but if you want more aggressive filtration, consider a dedicated ultra or HEPA filtration unit. These units can remove up to 90 percent of airborne particles such as mold spores, bacteria, and pollen. Passive filtration units use filters that must be replaced regularly. Electrostatic filtration units use a charged plate device which can be washed and reused indefinitely. Air filters can not only reduce asthma and allergic reactions, but can reduce the dusting ritual most people go through on a regular basis.

Humidifiers

Humidifiers are popular in dry areas and can improve breathing comfort especially in dry winter months. A properly humidified home is not only healthier, but more comfortable. A constant water supply can be attached for automatic water dispensing, making these units relatively maintenance free. Dehumidifiers — popular in the Southeast and other humid areas — draw water out of the air. The installation of a properly sized air conditioner unit should make these dedicated dehumidifiers unnecessary.

Heat Recovery Ventilators

As homes become better insulated and more airtight, indoor air quality becomes an issue. Fresh air exchange was guaranteed in older, leaky homes. Not so with modern, tightly sealed homes. A new and constant source of fresh air is often needed. But after spending all that money to insulate the home, who wants to waste all that expensive conditioned air by replacing it with fresh? Heat recovery ventilators use heat exchangers to heat or cool incoming fresh air, recapturing 60 to 80 percent of the conditioned temperature that would otherwise be lost. Models that exchange moisture between the two air streams are referred to as Energy Recovery Ventilators (ERVs). ERVs are especially recommended in climates where cooling loads place strong demands on HVAC systems.

The combination of super insulation and HRV's give homeowners the perfect combination of energy efficiency and indoor air quality. Because of their low air volume, HRV's may need a dedicated ducting system. Ventilators in kitchens and baths should also be connected to this system for maximum efficiency.

HVAC Subs

In a modern home, HVAC sub-contractors will be coordinating a sophisticated system of climate control. Here, experience and reputation is critical to get the best result. Most HVAC contractors will charge based on the size of

HEPA filtration unit.

Constant feed humidity control.

Typical energy recovery ventilator.

Ducting placed within I-joist framing.

Insulated ducting run between open web trusses.

your system. (BTU's and tonnage). So there is a tendency for them to overestimate your heating and cooling requirements for three reasons:

- More cost, more profit
- Nobody ever complains about having too much heating and cooling
- Habit

Ask your contractor to run a residential load calculation. This is based on an analysis of heat loss from the building through walls and ceilings, leaky ductwork, and infiltration through windows, doors, and other penetrations as well as heat gain into the building from sunlight, people, lights and appliances, doors, walls, and windows, and infiltration though wall penetrations. The Air Conditioning Contractors of America (ACCA) supplies guidelines for running this analysis. Your HVAC sub should also run a blower door test to calculate the rate of air infiltration. By analyzing these tests your sub should be able to install a properly sized unit, which will result in the greatest cost savings, both in installation and in operational costs. Shop carefully for your HVAC sub; a wrong decision here can cost you increased utility bills in the future.

Ductwork

The ductwork circulates conditioned air to specific points in the home and is split between air supplies that deliver conditioned air and air returns, which circulate the air back to the HVAC system. Ducts will usually be constructed of sheet metal or insulated fiberglass cut and shaped to fit at the site. Fiberglass ducts are rapidly gaining in popularity due to their ease of fabrication at the site. The duct itself is made of insulated fiberglass and can be cut with a knife. This extra insulation makes for a more efficient and quieter air system. Sheet metal ducts are notorious for developing mysterious noises caused by expansion and air movement that may be difficult or impossible to eliminate. If you plan to use foam-in insulation, ask your insulation contractor to coat metal ducting with a thin layer of foam insulation. This will not only reduce rattles, but increase energy efficiency.

Cold air for air conditioning is heavier than hot air, so it normally takes a higher speed fan. For this reason, look for units with two speed fans. The fan unit, normally a "squirrel cage" type, looks just like its name implies. If you have a two story home, a large split level or a long ranch, consider a zoned (split) system. Zoned systems work on the premise that two smaller, dedicated systems operate more efficiently than one larger one. For large homes (2500 sq. ft. or more), this becomes a STRONG consideration. Another less expensive alternative is to install a single, two speed unit with electronic duct vents to redirect air to different zones of the house as appropriate.

For efficiency reasons, your HVAC system should be located as close to the center of your home as possible, to minimize the length of ductwork required. Ductwork should have as few turns in it as possible. Air flow slows down every time a turn is made, decreasing the system's overall efficiency and effective output. 7"-diameter round ductwork is fairly standard.

Register Placement

Proper placement of supplies and returns is critical to making your total HVAC system efficient and economical to run. Most registers are placed near doors or under windows. This type of placement puts the conditioned air near the sources of greatest heat loss and promotes more even distribution of air. However, modern insulation practices and double pane windows have made this traditional method obsolete. Ask your sub to consider short run ducting if you are confident in the insulation quality of your home.

Returns should be located near the center of the home — low for warm air returns since the cooled air drops and high for cool air returns, to remove the air as it warms up. It is a good idea to have two supplies (high and low) and two returns (high and low) in critical rooms in order to handle either season. Your HVAC sub will be able to help you with the proper quantity and placement of your supplies and returns. As a general rule, you will need at least one supply and one return for each room.

Thermostats

Many different types of thermostats are available on the market today. Intelligent thermostats can save tremendous amounts of energy simply by varying indoor temperatures based on the time of day, day of week, or the season. Temperatures can be lowered at night in the winter, while the family is asleep, and raised in the morning just in time for sunrise. These thermostats can also be attached to electronic duct vents to direct heat or cooling to different zones of the house at different times. The latest models can be programmed through the Internet or via a home computer.

In any event, do not put the thermostat within 6 feet of any air supply register or facing one on an opposite wall. NEVER place a thermostat in a room which has a fireplace or is in direct sunlight. Consider placing the thermostat in a hallway.

Efficiency Ratings

Make sure to check the efficiency of your HVAC systems, as this will influence energy usage for years to come. Gas furnaces are rated by their annual fuel utilization efficiency.

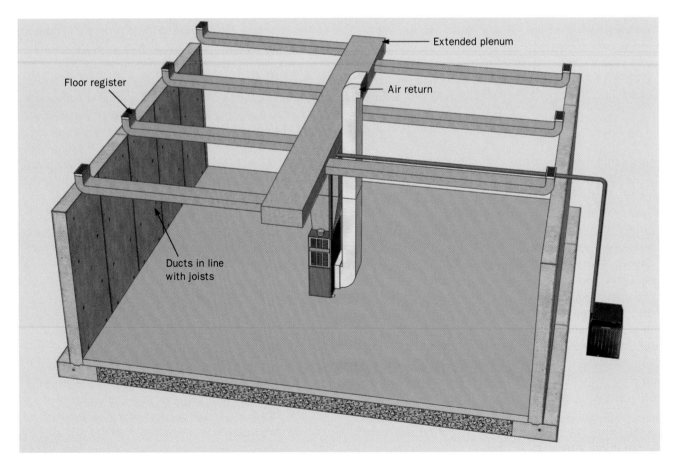

Extended plenum layout.

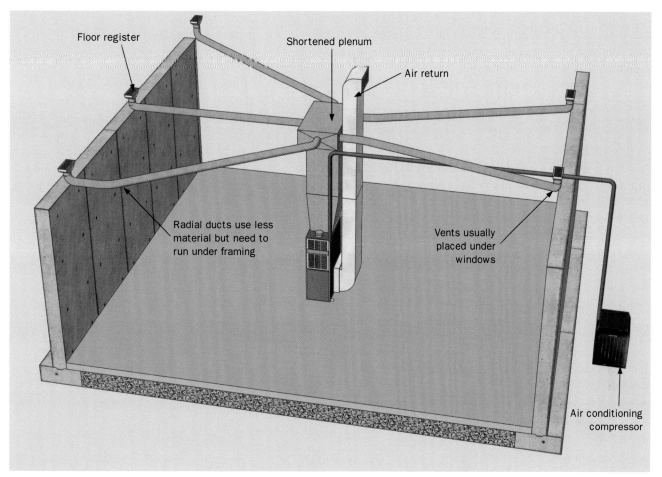

Floor register

Shortened plenum

Air return

Radial ducts use less
material but need to
run under framing

Vents usually
placed under
windows

Air conditioning
compressor

Perimeter radial ducting.

(AFUE) Air conditioners are rated by their seasonal efficiency rating (SEER). The most energy-efficient furnaces have an AFUE of 90% or higher. For air conditioners, the minimum requires a 13 SEER rating, however the most energy-efficient air conditioners top out at around a SEER rating of 20.

Get the Most Out of Your Subs

If you are building during extreme hot or cold months, subs may be hesitant to show up for work during the most unpleasant hours of the day. If you get your HVAC system approved early, normally by a quick test and inspection, you can turn the system on before the house is finished. Your sub-contractors will thank you and will tend to spend more time on site rather than at some other job with unconditioned space. Drywall, tile, and paint will also dry faster and wood floors will stabilize their moisture content sooner. Just make sure that a filter has been installed, since there will still be a lot of dust and dirt running through the system. Ask your HVAC sub to blow out the ducting system once all interior work has been completed. When you receive your occupancy permit, make sure to install a fresh, clean filter.

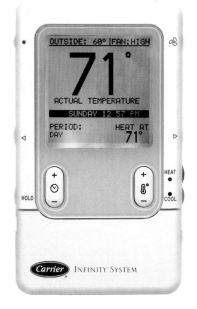

Intelligent thermostats can save energy costs by adjusting temperatures for different times of the day and week.

HVAC — STEPS

HV1	CONDUCT energy audit of home to determine HVAC requirements. Your local gas and electric companies can be very helpful with this, sometimes providing computer printouts at little or no cost. You may only have to drop off a set of blueprints. Decide whether you will have a gas or electric dryer and determine where it will be situated.
HV2	SHOP for best combination of cost, size and efficiency in heating and cooling systems. The higher the efficiency, the higher the price.
HV3	CONDUCT bidding process on complete HVAC job. Have them give a separate bid for the ductwork.
HV4	FINALIZE HVAC design. Have a representative from the local gas company or an inspector look over your plan to make sure you aren't doing any thing that could be a problem. Specify location of dryer exhaust vent.
HV5	ROUGH-IN heating and air conditioning system. Do not install any external fixtures such as compressors at this time. Compressors are a favorite theft item.
HV6	INSPECT heating and air conditioning rough-in. This should be done with your county inspector. Call him in advance.
HV7	CORRECT all deficiencies noted in the rough-in inspection.
HV8	PAY HVAC sub for rough-in work. Have him sign a receipt.

HVAC FINISH WORK — STEPS

HV9	INSTALL heating and air conditioning (finish work). This is where the thermostat and all other HVAC electrical work are hooked up. All registers are installed. The AC compressor is installed and charged with refrigerant. Discuss the placement of the AC compressor location with the HVAC sub. He may supply the concrete pad if you do not have one.
HV10	INSPECT heating and air conditioning final. Again, this should be done by your county inspector. Call him several days in advance so you can be present for the inspection. Some builders like to have their HVAC sub present, too. Make a note of any deficiencies. Be sure electrician has run service to the unit and that it is functional.
HV11	CORRECT all deficiencies noted in the final inspection.
HV12	CALL gas company to hook up gas lines.
HV13	DRAW gas line on plat diagram.
HV14	PAY HVAC sub for finish work. Get him to sign an affidavit.
HV15	PAY HVAC sub retainage after system is fully tested.

HVAC — SAMPLE SPECIFICATIONS

- Bid is to install heating, ventilation and air conditioning system (HVAC) as indicated on attached drawings.
- Install zoned forced air gas (FAG) fired heating unit. With five (5) year warranty. One in basement and one in attic.
- Install two AC compressors.
- Install gas range hook-up.
- Install 50-gallon gas water heater.
- Install downdraft line for island cooktop.
- Connect 1½" copper gas line to gas main, a minimum of 12" under ground.
- Install sloped PVC tubing for AC drainage.
- Install gas grill line and grill unit on patio.
- Install one fireplace log lighter and gas line.
- Install gas line to porch and mailbox gas lights.
- All work and materials to meet or exceed requirements of the local building code.
- Provide and install all necessary air filters.
- Opening to furnace large enough to permit removal and replacement.
- All refrigerant lines fully charged.

HVAC — INSPECTION

Rough-In

☐ All equipment UL-approved with warranties on file. Heating and air units installed in place and well anchored.

☐ Heating and air units are proper make, model and size.

☐ Zoned systems have proper units in proper locations.

☐ Air compressors firmly anchored to footings.

☐ All ductwork installed according to specifications.

☐ Proper number of returns and supplies have been installed.

☐ All ductwork meets local building codes.

☐ All ductwork joints sealed tightly and smoothly with duct tape or foam insulation.

☐ No return ducts in baths or kitchen.

☐ Bathrooms vent to outside if specified.

☐ All ductwork in walls flush with walls, not to interfere with drywall.

☐ All ductwork in ceiling tied off so as not to interfere with drywall or paneling.

☐ Attic furnace has a floor drain pan, if in attic.

☐ All heat exhaust vents isolated from wood or roofing by at least 1".

☐ No exhaust vents visible from the front of the home.

☐ Air conditioning condensate drain installed.

☐ All ductwork outlets framed properly to allow installation of vent covers.

☐ Vent hood or downdraft installed in kitchen according to plan.

☐ HVAC rough-in inspection approved and signed by local building inspectors.

☐ Gas meter in place.

☐ Gas line connected to gas main.

Finish

☐ Line to gas range hooked up. Range installed.

☐ Line to gas dryer hooked up. Dryer vent installed properly.

☐ Line to gas grill hooked up. Gas grill installed.

☐ HVAC electrical hookup completed.

☐ Thermostats operate properly and are installed near center of house. Thermostats located away from heat sources (fireplaces and registers) and doors and windows to provide accurate operation.

☐ Manual/Auto switch on thermostat works properly. A/C condensate pipe drains properly. This will be difficult or impossible to test in cold weather. The pipe should at least have a slight downward slope to it. You cannot test A/C if temperature is below 68 degrees.

☐ Fan noise not excessive.

☐ Furnace, A/C and electronic air filters installed.

☐ Water line to humidifier installed and operating.

☐ Water line from dehumidifier installed and operating.

☐ All vent covers on all duct openings and installed in proper direction.

☐ All holes and openings to exterior sealed with exterior grade caulk.

☐ Downdraft vent for cooktop installed and operating. All HVAC equipment papers filed away (warranty, maintenance, etc.).

☐ Strong air flow out of all air supplies.

☐ Air returns function properly.

ELECTRICAL

Electricians perform all the necessary wiring for interior and exterior fixtures and appliances. They often charge by the piece for providing and wiring each switch, outlet and fixture. Double switches will count as two switches and so on. You can save money by purchasing your own fixtures; the markup is incredible. Building codes will normally require at least one outlet per wall for each room and/or one for each specified number of running wall feet. Most counties require electricians to be licensed master electricians.

Electrical Outlets

If left to their own devices, electrical sub-contractors will make arbitrary decisions about where to place each switch and outlet, according to code. There are a few standards, such as minimum spacing of outlets and height of switches and outlets, which should be indicated in your local building code. That is why you should provide the sub-contractor with an electrical wiring diagram. Outlets seldom find their way to the exact location needed by your needs and furniture arrangements. In today's electronic world, most homeowners feel that their homes lack sufficient outlets. It is also best to walk through the house with the electrician and mark the locations of outlets and switches. A good electrician should make helpful suggestions on the location of fixtures.

You can save a little money by adding additional outlets yourself, but make sure to arrange this with your electrician ahead of time. The electrician needs to know the correct number of outlets and the expected electrical load in a room in order to place the properly sized circuit breaker in the service panel. If you have a room that you know will contain multi-media equipment, a large number of computers, or electric exercise equipment, make sure to point that out before wiring begins.

Consider multiple switches for a single light (such as at the top and bottom of stairs), floodlight switches in the master bedroom and photo-electric cells for driveway and porch lights. Other key areas for additional outlets:
- Plugs in breakfast areas
- Along kitchen counters
- Switched outlets in attic
- Foyer area

Ground Fault Circuit Interrupters

Ground Fault Circuit Interrupters (GFCI's) are required in all bathroom and kitchen outlets. GFCI's are ultra sensitive circuit breakers that provide an extra margin of safety in areas where the risk of shock is high. Hobby shops, garage areas, and outdoor circuits are also good sites for these devices.

GFCI's can also be placed in the service panel as long as kitchen and bath outlets are all wired to the GFCI circuit. This can save some money, but it's seldom done,

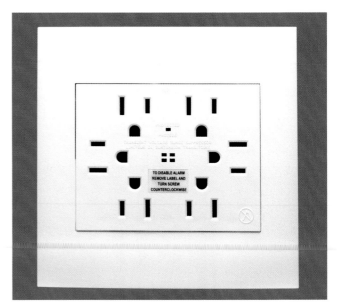

The increased demand for electrical outlets in homes has led to new innovations in electrical products and design.

Single-outlet ground fault circuit interrupter.

because most homeowners don't like the prospect of resetting these sensitive circuits at the panel. It's more convenient to use the reset button found on local outlets. Service panel GFCI's are critical for certain large switched devices like spas, whirlpool baths, and steam rooms.

Lighting

The electrician will be installing rough-in electrical wiring before drywall goes up. He will return later to install the finish lighting fixtures. Make sure to choose your lighting fixtures early. If possible, have the fixtures (or at least the installation instructions) available for the electrician at rough-in time. That way, he can make sure that electrical boxes are located in the precise location for the type and style of fixture being installed.

The lighting industry is going through a re-styling period as homeowners move away from incandescent lighting toward compact fluorescent and LED lighting options. Make sure your choice of fixtures is designed specifically for these bulbs. Resist the urge to buy closeout bargains unless you are sure that the (generally) larger compact fluorescent bulbs will fit properly into the fixture.

Most modern light fixtures are designed to hide compact fluorescent bulbs from view, since some people find their coiled elements to be a distraction.

Service Panel

The service panel is the main distribution area where outside utility lines meet inside wiring. Service panels are rated by amperage which also indicates the number of circuit breaker slots and electrical load the panel is rated to handle. Most codes require a minimum of a 100 amp service panel, but 150 amp panels are more common. It is much better to err on the side of extra capacity, since the amperage only indicates the maximum load and does not affect electrical usage. If your house contains a lot of electrical equipment, you may want to request a 200 amp service.

Wiring

Wire gauges used for circuits will vary. Electric stoves, ovens, refrigerators, washers, dryers and air compressors are each normally on separate circuits and use heavier gauges of wire. Your local building code will describe load limits on each circuit and the appropriate wire types and gauges to use.

After your home is dried in, the electrical sub will "rough in" all the electrical wiring, fuse box and electrical HVAC connections. Make sure to have the HVAC and plumbing sub-contractors complete the rough-in before calling in the electrician. This will prevent any wiring from having to be cut or rerouted in order to make room for plumbing or ducting. It is much easier for the electrician to work around obstructions. If all ducting is in place, the electrician will also know precisely where the furnace will be located and will be able to wire for it properly.

Additional Circuits

Electricians are highly trained contractors. Building codes require that a licensed electrical contractor sign off on any wiring done in a home, whether you choose to run the wiring or not. With the critical nature of electrical service, it's best to leave this job up to the experts. However, electricians are also accustomed to installing other wiring in the house. You may be able to work with the electrician to run the wiring for some of these circuits, but allow the electrician to make the final connections. Ask your electrician for bids on installing this other wiring, although his price may be quite high. Installing this wiring before drywall is installed is easy and saves considerable installation expense later.

Phone Lines

Use only professional grade four or six wire phone cable. This is usually available at most commercial building supply houses, or electrical supply houses. You can ask the electrician to purchase this for you, and then you can run the wiring. Phone lines are often used for high speed internet and some cable installations, so it's best to go with high quality wire.

Ethernet Cable

Homes have become Internet portals. While wireless Internet (WIFI) has become popular, you may prefer to "hard wire" your Ethernet system. It adds additional speed and privacy and can be used as a backbone for the latest internet based video delivery systems such as ATT u-verse service.

Security systems

Even if you're not considering a security system at this time, it's best to run the wiring just in case. Most security systems use standard double wire cable. Run these to all doors and windows and note their locations on a wiring diagram. If you decide to add a security system later, you'll want to be able to find the hidden connections.

Cable

This is another option. All these different cables can add up in cost, but are far cheaper to install now than later. Shielded cable can be used not only for cable TV service, but also for satellite TV and Ethernet systems. Choose the highest possible grade to insure compatibility with the latest high definition TV systems.

Speaker Wire

This simple installation can save a ton of irritation later. Run the wiring to wall mounted boxes near the intended location of speakers.

Media Center Access

Today's homes have become miniature theaters, with surround sound, big screen TV's, stereos, game consoles, computer media centers, etc. Wiring these systems behind media walls can be a big pain. If you know the likely location for your media center, consider providing access to the back of these areas. An example would be an access panel in a closet that backs up to the media wall. This can make wiring a pleasure later.

Home Automation

The building industry and the NAHB has been toying with the concept of the "smart house" for over two decades with many fits and starts. Finally, home automation appears to be on the horizon. These systems can control a home's sensors, wiring, lights, video cameras, and appliances from a single control center. The control center is similar to an electric service panel. Unlike electrical wiring, each multimedia outlet has dedicated cables that run from the central hub to the outlet and back again. The cables, multimedia outlets, and control center that make up the framework of a home automation system referred to as structured wiring or smart wiring. Wireless radio frequency systems will soon be available that do not require structured wiring.

These systems can do fancy things like monitor who's at the front door over the television set, turn the coffee pot on via the Internet, and time the switching on of lights while away from home. But perhaps their most useful purpose is to create flexible lighting systems. Want to place an overhead light switch in a new location? This is made easy with switches that can be programmed to turn on any light or outlet. Plan for the future by running the wiring ahead of time.

Structured wiring components add sophistication and flexibility to home automation. Porch and living room lights can easily be controlled from the bedroom by remote.

Backup power systems are becoming popular for homes that increasingly depend on electrical appliances.

STEPS — ELECTRICAL

EL1	DETERMINE electrical requirements. Unless your set of blueprints includes a wiring diagram, you will want to decide where to place lighting fixtures, outlets and switches. Make sure no switches are blocked by an open door. You may want to consider furniture placement while you are doing this. Even if your blueprint has an electrical diagram, don't assume it is correct. Most of the time, they can be improved. You may wish to investigate the use of low voltage and fluorescent lighting. Remember that you cannot dim fluorescent lights.
EL2	SELECT electrical fixtures and appliances. Visit lighting distributor showrooms. Keep an eye out for attractive fluorescent lighting which will be a long-term energy saver (but cannot use a dimmer switch). Special orders should be done now due to delivery times. You should listen to the door bell sound before purchasing.
EL3	DETERMINE if phone company charges to wire home for modular phone system. Find out what they charge for and how much. Even if you plan wireless phones, install phone jacks in several locations.
EL4	CONDUCT standard bidding process. Have subs give a price to install each outlet, switch and fixture. They will normally charge extra for wiring the service panel and special work. Use only a licensed electrician.
EL5	SCHEDULE to have phone company install modular phone wiring and jacks. If your phone company charges too much for this service, just have your electrician do the job or do it yourself. Also try to arrange to have your home phone number transferred to your new house later so that you can keep the same phone number.
EL6	APPLY for and obtain permission to hookup your temporary pole to public power system. You may need to place a deposit. Perform this early so that hook-up will take place by the time the framers are on site.
EL7	INSTALL temporary electric pole. Make sure the pole is within 100 feet of the center of the house foundation (preferably closer).
EL8	PERFORM rough-in electrical. This involves installing wiring through holes in wall studs and above ceiling joists. All the wiring for light switches and outlets will be run to the location of the service panel. To insure that your outlets and switches are placed where you want them, mark their locations with chalk or a marker. Otherwise, the electrician will place them wherever he desires. It is important to get lights, switches and outlets exactly where you plan to place furniture and fittings. Provide scrap 2×4s for blocking next to outlets.
EL9	INSTALL modular phone wiring and jacks and any other wiring desired. You may also wish to run speaker, cable TV, security and computer wiring throughout the house. It's much easier to do it now than later. Staple wires to studs where wire enters the room so the drywall crew will notice them. Many security system contractors will pre-wire your home for their security system at their cost if you decide to purchase their system.
EL10	SCHEDULE electrical inspection. The electrician will usually do this.
EL11	INSPECT rough-in electrical. The county inspector will sign off the wiring. Remember that this MUST be done before final drywall is in place.
EL12	CORRECT any problems noted during rough-in inspection.
EL13	PAY electrical sub for rough-in work. Get a receipt or use canceled check as a receipt.
EL14	INSTALL garage door(s). Typically this is not done by the electrician, but by a specialist. The garage door must be installed so that the opener can be hooked up by the electrician.
EL15	INSTALL electric garage door openers. Store remote control units in a safe place. Wire electrical connection.

Finish

EL16	PERFORM finish electrical work. This includes terminating all wiring appropriately (switches, outlets, etc.). Major electrical appliances such as refrigerators, washers, dryers, ovens, vent hoods, exhaust fans, garage door openers, doorbells and other appliances are installed at this time. The air compressor will also be wired.
EL17	CALL phone company to connect service.
EL18	INSPECT finish electrical. Again, your county inspector must be scheduled. Have your electrician present so he will know just what to fix if there is a problem. Store all appliance manuals and warranties in a safe place.
EL19	CORRECT electrical problems if any exist.
EL20	CALL electrical utility to connect service.
EL21	PAY electrical sub, final. Have him sign an affidavit.
EL22	PAY electrical sub retainage after power is turned on and all switches and outlets are tested.

ELECTRICAL — SPECIFICATIONS

- Bid is to perform complete electrical wiring per attached drawings.
- Bid to include all supplies except lighting fixtures and appliances. Bid to include installation of light fixtures.
- Bid to include furnishing temporary electric pole and temporary hook-up to power line.
- Bid to include brass switch plates in living room, dining room and foyer.
- Bid may include wiring and installation of the following items:
  - 200 amp service panel with circuit breakers
  - Light switches (1-way)
  - Light switches (2-way)
  - Furnace
  - Dishwasher
  - Garbage disposal
  - Microwave (built-in)
  - Door chime set
  - Door chime button(s) with lighted buttons
  - Bath vent fans
  - Bath heat lamp with timer switch
  - Jacuzzi pump motor with timer switch
  - Light fixtures (all rooms and exterior areas)
  - Three-prong (grounded) interior outlets
  - Three-prong (grounded) waterproof exterior outlets
  - Electric range
  - Double lamp exterior flood lights
  - Electric garage door openers with auto light switch
  - Washer
  - Dryer
  - Central air conditioning compressor
  - Climate control thermostats
  - Electric hot water heater
  - Hood fan
  - Central vacuum system
  - Intercom, radio units, speaker wires
  - Co-axial cable TV lines and computer cables
  - Sump pump
  - Electronic security system with sensors
  - Ceiling fans
  - Whole house attic fan with timer switch
  - Automatic closet switches
  - Humidifier/dehumidifier
  - Time-controlled heat lamps
  - Thermostatic roof vents
  - Phone lines in all rooms as shown
  - Time-controlled sprinkler system

- All work and materials to meet or exceed all requirements of the Electrical Building Code unless otherwise specified.
- Natural earth ground to be used.
- Dimmer switches to be installed on fixtures as shown on plan.
- Rheostat or remote to be installed on ceiling fans.

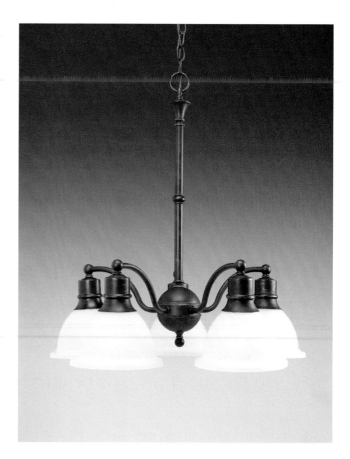

ELECTRICAL — INSPECTION

Rough-In

☐ Outlets and switch boxes offset to allow for drywall and base molding.

☐ All outlets placed at 11" and switches placed at 42" height as indicated on plans. Outlets and switch unobstructed by doors.

☐ All special outlets, switch boxes and fixtures are in place where intended. Refer to your specifications and blueprints.

☐ Electrical boxes for chandeliers and ceiling fans are adequately braced to hold weight of fixture.

☐ All lines are grounded.

☐ All visible splices have approved splice cap securely fastened.

☐ Bath ventilator fans installed.

☐ Attic power ventilators installed.

Finish

☐ All outlets and fixture wires measure 117 V with circuit tester or voltmeter. Test between the ground and each socket.

☐ Dimmer switches installed where specified (dining room, den, foyer and master bath).

☐ Air conditioner cut-off switch is installed at A/C unit.

☐ No scratches, dents or other damage to electrical appliances and fixtures.

☐ Furnace completely connected and operational.

☐ All bath vents connected and operational. Should be on separate circuit from light switch.

☐ Electric range connected and operational.

☐ Electric pilot light starter on gas range connected and operational.

☐ Range hood, downdraft and light connected and operational.

☐ Built-in microwave connected and operational.

☐ Garbage disposal connected and operational.

☐ Trash compactor connected and operational.

☐ Dishwasher connected and operational.

☐ Refrigerator outlet operational.

☐ Door chimes (all sets) connected and operational.

☐ Power/thermostatic ventilators on roof connected and operational.

☐ Washer and dryer outlet operational.

☐ Garage opener installed and operational.

☐ Electronic security system connected and operational. All sensors need to be tested one at a time.

☐ Phone outlets work at all locations.

☐ Service panel installed properly with sufficient load carrying capacity. Breakers are labeled properly.

☐ Exterior lighting fixtures installed and operational.

☐ All recessed lighting, ceiling lighting, heat lamps and other lighting fixtures connected as specified and operational.

☐ All intercom/radio units connected and operational.

☐ Jacuzzi pump motor connected and operational. For safety reasons, switch should not be reachable from the tub.

☐ Central vacuum system connected and operational.

☐ All other specified appliances and fixtures connected and operational.

☐ All switch plate and outlet covers are installed as specified (in rooms to be wallpapered, these will be temporarily removed).

☐ Finish electrical inspection approved and signed by local building inspector.

MASONRY AND STUCCO

Masonry

Masonry is the trade involved with the application of concrete blocks, clay or tile bricks and stone. Masonry is a true art; watch a bricklayer sometime and you will see why. A skilled bricklayer possesses a strong combination of good workmanship and efficiency — key ingredients to producing a strong, well built masonry structure on a schedule. Your brick mason and stone mason may be two different subs.

Concrete Blocks

Concrete blocks are normally used below ground as an alternative to a poured foundation. Although a poured foundation is 3 times stronger, concrete blocks are less expensive, while being of adequate strength. The standard size of a concrete block is $7\frac{5}{8}$" × $7\frac{5}{8}$" × $15\frac{5}{8}$" and are normally laid with $\frac{3}{8}$" flush mortar joints. Concrete blocks are applied on concrete footings.

While their use in foundations is falling out of favor, concrete blocks are still quite popular as structural backing for fences, stone posts, and retaining walls. Once the wall has been laid, a brick or stone façade will be installed.

Clay Bricks

Often called brick veneer, brick facing is normally comprised of brick $3\frac{5}{8}$" × $7\frac{5}{8}$" × $2\frac{1}{4}$" set in a $\frac{3}{8}$" mortar joint. For estimating purposes, brick veneer, laid in running bond (a typical pattern), requires seven bricks per square foot of wall. This does not apply to such items as chimneys or fancy brickwork. Your mason charges for laying units

Concrete block is still popular for above ground applications because of its low price and ease of installation. A good mason can lay an entire block wall in a day.

Nothing beats the durability and old world charm of brick and stone.

of 1,000 bricks (one skid). Additional expenses may be charged for multi-story or fancy brickwork where scaffolding and/or additional skill and labor is required. Bricks are anchored to walls with galvanized metal wall ties. Once bricks are laid, brick walls are self supporting.

In some styles of architecture, brick or stone veneer is used for all or part of the exterior finish. It is good practice, when possible, to delay applying the masonry finish over platform framing until the joists and other members reach moisture equilibrium. Waterproof paper backing or three dimensional rain screen should be used to protect against moisture buildup. A clear drainage path must be maintained behind the wall all the way to "weep holes" at the bottom of the wall. Three dimensional rain screen helps to maintain that channel by preventing excess mortar from blocking the flow of water. It is normal practice to install the masonry veneer with a 3" space between the veneer and the wall sheathing. This space provides room for the bricklayer's fingers when setting the brick.

Stone

Stone is laid much like brick but requires greater skill because stone masons must cut and fit the uneven surface. Because stonework has been around since the Middle Ages, there are many styles and patterns, some with mortar joints, and some without. Stone is generally more expensive than

Fancy brickwork over openings can add style and texture to brick walls, but requires a skilled mason.

brick, so it is often used as an accent piece or to add texture to small expanses of wall.

Cultured stone is a relatively new product that is manufactured to look like real stone. In some cases it IS real, just cut to thinner proportions. Others are made from colored concrete and the best brands are almost indistinguishable from the real thing. Special pieces are cast to wrap around corners and give the appearance of a

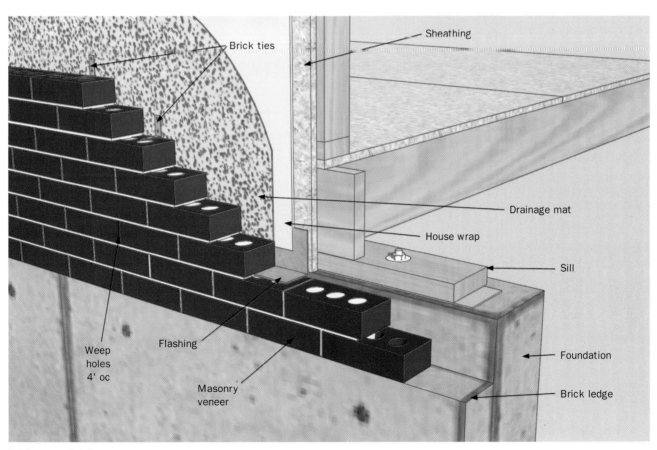

Brick veneer detail.

much thicker wall. Cultured stone is generally NOT less expensive than real stone, but has two major advantages: Lighter weight and ease of installation. Brick ledges and foundations are usually not required, which makes it easier to retrofit stone over an existing structure. Cultured stone is a natural surface for zero clearance fireplaces for this reason.

Stucco

Once popular and widely used throughout Western Europe, stucco brings an old world charm. Stucco is really just mortar applied over a metal lath or screen used for support and adhesion. Stucco finishes are applied over a coated, expanded metal lath and, usually, over some type of sheathing. In some areas where local building regulations permit, such a finish can be applied to metal lath fastened directly to the braced framework.

Some builders complain of maintenance problems (cracking) of traditional stucco after ten years or so of service. If stucco is applied properly with proper expansion joints, it should provide decades of sound performance. Expansion joints should be placed approximately every 300 square feet. Some stucco cracking should be expected. Stucco is not the best insulator. As such, you should go with heavier insulation in exterior walls. Higher grade sheathing is also in order.

Several brands of artificial stucco are on the market. Although they boast of higher insulating qualities and lower maintenance, they are also more expensive. It is easier to do fancier trim work like lintels, quoins and other work because it uses Styrofoam cutouts as a base. Artificial stucco has suffered from a bad reputation due to mistakes in application. The latest versions have solved most of these problems.

Cost

Stucco is charged by the installed square yard. Normally, area is not deducted for doors and windows except for the garage door. Rationale is that it is at least as much trouble to work around an opening as it is to cover the area. Extras will probably include:
- Smooth finish
- Decorative corner work (quoins)
- Decorative window and door work (keystones and bands)
- Corner bead — all edges

Stucco can be applied rough, smooth or anywhere in between. Subs will charge you more for a smooth finish because it takes longer and imperfections show up easier. If you use dyed stucco, it may not require painting.

For the Best Stucco Job
- Don't attempt to stucco foil or polyethylene backed sheathing due to poor adherence. Use either Styrofoam (⅝" to 2") or exterior gypsum (½"). Check to see what sheathing your stucco sub recommends.
- Avoid mixing pigments with stucco mortar. This can lead to uneven drying and coloration. Paint the stucco after it is dry.
- To avoid cracking, be liberal with expansion joints. They are hardly noticeable and do not detract from overall appearance.
- Don't attempt to apply stucco if there is ANY chance of rain in the forecast.

This cultured stone wall provides the visual appeal of real stacked stone but doesn't require a foundation or brick ledge. The Delta-dry rain screen behind the stone helps to drain and protect the wall from moisture.

Stucco and stone gives contrast to this Spanish revival home.

About Fireplaces

If you go with a custom masonry fireplace, the critical thing to get is a proper draft. If your fireplace doesn't "breathe" properly, you will always have smoke in your home during its use. If it "breathes" too well, you may not get too much effect from the fireplace. This is not a big problem with pre-fabricated fireplace units since they are pre-designed to operate properly. Below are a few options you may want to consider regardless of what type fireplace you have:

Fresh Air Vent—Allows the fire to get its combustion air from outside your home so that it doesn't rob all your warm air.

Ash Dump—Makes fireplace cleaning a lot easier.

Log Lighter—Easier log starting.

Raised Hearth—Easier access to working on the fire.

Blower—May provide more room heating.

Heat Recirculator—Increases heating and efficiency.

One important safety issue is the flue. Avoid burning pine whenever possible, especially if you are using a prefab fireplace. The resin buildup in the flue can cause problems later (fire hazard). If you plan to have a stucco home, consider a prefab fireplace. A masonry fireplace can look out of place unless covered over with stucco, which isn't very cost effective.

A fireplace adds that cozy feeling to a home.

MASONRY & STUCCO — STEPS

MA1	DETERMINE brick/stone pattern, color and coverage and joint tooling.
MA2	DETERMINE stucco pattern, color and coverage.
MA3	PERFORM standard bidding process. Contact material suppliers, local brick companies, masonry unions and other subs for the names of good masons. Do not select a mason until you have seen his work. Ask the suppliers if there is a penalty for returning unused bricks.
MA4	PERFORM standard bidding process for stucco subs. You really need to see their work, too.
MA5	APPLY flashing above all windows and doors if not already installed by roofer.
MA6	ORDER brick, stone, and angle irons.
MA7	LAY brick/stone veneer exterior, chimney, fireplace and hearth. This is the time to have the mason throw in a brick mailbox if you want one. Remember to provide a "postal approved" metal mailbox insert.
MA8	FINISH bricks and mortar.
MA9	INSPECT brickwork. Refer to specification, local code and checklist on following page.
MA10	CORRECT any brick/stone work requiring attention.
MA11	PAY mason. If you have extra bricks which either can't be returned or are not worth returning, turn them into a brick walkway or patio later.
MA12	CLEAN up excess bricks and dried mortar. Spread any extra sand in the driveway and sidewalk area. Masons charge too much by the hour to have them do this work.
MA13	INSTALL base for decorative stucco. This includes keys, bands and other trimwork. Use exterior grade lumber or Styrofoam if using synthetic stucco.
MA14	PREPARE stucco test area. This should be at least a 3' by 6' sheet of plywood covered with test stucco to make sure you agree with the actual color and texture. Make sub finish mailbox if desired. If you wish to protect the stoops from stucco, cover them with sand.
MA15	APPLY stucco lath and decorative formwork to sheathing.
MA16	APPLY first coat of stucco. The color of the first coat does not matter.
MA17	APPLY second coat of stucco. You need to wait a few days for the first coat to dry completely. Place a thick mat of straw at base of house to prevent mud from staining stucco. While the scaffolding is set up, have the painter work on cornice and second story windows.
MA18	INSPECT stucco work.
MA19	CORRECT any stucco work requiring attention.
MA20	PAY stucco subs.
MA21	PAY mason's retainage.
MA22	PAY stucco sub's retainage.

MASONRY — SPECIFICATIONS

- Brick veneer coverage to be as indicated on attached drawings.
- All brick, sand and mortar mix to be furnished by builder.
- All necessary scaffolding to be supplied and erected by mason.
- Brick entrance steps to be laid as indicated in attached drawings.
- Brick patio to be laid as indicated in attached drawings.
- All excess bricks to remain stacked on site.
- Wrought iron railings to be installed on entrance steps as indicated in attached drawings. Wrought iron to be anchored into surface with sufficient bolts.
- Bricks are to be moistened before laying to provide superior bonding.
- Window, door, fascia and other special brick patterns are to be done as indicated in attached detail drawings.
- Moistened brick is to be laid in running bond with ⅜" mortar joints.
- Mortar is to be medium gray in color; Type "C" mortar is to be used throughout job.
- Mortar joints are to be tooled concave.
- Steel lintels are to be installed above all door and window openings.
- Masonry walls are to be reasonably free of mortar stains as determined by builder. Mortar stains are to be removed with a solution of one part commercial muriatic acid and nine parts water to sections of 15 square feet of moistened brick and then washed with water immediately. Door and window frames are to be protected from the cleaning solution.
- Flashing to be installed at the head and sill of all window openings.
- Flashing to be installed above foundation sill and below all masonry work.
- Bid to include all necessary touch-up work.

STUCCO — SPECIFICATIONS

- Bid is to provide all material, labor and tools to install lath and stucco per attached drawings.
- Stucco to be finished smooth, sand texture. (Specify color.)
- All trim areas to be 2" wide unless otherwise specified.
- Quoins to have beveled edges; alternating 12" and 8" wide.
- Two coats of stucco to be applied; pigment in second coat.
- Stucco work to come with a twenty year warranty for material and labor against defects, chipping and cracks.
- Expansion joints to be as indicated on blueprint or as agreed upon.

MASONRY & STUCCO — INSPECTION

- ☐ Brick and stone style and color are as specified.
- ☐ Mortar color and tooling are as specified.
- ☐ All wall sections plumb, checking with plumb bob and string. No course should be out of line more than ⅛". Overhang distance between top course and bottom course should not exceed ⅛".
- ☐ Flashing installed above all windows, exterior doors and foundation sill.
- ☐ No significant mortar stains visible from up close.
- ☐ Stoops and patio steps drain properly.
- ☐ No loose brick or stones.
- ☐ All quoins smooth, even and as specified.
- ☐ Touch-up brickwork finished.
- ☐ All wrought iron railing is complete per drawings and is sturdy.

Stucco
- ☐ Stucco coverage, colors and materials are as specified.
- ☐ Stucco is finished as specified.
- ☐ Stucco finish is even with no adverse patterns in surface. Surface level with no dips or bumps. Sight down surface from corners.
- ☐ Stucco is applied to proper depth.
- ☐ All decorative stucco work such as quoins, keys and special trim are in place, complete as specified and well finished with sharp edges and corners.
- ☐ Adequate number of expansion joints (one approximately every 300 square feet). Expansion symmetrical between window openings.
- ☐ Openings for exterior outlets and fixtures not covered up.
- ☐ No large cracks.

SIDING AND CORNICE

The siding and cornice sub will handle all siding application and cornice work (soffit, fascia, frieze and eave vents). This sub can also install windows and exterior doors, if needed. Usually, the framing sub will have already installed the sheathing, windows, and doors. The siding sub should make sure that the house wrap and water-proofing around windows and doors has been properly installed before any siding goes up.

Siding

Siding is the most economical exterior covering. The most popular materials include cedar, vinyl, aluminum and fiber cement. There are advantages and disadvantages to each:

Type	Advantages	Disadvantages
Cedar	Natural appearance Moderate cost	High maintenance. May raise fire insurance premium. May warp or crack
Texture 111®	Inexpensive	Poor appearance
Masonite®	Inexpensive	High maintenance. Requires painting. Prone to swelling and delamination
Vinyl	Washes easily. Low maintenance	Expensive
Aluminum	Washes easily. Low maintenance	Expensive
Fiber Cement	Durable, fireproof, and pest proof. Low maintenance	Expensive, can be difficult to cut

Wood Siding

Wood siding provides the greatest variety in styles and textures while remaining cost effective. It can be installed horizontally, vertically, diagonally or in a combination to provide a creative and varied exterior. Some of the wood types available include pine, cedar, cypress and redwood. The full range of types and styles exceeds the scope of this book. Check with your local supplier for the styles available in your area.

Regardless of the type of wood siding you choose, the important properties required for quality are freedom from warping, knots and imperfections. Wood siding is prone to shrinkage after installation as the moisture level begins to drop. If painted right after installation, lap siding will continue to shrink and leave an unpainted strip of wood where the pieces overlap. Try to delay the final coat of paint until the siding has had an opportunity to stabilize. Drying will also cause boards to warp and crack as they shrink. Always retain a portion of your siding sub-contractor's payment so that you can get him to return later to replace these defective pieces.

The introduction of several plywood siding types such as Texture 111® has simplified siding installation. These plywood sheets simulate several styles of wood siding such as board and batten, but are much easier to install and do not shrink. Labor costs to install are also much lower than traditional wood siding.

Wood siding is prone to leakage around windows and doors, especially on contemporary designs that have many angled walls. Therefore, it is important that your siding sub or painter caulk all butt joints. Specify a tinted caulk that matches the final color of the siding. Ask the siding sub to hand-pick the best siding pieces for the front and back of the house. Areas bordering decks and porches are the most visible; where visitors will be looking straight-on at the wood pieces.

Fiber Board Siding

Fiber board siding was very popular in the seventies and eighties. It has many of the aesthetic qualities of wood. It is usually free of imperfections and will not shrink. It also has the added advantage of being pre-primed, which may eliminate one paint coat. Its popularity plummeted when quality control issues cropped up, leading to several class action lawsuits. The bottom edges were prone to swelling if not primed and painted properly. When installation instructions were followed to the letter, the siding was usually quite durable and acceptable. This situation was a perfect example of the need for best practices when installing a product.

Vinyl and Aluminum Siding

This is usually installed by a sub-contractor who supplies labor and materials for the job. Although expensive, these siding types will last for many years with little or no maintenance. Many come with a 20 year unconditional guarantee. Early vinyl products suffered from a flimsy appearance, but recent products are often backed by foam insulation to increase its apparent stiffness. It also provides additional insulation.

Fiber Cement

Fiber cement siding has replaced fiber board as the siding of choice for most installations. It doesn't shrink or warp and is very fire resistant. It is manufactured from Portland cement with fibers added and comes in several different styles, including lap, board and batten, clapboard, and cedar plank. The fiber cement is cast in forms that give it a wide variety of wood grains. When painted and stained

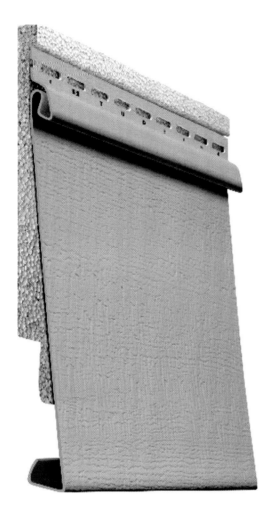

Foam backed, interlocking vinyl siding.

properly, fiber cement can be virtually indistinguishable from real wood. At first, siding contractors were not crazy about the material because the fiber cement was difficult to cut and kicked up an irritating dust. However, over time, contractors learned to adapt and now it's one of the most common installations.

House Wrap and Waterproofing

House wrap, window, and door waterproofing may be installed by either the framing contractor or the siding sub or a combination thereof. Whoever does the job needs to prepare the wall properly before siding is attached or water problems may crop up later. All siding is designed to drain water away by gravity, but wind driven rain can still migrate into unforeseen areas through cracks, seams, and voids in the wood. It's critical that the sheathing provide a drainage path for trapped water.

Structural sheathing like OSB board and porous sheathing is usually covered with a house wrap, which is designed to allow water vapor from inside the home to escape, but has small pores that stop the migration of liquid water from the outside, inward. Siding that is nailed to structural sheathing that has house wrap properly installed will have natural voids and air pockets that allow wind driven rain to drain down the backside of the siding.

Foam sheathing is usually impervious to water and so eliminates the need for house wrap. However, its soft texture creates a problem. Foam sheathing tends to conform to the shape of siding that has been nailed to it, functioning much like a gasket. This can stop the natural drainage of water and trap moisture against the backside of the siding. This type of installation can cause warping of wood and blistering of paint as moisture tries to migrate its way through the wood.

Fiber cement siding, when properly stained, can attain the rich look of real wood.

House wrap is applied from the bottom up, so that upper layers overlap lower layers by several inches.

This has led to a new category of house wraps known as "drain wrap," "three-dimensional membrane," or "ventilated rainscreen." These products usually have a three-dimensional texture that creates air pockets behind siding. This not only allows for ventilation, but creates a drainage route for any trapped water. Other two-dimensional wraps have vertical ridges or textures pressed into the material to provide a continuous drainage channel.

Green building technology has taken this to the next logical step by creating "rain walls." The rainscreen material is first applied to foam sheathing and then a set of vertical furring strips attached. Siding is then nailed to these strips, providing a continuous ventilation channel behind the siding. In summer, heat is carried away from the wall through these channels while water drains downward. In winter, the channels are blocked at the top and the air gap acts as an additional insulation layer. Rain walls can extend the life of siding and paint while reducing energy usage significantly.

Delta-Dry three dimensional membrane.

Waterproofing

The gaps around windows and doors provide ample opportunity for liquid water to migrate into the framed wall. The siding sub-contractor must make sure that all waterproofing sealants have been properly installed around windows and doors before attaching the siding. Most waterproofing methods must be done as the windows are installed, so this job has usually been done by the framer before the siding sub-contractor shows up. Make sure to have a joint meeting with the framer and the siding sub-contractor early in the project so that they can coordinate their activities. If the siding sub discovers an improper waterproofing job by the framers, his job will be delayed while the framers return to re-install windows and doors.

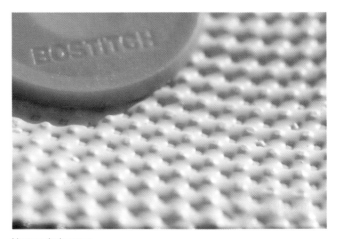

Vortec drain wrap.

There are several sealing technologies available on the market and each has their own detailed installation instructions. Make sure that all house wrap or drain wrap has been folded over the rough openings of doors and windows before they are installed. Once doors and windows are in place, flashing and waterproof sealants must be applied in the proper order to insure that any liquid drains away from the framed wall, not toward it.

Once siding is installed it is best to allow some time for moisture content to stabilize before painting. Painting will seal the wood and slow drying. The paint sub-contractor will than caulk around siding and corner boards. If the wood has not finished shrinking, it can pull away from the caulk, breaking the seal.

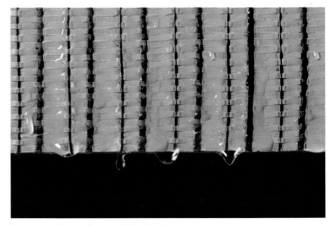

Greenguard Raindrop ventilated rainscreen.

Cornice

The cornice is the finishing applied to the overhang of the roof called the eave. The overhang of the roof serves a decorative function as well as protecting the siding from water stains and leakage. Generally, the greater the overhang, the more expensive the look of the house.

Proper overlap between house wrap and window sealant is essential before siding can be installed reliably. Note how the house wrap overlaps the waterproofing at the top of the window.

Window flashing and waterproofing membrane can be installed either before or after house wrap. Each method has a precise sequence. Note how house wrap has been folded over corner flashing to direct water down and away from window openings.

Window flashing

Waterproof membrane over house wrap

House wrap

Siding flush with bottom of sill

8" min. clearance

Furring strip

Installation of lap siding.

Waterproof membrane can also be installed before house wrap if necessary. Refer to manufacturer's installation instructions for the proper method and sequence of steps.

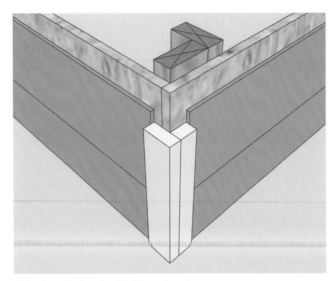

Siding installation details: Corner boards.

The cornice will be installed by your siding or trim sub. If you plan to use fancy trim around your cornice such as dentil molding, you may want to trust the job to your trim sub. He is more used to working with decorative trim.

The two types of cornices most commonly used are the open cornice and the box cornice. The open cornice is the simplest type and is more appropriate for contemporary styles. The open cornice looks just like it sounds. The rafter overhang is left unfinished and open and requires care in the choice of roof sheathing, since it will be visible from below. The box cornice has the overhang boxed in with plywood and is finished off with a variety of trim types and styles. This style is more commonly used on traditional styling and has the added advantage of providing a place to install eave ventilation, which will improve the energy efficiency of the house.

Polymer trim boards eliminate one of the key sources of rot and decay on cornices and soffits. They are more expensive, but have a long life with low maintenance.

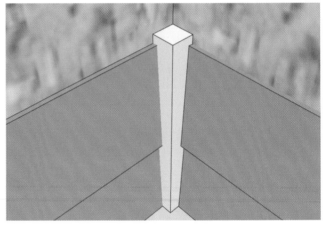

Siding installation details: Inside corner strip.

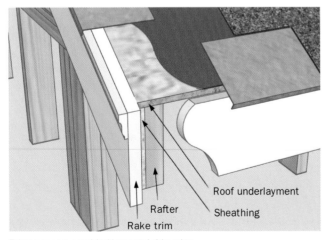

Roof underlayment

Rafter

Sheathing

Rake trim

Edge support provided by metal drip edge.

Cornice and soffit installation required meticulous installation and attention to detail.

The style and quality of the cornice can have a big effect on the appearance of your house, so take care in the design. If you use a draftsman or an architect, ask him to provide a cornice detail drawing so that its design will be clear to the cornice sub. This will also help you in estimating materials for the job.

Cornice work is installed before brick or stucco; so if you have that type of finish, be sure that the cornice sub builds the cornice out (usually with 1x6 or 2x6) so that the cornice sits out beyond the brick or stucco. Once installed, the cornice should be primed with exterior paint as soon as possible. For convenience you may wish to have your painters prime the cornice material prior to installation.

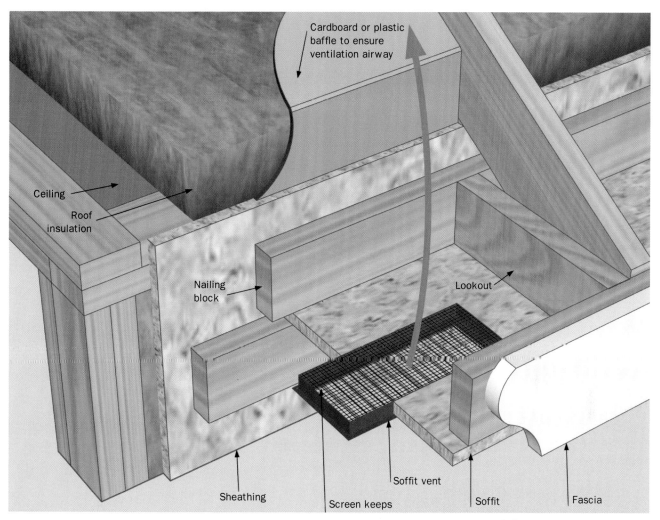

Cardboard or plastic baffle to ensure ventilation airway

Ceiling

Roof insulation

Nailing block

Lookout

Sheathing

Screen keeps

Soffit vent

Soffit

Fascia

Soffit framing details facing page.

SIDING AND CORNICE — SPECIFICATIONS

Siding
- Exterior hot dipped galvanized finish nails to be used exclusively.
- All nails are to be flush or counter sunk.
- All laps of siding to be parallel.
- Joints to be staggered between courses.
- All work and materials are to conform to the local building code.
- All siding edges to be terminated in a finished manner and caulked.
- All openings and trim caulked and flashed.

Cornice
- All soffit and fascia joints to be trimmed smooth and fit tight.
- All fascia to be straight and true.
- Soffit vents to be installed.
- All soffit, fascia, frieze and trim material to be paint grade fir or equivalent.

SIDING AND CORNICE — STEPS

SC1	SELECT siding material, color, style and coverage.
SC2	CONDUCT standard bidding process. Most often the siding and cornice will be bid as a package deal.
SC3	ORDER windows and doors. Before ordering, have the window supplier walk through the site to check openings. All window and door brands differ in their real dimensions. The supplier can help to provide more exact dimension information. Mark the proper dimensions on the openings if adjustments need to be made by the framing or siding sub.
SC4	INSTALL windows and doors. Inspect windows and doors for bad millwork BEFORE installing. Have wood shims available to adjust openings properly. If installing brick veneer, nail a 2×4 under exterior door thresholds for additional support until brick is installed. Inspect window and door positioning. They should be plumb and secure. Galvanized finish nails should be used, with the heads of the nails set below the surface of the trim. Install locks and hardware on doors and windows.
SC5	INSTALL flashing above windows, exterior door openings and around bottom perimeter of house.
SC6	INSTALL siding.
SC7	INSTALL siding trim around corners, windows and doors.
SC8	CAULK all areas where siding butts against trim or another piece of siding.
SC9	INSPECT siding. Refer to related checklist and specifications.
SC10	PAY sub for siding work. Make sure to retain a portion of payment for call backs.
SC11	INSTALL cornice. This involves installing the cornice, fascia, frieze, soffit and eave vents. If you have a traditional front, you may want to use a dentil molding. Stucco fronts normally use an exterior grade of crown molding ranging from 3" to 6" thick. Frieze bed mold runs in size from 4" to 12" and should be made of clear wood. Soffit should be made of an "AB" grade of ½" plywood.
SC12	INSPECT cornice work.
SC13	CORRECT any problems noted with cornice work.
SC14	PAY for cornice work.
SC15	CORRECT any problems noted with siding after it has had time to shrink.
SC16	ARRANGE for painter to paint and caulk trim.
SC17	PAY siding retainage.
SC18	PAY cornice retainage.

SIDING — CHECKLIST

- [ ] Foam sheathing is not torn, split, or dented. Torn areas should be sealed with waterproofing tape.
- [ ] Flashing and waterproofing around openings properly installed before siding.
- [ ] No splitting, warping, or cupping of lap siding.
- [ ] No large, unpatched knots. Large knots should be trimmed off boards or patched with Bondo.
- [ ] Siding joints staggered for more random look.
- [ ] No double nailing of lap siding (where bottom nail goes through top of siding underneath). This restricts movement and causes splitting as wood shrinks.
- [ ] Nails heads are flush with siding and not overdriven.
- [ ] Nailing pattern hits studs, not just structural sheathing (have siding sub pull chalk lines to use as guides).
- [ ] House wrap overlaps 4" or more from bottom to top.
- [ ] Vertical house wrap joints taped with waterproofing tape.
- [ ] Wood siding is back-primed before installation to equalize moisture absorption.

CORNICE — CHECKLIST

- [ ] Baffles installed in soffit to allow unobstructed air flow to roof vents.
- [ ] Soffit vents installed with rodent screens.
- [ ] Open cornice shows no nails from roof sheathing that have missed rafters.
- [ ] Cornice overhang properly flashed with shingles overlapping flashing.
- [ ] Frieze board notched to overlap last siding board.
- [ ] No large knots in fascia or soffit board that would allow entrance of rodents.
- [ ] Small knots patched to prevent insect infestation.
- [ ] Cornice fascia beveled to fit flush against underside of roof sheathing.
- [ ] Fascia boards scarfed or mitered (no butt joints) to allow for shrinkage.

INSULATION AND SOUNDPROOFING

Green Building

The increasing cost of energy and the threat of global warming has resulted in an explosion of interest in environmentally sensitive construction. "Building green" covers a wide range of concepts that goes beyond the scope of this book, although you will find references to it throughout the chapters. One of the key elements of any green project is energy efficiency, and that means insulation and lots of it.

To learn more about green construction, check out these excellent books on the subject:

- Building Today's Green Home, by Art Smith
  Rocky Ridge Designs
  www.rockyridgedesigns.com
- Mascord Efficient Living, by Alan Mascord
  Alan Mascord Design Associates
  www.mascord.com
  Green construction consists of several strategies:
- The use of sustainable or recycled materials that are easily renewed without increasing the home's carbon footprint (the amount of carbon dioxide added to the atmosphere by the construction of the home)
- The reduction of VOC's (volatile organic compounds) that are off-gassed from materials like flooring and insulation. These compounds reduce indoor air quality.
- The use of energy efficient appliances
- An increase in insulation and air-tightness which can greatly reduce energy bills

Green Building Initiatives

Energy Star Program

ENERGY STAR is a joint program of the U.S. Environmental Protection Agency and the U.S. Department of Energy which promotes energy efficient products and practices. Look for the Energy Star label on appliances to gain significant energy savings. The higher the rating, the more energy you will save. Energy Star rated appliances must meet strict energy efficiency guidelines set by the EPA and US Department of Energy. You can also seek certification as an Energy Star rated home, which can often garner discounts from local utilities. Energy Star homes are at least 15% more energy efficient than homes built to the 2004 International Residential Code (IRC), and include additional energy-saving features that typically make them 20–30% more efficient than standard homes. Check the Energy Star website at www.energystar.gov for information on numerous government sponsored tax credits that may be available.

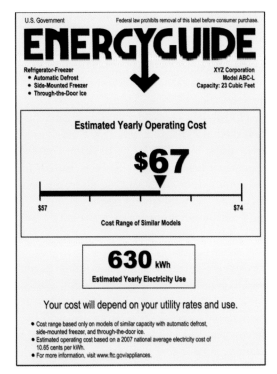

Look for the EnergyGuide label on all energy star rated appliances.

LEED Certification

LEED is a green building certification initiative developed by the U.S. Green Building Council (USGBC). LEED certification provides third-party verification that the home is built using sound techniques to reduce its environmental impact. These techniques include: energy and water saving technologies, carbon footprint reduction, higher indoor air quality, and a minimal impact on the environment. LEED certification can result in significant energy savings and a higher resale value for the home.

Forest Stewardship Council

The Forest Stewardship Council was created to coordinate the development of forest management standards and to promote sustainable lumber products through a certification program. FSC accredited, independent, "third-party" bodies assess forest management principles, criteria, and standards. The FSC standards for forest management have now been applied in over 57 countries around the world. If you would like to use lumber products bearing the FSC logo in your construction project, visit their website at www.fsc.org to find a list of certified retailers and manufacturers.

NAHB National Green Building Program

The National Association of Home Builders (NAHB) is helping its members move green building into the mainstream. Energy efficiency, water and resource conservation, sustainable or recycled products, and indoor air quality are increasingly incorporated into the everyday process of home building. The website can be found at www.nahbgreen.org. The NAHB provides several resources to builders:

- The ANSI approved ICC-700-2008 National Green Building Standard (NGBS) establishes a much-needed and nationally-recognizable standard definition of green building.
- NAHB Model Green Home Building Guidelines. Published in 2005, the voluntary Guidelines cover seven areas, including lot preparation and design, resource efficiency, energy efficiency, water efficiency and conservation, occupancy comfort and indoor environmental quality, and operation, maintenance and homeowner education.

Regional Green Building Organizations

There are many local and regional non-profit organizations dedicated to green construction. One example is Southface in Atlanta, Georgia (www.southface.org.) They provide training materials, certification, and outreach programs for homeowners and builders. Search the Internet for organizations in your area.

INSULATION

Proper insulation is essential to any energy efficient or green building construction strategy. Improvements in insulation and air infiltration can greatly improve protection against cold, heat, drafts, moisture, pollution, and noise. An energy-efficient home also helps to ensure consistent temperatures between rooms, improved indoor air quality, greater durability, and a significant reduction in energy costs.

The insulation ability of a material is measured as R-factor or R-value. R-value is a measure of thermal resistance to the transmission of heat. The thickness of a material has little to do with its R rating. Density and resistance to air motion play a significant role here. Don't assume thick insulation has the highest R-value.

Insulation Materials

There are many new insulation technologies available to homeowners in a wide range of costs and efficiencies.

Fiberglass

Fiberglass is the most common and cost effective insulation available and is probably found in 80% of homes built in the past fifty years. It has high durability, fire resistance, pest resistance, and a moderate R-factor. Sometimes the R-factor of fiberglass is overrated unless the batts are perfectly installed. They have a tendency to leave gaps where air infiltration can negate the insulation's natural thermal resistance. Moisture infiltration can also reduce its effectiveness. While usually sold in faced and unfaced batts for wall installation, fiberglass is also available as a blown-in product for attics. Blown-in materials are very popular as attic insulation because of ease of installation and the

ability to fill in around odd shaped areas and framing. It is usually not suitable as wall insulation, however, because of its tendency to settle. Fiberglass, rock wool and cellulose can be blown in (as a loose material). This type of insulation is normally applied with depths ranging from 6" to 12".

Fiberglass is also famous for being unpleasant to install. Fiberglass fibers are itchy, and require installers to wear long sleeved clothing and dust masks. Fiberglass contains formaldehyde and can outgas volatile organic compounds. For this reason, it is not considered a good green product, although proper installation of vapor barriers can alleviate some out-gassing.

Cellulose

Cellulose is the second most common and cost effective insulation. Cellulose is usually made from recycled newspaper and commonly blown into walls and attics. For wall installation, this is either accomplished by installing a netting or vapor barrier on the wall to hold the insulation in place, or by impregnating the cellulose with a binder that holds the fibers in place after blowing. Because it is

The effectiveness of fiberglass insulation depends on the quality of installation. Gaps should be filled carefully.

Blown-in wall insulation.

blown in place, cellulose tends to fill gaps and air infiltration areas better than fiberglass. The R-factor is moderate to good and pest resistance is high when the cellulose is properly treated with boron. Beware of cellulose that does not conform to local fire and pest treatment standards. Like fiberglass, the R-factor of cellulose is reduced by moisture. Cellulose is also susceptible to mold in high moisture areas.

Recycled Denim

The color of green is blue. Recycled denim is usually produced from old blue jeans, hence its color and has become a popular green building alternative. It has a high R-factor and has virtually no chemical irritants. Being a recycled product, its impact on the environment is negligible. Like cellulose, it is commonly treated with borate, which is a safe mold, fire, and pest resistant coating.

Foam-in-Place Insulation

This category covers a wide range of new products with their own advantages and disadvantages. The main advantage of foam-in insulation is its excellent thermal properties. The foam expands and fills every nook and cranny, and most foam products double as vapor barriers, making installation simpler. Their biggest disadvantage is cost. However, remember that sometimes, more is less. Because of its high R-factor per inch, foam-in insulation can provide higher thermal resistance in the same area as fiberglass. A highly insulated home with fiberglass would probably require a 2×6 frame wall. You could achieve the same insulation with a 2×4 wall when using foam-in insulation. Vapor barriers can often be eliminated. A little extra investment in cost can reap benefits for years to come. Foam-in insulation is highly moisture, pest, and mold resistant.

There are two types of foam-in insulation: Open-cell (isocyanurate) and closed cell (polyurethane). The closed cell foams typically have a higher R-value than open-cell foam. They are also denser (and harder) and can add

significant structural strength to a building, literally gluing the sheathing to the studs. Its cost is also higher, prompting some builders to use a hybrid insulation approach. High density foam is sprayed in as a thin layer to seal and strengthen the wall. Later, after electrical and plumbing has been installed, open cell foam or cellulose is then blown in to finish filling the wall cavity. This approach can save money and provide an extremely tight and vapor resistant structure.

Sprayed foam insulation is applied as a liquid which contains a polymer (such as polyurethane or modified urethane) and a foaming agent. The liquid is sprayed through a nozzle into wall, ceiling, and floor cavities where it expands. It requires an experienced installer. As the foam begins to dry, excess material will be trimmed away to leave a straight wall. In attics, the excess foam

Unvented attic insulated with foam directly on the backside of the roof sheathing. HVAC vents in the conditioned attic will save considerable energy.

Foam-in-place Icynene insulation seals all gaps, even around electrical outlets. Its water resistance eliminates the need for interior vapor barriers.

is simply left in place. Foam insulation works well in unvented "conditioned attics." In this approach, the foam is sprayed directly on the roof sheathing, leaving the rest of the attic as part of the house's "insulated envelope." By sealing and insulating the roof directly, the need for attic ventilation is eliminated. Any HVAC installed in a conditioned, unvented attic will be far more energy efficient as a result.

By acting as a wind and air barrier, it often eliminates the need for separate air-tightness detailing which can increase energy efficiency and allow downsizing of the heating and cooling system equipment. Sprayed foam insulation does not shrink, sag, settle, or biodegrade. There are also bio-based foams made from soy products that are environmentally friendly, renewable, and highly efficient.

Foamed Concrete

Otherwise known as cementitious foam insulation, this new technology is gaining in popularity. It consists of air entrained concrete with a consistency of shaving cream. After a few days, the foam hardens into a stable insulation that is fireproof, mold-proof, pest-proof and completely non-toxic. While concrete in its normal form is not considered particularly "green," cementitious foam insulation is considered very green because of its high insulation value and permanence. When the house reaches its end of life, the concrete foam can be recycled. Because of its strong fireproof properties, it is particularly appropriate for areas at great risk for fire, such as California. It also has the additional benefit of having excellent sound deadening properties.

Installation

Insulation requirements vary across the country. Contact your local power company and they will assist you with recommended R ratings for ceilings, floors, walls, attics, etc. Observe that their recommendations may exceed your minimum local code requirements. Don't worry. You can pay for insulation now or pay for energy later. R-factor requirements are not necessarily the same for all walls. Exposures which receive heavy sunlight or wind should be more heavily insulated.

Sheathing

Sheathing is a lightweight, rigid panel nailed to the outer side of the exterior walls. Foil covered sheathing has several advantages including higher insulation values, a finish more resistant to abrasion during installation and greater airtightness. When manufactured as a rigid foam material, it can add to the R-factor of the insulation, so research your insulation needs carefully before deciding which type to use. Foil backed urethane insulation was a popular alternative, but it gained a reputation for causing moisture and siding problems in humid climates. The foil backing acted as a

Sealing behind the rim joist eliminates a key source of air infiltration and moisture.

vapor barrier, trapping interior moisture in the walls. The foil backing also caused paint and siding issues where moisture would accumulate at the foil base directly behind siding, causing paint blistering and warping. A better alternative is foam sheathing.

Air Tightness and Moisture Control

These two subjects are discussed together because they are so closely related. Most moisture problems in a house are caused by the movement of air from one temperature zone to another. If you eliminate air migration, you eliminate most moisture issues. That is why new foam-in-place insulation methods are good choices not only for insulation quality but to eliminate air movement and moisture. They tend to seal vulnerable areas, but you'll also want to look for a few classic areas of vulnerability.

Non-structural foam sheathing adds insulation and an exterior vapor barrier.

Moisture

Most moisture problems, with the exception of flood damage and leaks, result when warm, moist air moves to a cooler area. In northern climates in the winter, the warm moist air is migrating from the inside toward the outside, where it then condenses in the cooler air pockets of the exterior wall. In southern climates in the summer, the migration is usually from the outside in, where warm exterior air condenses on cooler interior walls. You will need to analyze your climate situation when deciding on the proper installation of vapor barriers. Check with your local code inspectors for suggestions. They are usually quite familiar with the climate related problems in your part of the country.

When combined with leaks from roof breaches and poor moisture control behind siding and windows, moisture accumulation can create a serious toxic mold problem. Attack these as a single problem and you'll be able to breathe easier at night knowing that your house will remain warm, dry, and healthy.

Air Leaks

Homeowners can spend thousands of dollars on energy efficient windows, thick walls, and copious insulation, only to see their investment (and energy) drain away in the wind. If you have zones of air leakage, then all the insulation in the world will not stop the loss of your expensive conditioned air. In many cases, air leakage and poorly insulated pockets result in more heat (and cold)

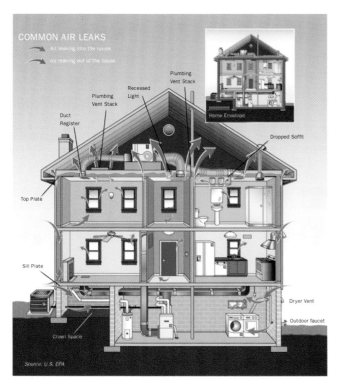

Plug air leaks to reduce air infiltration and moisture issues.

loss than poor insulation. To defend against leaks, do the following:

- Keep and can of foam insulation with you during the framing phase and plug leaks around windows, electrical outlets, plumbing pipes and vents, HVAC, corners, and wall and ceiling junctions.
- Install the proper vapor and moisture barriers for your area of the country. These include poly film behind drywall and house wrap over exterior sheathing.
- Seal windows properly against wind and moisture.
- Look for pockets of thin or no insulation where floor and roof joists intersect with walls.
- Make sure cantilevers have been insulated underneath.
- Consider foam-in insulation due to its air tightness and vapor barrier properties.

Soundproofing

This is a refined touch. Consider soundproofing bathrooms and the HVAC area to quiet down the system. Rooms where you plan to have stereo systems are also good candidates. Soundproofing material comes in 4×8 sheets and can be applied between the drywall and studs. The addition of insulation will also add to the sound deadening affect. If you plan to use foam-in insulation, ask your insulation sub-contractor to spray around interior plumbing and waste vents, especially where vents breach the roof. HVAC ducts can rattle during use if not installed tightly. A generous coating of foam insulation not only eliminates vibration, but seals air leaks as well.

There are many different techniques for reducing sound transmission but they all depend to two physical properties: absorption and transmission. Absorption uses a porous material like insulation to capture sound waves traveling through the air, as in the cavities of interior walls. Another strategy is to reduce the transmission of vibrations through solid structures. This is done by isolating one structure from another. Drywall nailed to wood studs acts like a huge speaker diaphragm. Sound on one side moves through the wall and stud, then causes the drywall on the other side to vibrate.

Sound deadening strategies combine these two factors to absorb and isolate sound. Use a combination of the following techniques to isolate sounds:

- Fill wall cavities with fiberglass or foam insulation. Foam concrete is particularly good at absorbing air based sound.
- Place a sound deadening board between drywall and stud. This not only absorbs sound but breaks the direct route of vibration from one room to another. Dense rubber sheets work quite well for this purpose, but are fairly expensive.
- Double up drywall on both sides of the wall. This adds to the wall's density and helps to resist vibration.

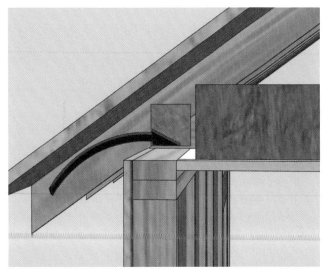

Roof insulation doesn't extend far enough past the wall, allowing cool, moist air to migrate through the ceiling, causing moisture buildup.

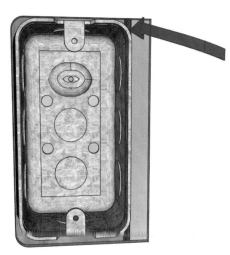

Air can flow unimpeded around electrical outlets and switches if the insulation and vapor barriers are not sealed properly with caulk or tape.

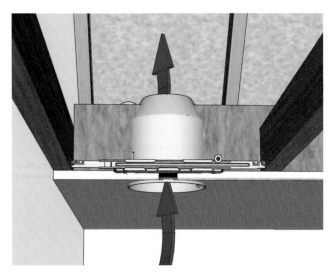

Tall, recessed light cans can provide a direct route for warm and humid air to enter the attic, causing condensation and resulting loss of insulation.

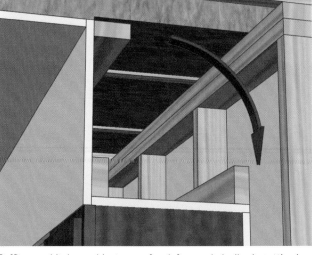

Soffits over kitchen cabinets are often left unsealed, allowing attic air to infiltrate into uninsulated inner walls. Moisture from cooking can accumulated there, encouraging mold growth.

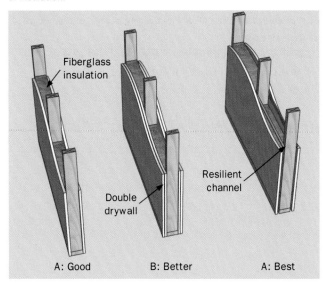

Fiberglass insulation

Double drywall

Resilient channel

A: Good   B: Better   A: Best

Soundproofing methods using traditional material. Resilient channels isolate drywall and stop sound transmission.

- Mount drywall to studs using a resilient metal channel. This isolates the drywall from the studs completely.
- Build two stud walls back to back or 2×4 studs staggered on a 2×6 plate. Make sure adjacent studs do not touch. This will create a break for vibrations and reduce sound transmission.

INSULATION — SPECIFICATIONS

- Bid is to provide and install insulation with minimum R-values as indicated below:

Location	R Factor	Type
Walls	R13	Fiberglass batts
Ceiling	R19	Blown insulation

- Bid is to include labor and materials for wall and ceiling insulation.
- All insulation is to include a vapor barrier on the warm side of the insulation, free of rips and reasonably sealed against air infiltration.
- All joints and gaps around door frames, window frames and electrical outlets to be packed with fiberglass insulation.
- All insulation used will be of thickness sufficient to meet R values specified after insulation has settled.

INSULATION — INSPECTION

- ☐ All specified insulation has been installed per specifications. Check labels on insulation.
- ☐ All insulation is installed tightly with no air gaps.
- ☐ No insulation packed down. This reduces the effective R value.
- ☐ Vapor barrier faces the INTERIOR of the home (the side heated in the winter).
- ☐ Any rips or tears have been repaired.
- ☐ No recessed lighting fixtures covered with insulation. Need to allow heat to flow.
- ☐ Pull-down attic staircase weather-stripped and well insulated.
- ☐ No punctures have been made in vapor barrier.
- ☐ Insulation adheres firmly to all adjoining surfaces.
- ☐ No eave vents have been covered by insulation.
- ☐ Ductwork insulated in basement.
- ☐ Plumbing in basement insulated.
- ☐ All gaps around doors and windows stuffed with insulation.
- ☐ Gap between wall and floor caulked completely.
- ☐ All openings to the outside for plumbing, wiring and gas lines sealed with spray foam insulation.
- ☐ All gaps in siding around windows and corners caulked thoroughly.
- ☐ All fireplaces properly insulated.

INSULATION — STEPS

IN1	DETERMINE insulation requirements. Your local energy company can help give you guidelines and suggestions.
IN2	PERFORM standard bidding process.
IN3	INSTALL wall insulation. Hand-pack insulation in small nooks and crannies first. Then around chimneys, where framing offsets occur, and where pipes come through walls. Make sure the sub uses plenty of staples when attaching vapor barrier to avoid gaps.
IN4	INSTALL soundproofing. This is a nice touch to deaden noise. Use in bathroom walls to deaden sound.
IN5	INSTALL floor insulation. Floor insulation will be installed on crawl space and basement foundations. Metal wires cut to the length of joist spacing are used to hold fiberglass insulation in place. Foam-in insulation can just be sprayed in place.
IN6	INSTALL attic insulation. This normally consists of batts or blown-in insulation. If blown-in, attic insulation will be installed after drywall has been nailed into place on the ceiling. If you use two layers of fiberglass batts, place the second layer perpendicular to the first to cut off any major air leaks. The second layer should have no vapor barrier. Plug up all HVAC vent openings with leftover insulation. This keeps drywall dust out of the HVAC vents during the dusty drywall installation and sanding process.
IN7	INSPECT insulation. A kraft paper or poly vapor barrier should be installed on the warm side of the insulation. All areas around plumbing, electrical fixtures, doors and windows should be stuffed with insulation to prevent air infiltration.
IN8	CORRECT insulation work as needed.
IN9	PAY insulation sub and have him sign an affidavit.
IN10	PAY insulation sub retainage.

DRYWALL

Drywall

Drywall is composed of gypsum sandwiched between two layers of heavy gauge paper. In recent years, manufacturers have offered a mold resistant drywall that is sandwiched with fiberglass instead of paper. A third type of drywall is "greenboard" which is made with water resistant paper and is used in bathrooms and other moisture laden environments. Drywall comes in 4' × 8', 4' × 10' and 4' × 12' sizes. When planning your home, it may be helpful to plan around these dimensions to some extent to minimize waste. Whenever possible, plan for butt joints to intersect above and below windows and doors. This minimizes the size of the butt joint, which requires extra labor to finish properly.

Drywall comes in three common thicknesses: ⅜", ½", and ⅝". One half inch thickness is most common on walls with five-eighths inch being popular for ceilings because it does not sag or warp as easily. If you are framing 24" on center (O.C.), you may want to upgrade drywall thickness for the walls to five eighths inch. Drywall sheets have the long edges tapered and the short edges full thickness. This is to allow room for the drywall mud and tape. Sheets should be installed so that the full edges butt against studs. After drywall is applied to studs, the joints between sheets of drywall are smoothed using special drywall tape. Moisture affects drywall adversely, so always store it in a dry place until used.

Drywall installation normally involves the following steps:
- Apply drywall to studs with either nails or screws, and glue.
- Apply fill coat of joint compound and drywall tape.
- Sand all joints smooth (if necessary).
- Apply joint compound (second coat).
- Sand all joints smooth (if necessary).
- Apply finish coat of joint compound (third coat).
- Sand all joints smooth.

Application should not be done in extreme heat or cold. Ceilings are treated just as walls, unless you decide to stipple them. Stipple is a coarse-textured compound which eliminates the need for repeated sanding and joint compound application. This is a time and money saver.

Drywall Subs

Your drywall sub is a key player in the final, visible quality of most surfaces in the interior. Drywall sub-contractors normally bid and charge for drywall work by the square foot, although they install it by the sheet. Drywall subs charge extra for special projects such as:
- tray ceilings
- vaulted ceilings
- open foyers
- curved walls and openings
- high ceilings
- smooth ceilings
- water-resistant drywall
- mold resistant drywall

Of all your sub-contractors, quality here is important. Drywall is one of the most visible parts of your home. Get the best drywall sub you can afford. Below are signs of a good drywall sub:
- Finishing coats are very thin and feathered.
- No need to sand first two coats because they are so smooth.
- Pieces are cut in place to assure proper fit.

Your drywall sub can quickly tell you the quality of the framing job. Knots, warped lumber and out-of-square walls will make the drywall sub's job more difficult. If the problems are major, have the framer fix them before continuing. Your drywall sub can't fix these underlying problems effectively.

Warning: Your drywall sub may use stilts to reach ceilings and high walls. Stilts are dangerous, but many times virtually unavoidable. Some states will not award worker's compensation to subs injured while using stilts. If this is true in your state, have your sub sign a waiver of liability to protect you from such accidents. Scaffolding should be used for stairway work. If stilts are used, the floor should be cleared of scrap wood and trash to avoid tripping.

Nails vs. Screws

Most drywall subs prefer nailing as it goes fast. However, nailed drywall is impossible to remove without tearing up the board completely. Nailed drywall is also prone to "nail pops." These usually occur a few years after construction and show up as bumps and cracks in the smooth drywall finish. Nail pops occur when wood studs shrink over time due to climatic changes. The sagging or settling of ceilings and roofs are another common cause of nail pops.

Make sure to inspect installed drywall before the first coat of drywall compound is applied. Overzealous installers may hit the drywall nails with too much force, causing the paper around the nail to rip. When this occurs, the structural integrity of the drywall at that attachment point is destroyed. If you see torn drywall paper, ask the drywall sub to place a few additional nails nearby.

Drywall screws help to alleviate nail pops and provide a much stronger bond between drywall and stud. Screws are particularly valuable for attaching ceilings, due to the ceiling's weight. Commercial power tools are available with drywall screws dispensed from long rolls, allowing screwed walls to go up almost as fast as with hammer and nail. These drywall drills are designed to seat the screw head at the perfect depth without ripping the drywall paper.

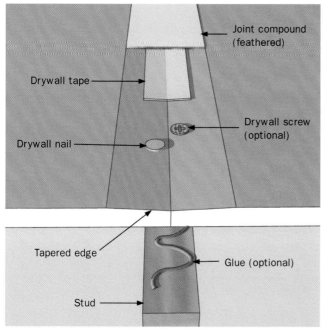

Joint compound
(feathered)

Drywall tape

Drywall screw
(optional)

Drywall nail

Tapered edge

Glue (optional)

Stud

Tapered drywall edge allows for flush joint channels that isolate drywall and stop sound transmission.

Joint compound
arched & tapered

Drywall tape

Butt joint

Stud

Tapered edge

Non-tapered butt joints require more layers of joint compound and labor to feather the joint.

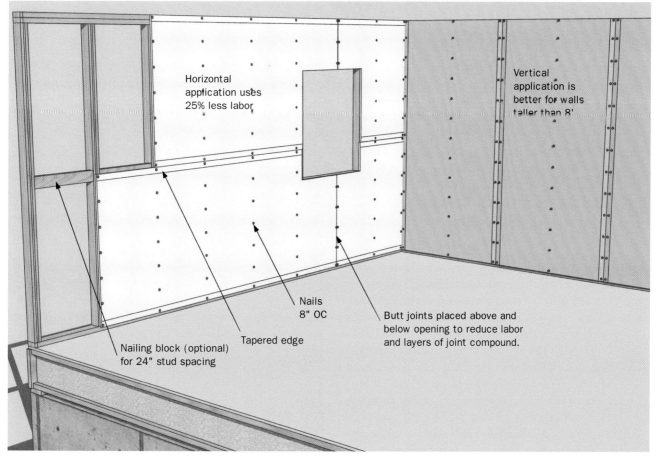

Horizontal application uses 25% less labor

Vertical application is better for walls taller than 8'

Nails 8" OC

Tapered edge

Butt joints placed above and below opening to reduce labor and layers of joint compound.

Nailing block (optional) for 24" stud spacing

Application of gypsum wallboard.

Scheduling

Drywall is extremely heavy and bulky to manage. Imagine trying to carry drywall up a flight of stairs to the second floor, or down to the basement. For this reason, drywall trucks are usually equipped with cranes so that drywall can be lifted and placed on upper floors through openings like windows and sliding doors. It's best to schedule the delivery of drywall before the framers have installed the windows. Most roughed-in window openings are barely large enough to fit drywall through and you don't want to risk damaging or breaking newly installed windows. If you don't have a second floor window or door large enough to accommodate drywall, then ask your framer to leave one or two sheets of sheathing unattached to provide an opening.

DRYWALL — STEPS

DR1	PERFORM drywall bidding process (refer to standard bidding process). Refer to drywall contract specifications for ideas. You may wish to ask painters for names of good drywall subs; they know good finishing work — they paint over it for a living!
DR2	ORDER and receive drywall materials if not supplied by drywall sub.
	NOTE: Before drywall is installed, mark the location of all studs on the floor with a builder's pencil. This will make it easier for the trim sub to locate studs when nailing trim to the wall.
DR3	HANG drywall on all walls. Outside corners must be protected with metal edging. Inside corners must also be taped.
DR4	FINISH drywall. Drywall must be finished in a series of steps as outlined below: • Spackle all nail dimples. • Sand nail dimples smooth. • Apply tape to smoothed nailed joints. • Spackle tape joints (fill coat). • Sand tape joints (as needed). • Spackle tape joints (second coat). • Sand tape joints (as needed). • Spackle tape joints (finish coat). • Sand tape joints (third time).
DR5	INSPECT drywall. Refer to specifications and inspection guidelines. To do this properly, turn out all lights and look at the wall while shining a light on it from an angle. Slight shadows will appear if there are imperfections on the surface. Mark them with a pencil so the drywall sub will know where to fix walls. Don't use a pen, as it will leave a mark that will show through paint.
DR6	TOUCH up and repair imperfect drywall areas. There are always a few. Place large drywall scraps in basement if you want to finish the basement later.
DR7	PAY drywall sub. Get him to sign an affidavit.
DR8	PAY drywall sub retainage. You may want to wait until the first coat of paint has been applied so that you can get a good look at the finished product.

DRYWALL — SAMPLE SPECIFICATIONS

- Bid is to provide all material, labor and equipment to perform complete job per specifications of attached plans. This includes:
  - Drywall, tape and drywall compound
  - Metal corner bead
  - All nails
  - Sandpaper
  - Ladders and scaffolding
- Apply 5/8" gypsum board (drywall), double nailed at top. Four nails per stud.
- All joints to be taped with 3 separate coats of joint compound each sanded smooth.
- All outside corners to be reinforced with metal corner bead.
- All inside corners to be reinforced with joint tape.
- All necessary electrical outlet, switch and fixture cutouts to be made.
- All necessary HVAC ductwork cutouts to be made.
- All ceilings to be stippled as specified.
- All ceiling sheets to be glued and screwed in place.
- Wall adhesive to be applied to all studs prior to applying drywall.
- Moisture resistant gypsum to be used along all wall areas around shower stalls and bathtubs.
- Use drywall stilts at your own risk.

DRYWALL — INSPECTION

Before Taping

- [ ] No more than a 3/8" gap between sheets.
- [ ] Nails driven in pairs (2" apart). Nail heads dimpled below the surface of the drywall.
- [ ] Nails not hit so hard as to break the surface paper. All joints shall be double nailed or glued.
- [ ] No nail heads exposed to interfere with drywall.
- [ ] No sheet warping, bowing or damage. Sheets are easier to replace before taping.
- [ ] Rough cuts around door and window openings cut close so that trim will fit properly.
- [ ] Waterproof drywall, or wonderboard, installed in shower stalls and around bathtubs. No taping is necessary here.
- [ ] Metal bead installed flush on all outside corners.

During Finishing

- [ ] Three separate coats of mud are applied to all joints. Stippling will hide any imperfections. Each successive coat should leave a wider track and a smoother finish.

After Finishing

- [ ] Look down the length of installed drywall; there is no warping or bumps. If found, circle the area with a pencil (ink may show through when painted). Have the drywall sub-contractor fix all marked areas before final payment.
- [ ] All joints feathered smooth. No noticeable bumps, either by sight or touch.
- [ ] All electrical wiring remains exposed including bath and kitchen vent fans, garage door opener switches and doorbell in garage.
- [ ] Proper ceilings smooth and stippled as specified.
- [ ] Clean cuts around register openings, switches and outlets so that covers will cover exposed area.
- [ ] Nap of paper not raised or roughened by excessive or improper sanding.
- [ ] All touch-up work completed and satisfactory.
- [ ] No exposed corner bead.

Note: Some drywall imperfections will not appear until the first coat of paint has been applied. This is the first time you will see the wall as a single, uniform color. If you can, wait until this point to pay retainage.

TRIM

Good trim work requires a skilled carpenter. If you are not very experienced in this type of work, you may not want to tackle this job. Interior trim sub-contractors usually will do the following:

- Set interior/exterior doors and door sills
- Set windows and window sills
- Install base, crown, chair rail, picture and other moulding
- Install paneling, raised paneling and wainscoting
- Install stairway trim
- Install fireplace and main entry door mantels
- Install closet shelves and hanger rods
- Install any other special trim
- Install door and window hardware, door stops, etc.

Trim subs charge for work in many ways:
- By the opening (for doors, windows and entryways).
- By the cut. The more cuts involved, the greater the cost.
- By the hour or day.
- By the job.

Decorative trim is one of the most visible items in your house and can make the difference between an ordinary interior and one that stands out. The workmanship of your trim sub will be very visible and hard to repair if not done properly. Make sure your trim sub is very quality conscious and watch his work closely. If you want to keep a close eye on his work, volunteer to assist him.

The critical part of door installation is in making sure that the door is set level and square so that doorknob hardware will work properly and the door will operate without binding. Special care should also be taken with bi-fold and sliding track doors. Improper installation is hard to fix and results in binding.

Weather stripping is essential in all exterior doors. This has become a specialty trade in itself. As such, your trim sub may not necessarily be the one who does this small, but critical job. Trim work has become somewhat easier with the advent of preset doors and window trim kits. To save money, consider using pre hung hollow core interior doors. These doors install easily and look as good as or better than custom hung doors.

In choosing your trim material, you must first decide whether you plan to paint or stain the trim. You may be able to use lesser grades of millwork if they will be painted. Finger joint trim consisting of scraps glued together is fine if you plan to paint. Chipboard panel doors are becoming very popular and are inexpensive. If you plan to stain, you must use solid wood trim and birch paneled wood doors; but the fine appearance and low maintenance of stained trim may be worth the expense.

Protect trim from abuse prior to installation to avoid dents, cracks, scratches and excessive waste. Lay trim on the floor in neat stacks. Don't purchase trim from different suppliers — the different pieces may not match.

When installing base moulding, consider whether trim will go around wall registers or merely be interrupted by the register. The first example is more expensive but yields a nicer appearance. If you use a tall base moulding, you can build your registers and receptacles into the base moulding.

Modern houses are making use of decorative trim extensively, such as paneling and wainscoting. Ask your trim sub for suggestions of decorative approaches. It is amazing what can be done with a minimum of materials.

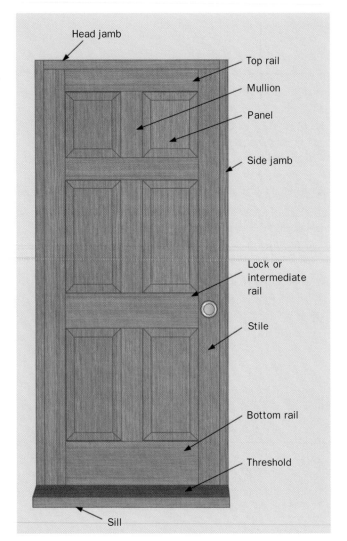

Panel door components.

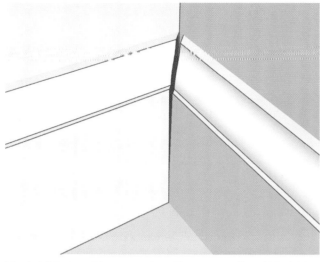

Wood shrinkage and/or uneven drywall can cause mitered trim joints to separate.

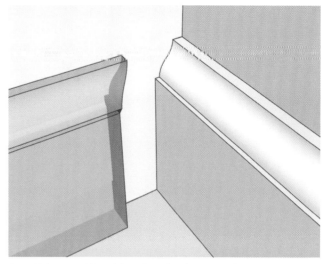

Coped joints provide a tighter fit and resist separation.

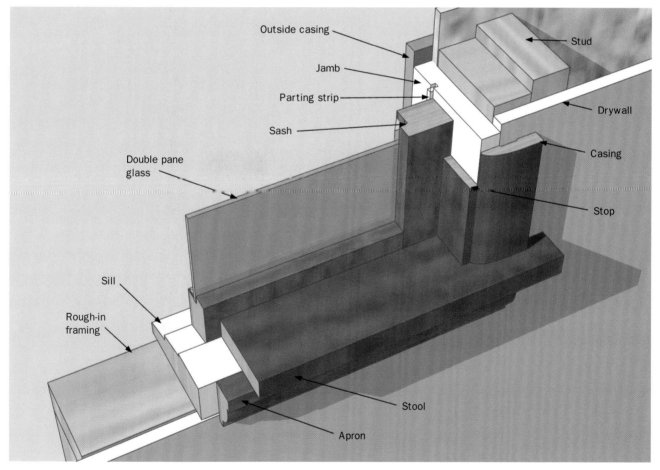

Window framing details.

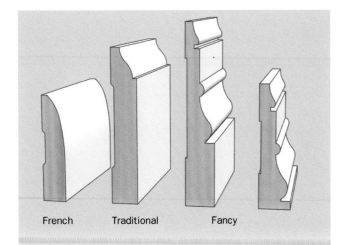

Base moulding styles.

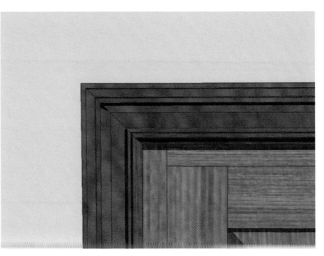

Traditional miter joint casing.

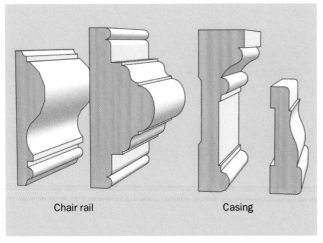

Other moulding styles.

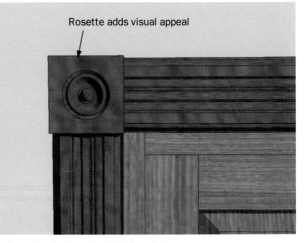

Use of rosette simplifies butt joint casing.

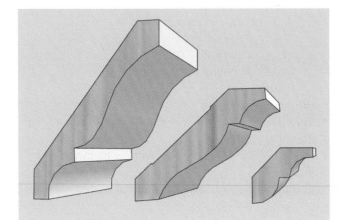

Crown moulding styles.

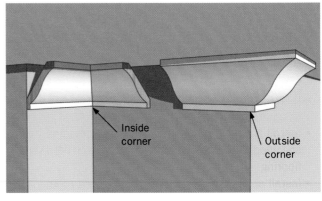

Crown moulding requires precise, multi-angle cuts..

TRIM — STEPS

TR1	DETERMINE trim requirements. Millwork samples can help. Mark all walls that get different trim treatments.
TR2	SELECT moulding, window frames and door frames.
TR3	INSTALL windows. Windows and exterior doors should be installed as soon after framing as possible. Don't have windows and doors delivered until just before installation. Remind the trim sub to use the best trim pieces around the openings in the most visible locations.
TR4	CONDUCT standard bidding process.
TR5	INSTALL interior doors. If pre-hung doors, install and adjust door stops and doors.
TR6	INSTALL window casing and aprons.
TR7	INSTALL trim around cased openings.
TR8	INSTALL staircase moulding. Treads, risers, railings, newels, baluster, goosenecks, etc.
TR9	INSTALL crown moulding. This includes special-made inside and outside corners if specified. This is installed first so that the ladder legs will not scratch the base moulding.
TR10	INSTALL base and base cap moulding. Where the final floor level will be higher than the subfloor (when hardwood is installed) the base should be installed a little higher.
TR11	INSTALL chair rail moulding.
TR12	INSTALL picture moulding.
TR13	INSTALL, sand and stain paneling.
TR14	CLEAN all sliding door tracks.
TR15	INSTALL thresholds and weather stripping on exterior doors and windows.
TR16	INSTALL shoe moulding after flooring is installed. Painting the shoe moulding before installation will reduce time and touch-up work.
TR17	INSTALL door knobs, deadbolts, door stops and window hardware. Re-key locks if necessary. Consider keying all the locks to just one key. Install a deadbolt-type lock on sliding glass doors for improved security.
TR18	INSPECT trim work. Refer to your specifications and the checklist that follows.
TR19	CORRECT any imperfect trim and stain work.
TR20	PAY trim sub.

TRIM — SPECIFICATIONS

- All materials except tools, to be furnished by builder.
- Interior doors will be pre-hung.
- Finish nails to be used exclusively and set below the surface, puttied and sanded over smooth.

Bid is to include the following:
- Hang all interior doors. All doors to be right- and left-handed as indicated on attached drawings.
- Install all door and window trim according to attached schedule.
- Install base moulding and base cap moulding as specified. Raise base moulding in areas of hardwood flooring.
- Install apron and crown moulding in living room, dining room and foyer area.
- Install two-piece chair rail moulding in dining room as specified.

- Install 1" shoe moulding in all rooms not carpeted.
- Install den bookshelves as indicated on attached drawings.
- Install staircase trim as indicated below:
  - 12 oak treads
  - 12 pine risers
  - solid oak railing with involute
  - gooseneck
- 36 spindles.
- Build fireplace mantel as indicated on attached drawings.
- Install door locks, door knobs, window hardware and dead bolts.
- All trim paint grade unless otherwise stated.
- Install all closet trim, shelves and closet rods.

Doors

- [ ] Proper doors installed. Correct style, size, type, etc.
- [ ] Doors open and close smoothly and quietly. Hinges do not bind or squeak. Doors should swing freely with no noise or friction against adjoining surfaces. Open doors at 30, 45 and 60 degree angles. Doors should remain where positioned. If not, remove center pin, bend it slightly with a hammer and replace it.
- [ ] Door knobs and latches align with latch insets. All deadbolt locks should align properly. Privacy locks installed on proper side of doors. Passage locks on proper doors.
- [ ] All exterior doors lock and unlock properly. Locks should function freely.
- [ ] All keys available. Have all locks keyed the same for ease of use.
- [ ] All doors open in proper direction. Door latch faces the proper direction.
- [ ] All doors are plumb against door jambs. With door slightly open, check for alignment and evenness of opening. Two screws in each strike plate. A pin in each door hinge.
- [ ] All door casing nails set below surface and filled with putty.
- [ ] Door knobs, door locks and dead bolts work properly without sticking.
- [ ] No hammer dents in door or door casing.
- [ ] Thresholds in place and properly adjusted.
- [ ] Weatherstripping in place if specified.
- [ ] Proper clearance from floor (about 1/2" above carpet level). Consider height of carpet and pad, tile, wood floor, etc.
- [ ] Door stops in proper places.

Windows

- [ ] Window frame secure in place with leveling wedges or chips.
- [ ] Windows installed with less than 1/2" gap between wall and frame.
- [ ] Windows open and close smoothly and easily. Windows should glide easily along tracks. Windows should close evenly and completely.
- [ ] Window sash locks installed and operating properly. No hammer dents in window or window casing.
- [ ] All window casing nails set below surface and sealed with putty.
- [ ] Weatherstripping in place if specified.
- [ ] Screens installed if specified.
- [ ] All window latches function properly.
- [ ] All window pulls in place and secure.
- [ ] No excessive damage to millwork.

Trim and Paneling

- [ ] All crown molding installed and finished as specified.
- [ ] All base molding installed and finished as specified.
- [ ] All chair rail molding installed and finished as specified.
- [ ] All wainscoting installed and finished as specified.
- [ ] Trim joints are caulked, sanded smooth and undetectable, both by sight and feel.
- [ ] Trim intersects with walls, ceilings and floors evenly with no gaps or other irregularities.
- [ ] All trim is void of major material defects.
- [ ] All finishing nails set below surface and sealed with wood putty.
- [ ] All den paneling and shelving installed and finished as specified.
- [ ] Fireplace mantel installed and finished as specified. Will support heavy load.
- [ ] Fireplace surround installed and finished as specified.
- [ ] All closet shelving and coat racks installed at proper height and level.

Staircase

- [ ] Proper treads used.
- [ ] Finished and curved treads installed correctly. All railings in place, secure, and angled parallel to stairs.
- [ ] Spindles and newel posts secure in place and properly finished.
- [ ] Balusters secure in place and properly finished.
- [ ] No squeaks in staircase steps or elsewhere on staircase.

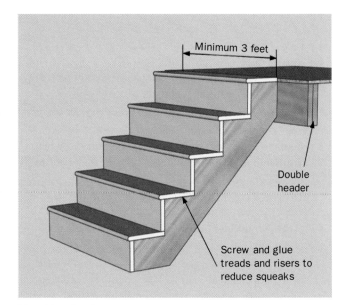

Closed tread stairs.

PAINTING AND WALLCOVERING

Paint

If there is any job you can feel safe about doing yourself, it is painting and wallpapering. It may be worth several thousand dollars to you. Although painting can be easy, preparation is essential for a professional look. Even the smallest defect will show up on a painted wall or piece of millwork. The goal of preparation is to make every surface smooth; getting rid of all cracks, bumps, nicks, scratches and rough areas.

Paints can be purchased in standard colors or custom mixed. Custom mixed paints can't be returned, so be careful with color schemes and order quantities. If you decide to paint yourself, purchase the highest quality paints you can find — but insist on a builder discount.

New drywall should always receive two coats of paint, but the one-coat paints will guarantee the best coverage on the second coat. A good primer should be used for the first coat to provide the best adhesion surface. Make sure to paint ALL walls, including ones that will be covered with wallpaper. If wallpaper is applied to bare drywall, it will be difficult or impossible to remove later without tearing off the top layer of drywall paper.

Colors on small samples in a store will appear darker when they are applied to a room. If you plan to stipple ceilings, this should be done before any painting has begun. Although roller application is quick and yields good results, consider renting an electric paint sprayer. You may need to thin the paint if you go this route, and you will definitely need some form of eye and nose protection. Keep paint off areas to be stained or it will show up in the final work. When enamel thickens, thin it with paint thinner.

Hiring a Professional

Painting your home can be a very rewarding experience, but it is time consuming and tedious. Before you decide to tackle the painting, ask yourself the following questions:

1. Do you have the time and patience for the job? If you are working full-time, your time is worth money. Is your time worth more than the painting contractor?
2. Do you have the skills to do the quality job you demand? Just because you know a good job when you see one -- doesn't mean you have the talent to do the job yourself.
3. Do you have the physical stamina? Painting itself is not hard, but setting up scaffolding, climbing ladders and scraping paint can be backbreaking work.

If you hire professional painters, make sure they carry worker's compensation and liability insurance. Since painting can be more of an "amateur" trade than most, anyone can call themselves a painter. They are less likely to carry professional liability coverage. Ask for references from other jobs they have done and specify the masking and drop cloth requirements to protect your property.

Types of Paint

Paint formulations and technology have changed tremendously in the past few years. Before choosing the correct paint for your project you need to know the nature of the material underneath the paint, the composition of the existing paint finish, and the properties of the paint you plan to use. The many types of paint fall into three main categories:

1. Oil based pigments which use linseed oil as the carrier. This class of paint adheres well to most surfaces, provides a tough, durable surface and creates a water repellent surface.
2. Latex paints which use water as the carrier for the pigment. These paints produce little or no fumes, dry fast, are easy to use, and produce a breathable finish that allows water vapor to escape. Latex is not quite as durable as oil based paints, but recent improvements have narrowed the gap considerably.
3. Varnishes and other solvent based finishes. These compounds use solvents such as mineral spirits, alcohol, and acetone as the carrier. These finishes provide a myriad of uses as penetrating finishes, floor finishes, epoxies, elastomeric finishes, stains, and other finishes.
4. Epoxy paints. These are two-part, self curing paints most commonly used to coat concrete, garage floors, or countertops. They come in two formulations that must be mixed right before being used.

All paints are made up of similar compounds that determine the quality of the paint. The carrier in paint is the liquid that suspends the pigments and additives and allows the material to penetrate the painted surface. As the carrier evaporates, the pigments harden into a solid surface that protects the surface against moisture and wear. Additional solvents also serve as carriers and are added to thin the paint to the proper consistency.

The pigments in the paint supply the "holdout" or covering ability of the paint. Pigments provide both opacity and tint to the paint. The most common pigments found in almost all paints today are the oxides: zinc oxide, and titanium oxide. Lead oxides used to be quite popular twenty years ago until it was discovered to be a health hazard, especially to children. Removal of old paint that may contain lead oxide is a serious consideration when stripping old finishes. Lead oxide paints have been banned for many years. Even with new paints, you should take precautions to avoid inhaling the paint dust or fumes. There is a new category of paint called "low VOC" paint. This formulation is designed to have little or no volatile organic compounds that can outgas and affect air quality. This can be a big factor if you have young children in the family.

Sherwin-Williams low VOC paint.

The amount and quality of pigments in your paint will determine the total quality of the paint grade. Certain types of clay are used as "budget" pigments in less expensive paints to reduce the need for the oxide pigments. Additives make up a diverse selection of compounds that serve many purposes in special paint formulations. Some examples are: rust inhibitors, drying agents, fungicides, bonding agents, antifreeze, emulsifiers, and thickening agents. The specific use of the paint will determine which and how much ingredients are present.

Most paint is sold in at least three quality grades: premium, budget and professional. For maximum quality, durability, and coverage always choose the premium grades of paint. These may cost a little more, but they last much longer. Surprisingly, the "professional" or "contractor" grades of paint are actually the cheapest grades. These paints are sold in large quantities to contractors at low prices. Many contractors are mostly concerned with the price and coverage of the paint; not the durability. These grades usually contain more clays and chalk pigment, which cover very well but don't last as long as the oxide pigments. Don't assume that your "professional" contractor is using the best quality paint. Specify in your contract the grade of paint that you want for your project.

Oil Based Paint

Until recently, the most common paints were oil based. These paints are still popular but have slowly been edged out in popularity by latex paints. Oil paints are still the most durable paints and are most appropriate for exterior finishes and trim. Oil paint bonds strongly to wood and metals and also bonds better on sub-standard finishes, such as chalky, dirty, or oily surfaces. Because oil repels water, however, it does NOT adhere to damp surfaces at all. Oil paints must be applied to a dry surface. Cleanup also requires the use of smelly solvents such as mineral spirits. This makes oil paints less attractive for inside projects.

Oil Based Stains

Oil based stains are similar to the paints but contain much less pigment. Since they do not cover well, they are designed to be used on new unfinished wood as a sealer and water-proofer. They allow the natural beauty of the wood to show through. Stains come in transparent, semi-transparent, and opaque formulations. Penetrating stains are effective and economical finishes for all kinds of lumber and plywood surfaces. They are especially well suited for rough-sawn, weathered and textured wood and plywood. Knotty wood boards and other lower quality grades of wood which would be difficult to paint can be finished successfully with penetrating stains.

These stains penetrate into the wood without forming a continuous film on the surface. Because there is no film or coating, there can be no failure by cracking, peeling and blistering. Stain finishes are easily prepared for refinishing and are easily maintained.

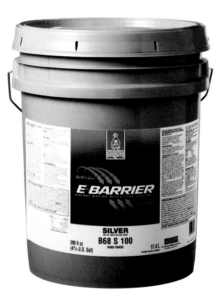

Buy paint in five gallon containers to save money.

Penetrating semi-transparent stains form a flat and semi-transparent finish, which allow only part of the wood-grain pattern to show through. A variety of colors is available, including shades of brown, green, red and gray. The only color which is not available is white — it can be provided only through the use of white paint. The opaque stains contain enough pigment to behave much like paint and can be used over previous stain colors.

Stains are quite inexpensive and easy to apply. To avoid the formation of lap marks, the entire length of a course of siding should be finished without stopping. Only one coat is recommended on smoothly planed surfaces — it will last 2 to 3 years. After refinishing, however, the second coat will last 6 to 7 years because the weathered surface has absorbed more of the stain than the smoothly planed surface.

Water Repellent Finishes

These finishes contain little or no pigments and are designed only to penetrate and seal the wood grain. These are used mainly on patios and decks where foot traffic would damage a surface finish. A simple treatment of an exterior wood surface with a water repellent finish greatly reduces the natural weathering process. Staining seals the wood and promotes uniform natural tan color in the early stages of weathering and a reduction of uneven graying which is produced by the growth of mildew on the surface.

Water-repellent finishes generally contain a preservative, a small amount of resin and a very small amount of a water repellent that is frequently wax-like in nature. The water-repellency greatly reduces warping, excessive shrinking and swelling which can lead to splitting. It also retards the leaching of chemicals from the wood and the staining from water at the ends of boards.

This type of finish is quite inexpensive, easily applied and very easily refinished. Water repellent finishes can be applied by brushing, dipping and spraying. Rough surfaces will absorb more solution than smoothly planed surfaces — the treatment will also be more durable. It is important to thoroughly treat all lap and butt joints and the ends of all boards.

Two-coat application is possible on rough sawn or weathered surfaces, but both coats should be applied within a few hours of each other. When using a two-coat system, the first coat should never be allowed to dry before the second is applied. If it does, the surface will be sealed, preventing the second coat from penetrating.

Initial applications may be short lived (1 year), especially in humid climates and on species that are susceptible to mildew, such as sapwood and certain hardwoods. Under more favorable conditions, such as on rough cedar surfaces which will absorb large quantities of the solution, the finish will last more than 2 years.

Latex

Latex has become the paint of choice in recent years. It is durable, safe, and easy to clean up with water. It dries quickly and produces few fumes, making it ideal for interior application in living areas. They can also be used for exterior painting.

Latex paints are called "breather" paints and are more porous than conventional oil paints. If these are used on new wood without a good oil primer, or if any paint is applied too thinly on new wood (a skimpy two-coat paint job, for example), rain or even heavy dew can penetrate the coating and reach the wood. When the water dries from the wood, the wood chemicals leach to the surface of the paint. This happens most often with red cedar and redwood. For those surfaces, consider using an oil primer coat first and then finish with latex.

Latex also works well on stucco and masonry finishes. It is more resistant to the alkali by-products that can leach to the surface of masonry. Latex is thick and easily applied to stucco with a paint roller.

Interior vs. Exterior Paints

Make sure to use the proper formulation for the job. Interior and exterior paints are formulated quite differently and are inappropriate in situations for which they were not designed. Interior paints are formulated for hardness and durability, so that marks and stains can be easily scrubbed away without damaging the finish. Exterior paints contain far more elastomers, which make it far softer than interior paints. Exterior paints are designed to stretch and move as exterior walls heat up and cool down with the sun. If you use an interior paint outside, it will crack and blister in no time. Exterior paint used inside will be so soft that any scrubbing will wear right through the finish.

Varnishes

Clear finishes based on varnish, which forms a transparent coating or film on the surface, are used mainly for coating interior finishes such as floors and trim. The polyurethane type varnishes are very durable and resistant to water, alcohol, and oil.

Regular varnishes should not be used on exterior wood exposed fully to the sun. These finishes deteriorate and often begin to disintegrate within one year. A special exterior varnish called "spar varnish" is appropriate for outside use. It contains "UV blockers" that filter out the ultraviolet rays from the sun that damage regular varnish (and give you a sunburn.).

Special Purpose Paints

Several other types of paints are available for special painting needs. New water-based varnishes are available for interior finishing. These unique formulations combine the hardness and durability of polyurethane with the easy to clean properties of latex. They also create no fumes, which can make an interior paint job much more pleasant.

Porches and decks receive too much foot traffic to use standard exterior paints. Porch and deck paint dries to a much harder finish than standard paint. It comes in a limited number of colors and should be applied in two coats with no primer.

Steel or wrought iron fixtures are very prone to rusting. If painted with standard paint, any moisture that seeps through can cause a rust spot to develop. This rust spot will continue to grow and will eventually push the paint off the surface. Rust inhibitor paints and primers are available that contain the rust inhibiting chemicals. They bond tightly to the metal and create a chemical bond that prevents rust from starting. Since they come in a limited number of colors, they are better used as primer coats instead of final finishes.

Concrete sealants are highly elastic paints designed to maintain a waterproof seal even if cracking of masonry develops. They are so elastic that the surface will stretch and span any small cracks. They usually come in white only and are not very durable since the finish is soft. They will not hold back a large amount of moisture and can blister badly in extremely wet conditions. You can paint over them with latex however, if you want a more versatile color selection.

Primers

Primers are designed to seal the surface so that subsequent layers of paint will be absorbed evenly. Primers are specially formulated to provide an optimum surface for final coats and come in most of the same formulations as finish coats. The difference is in the bonding and sealing agents. Primers are usually thinner so they can soak in and seal the surface better. They usually dry very fast. Oil based primers can be used under latex paints. A combination of oil primer and latex finish coat can provide the best of both worlds. The big disadvantage however, is that two different clean ups are needed -- one for the primer and one for the latex. Use primer coats under your final finish whenever possible. The final finish will usually be superior to just two coats of regular paint.

Materials and Equipment

Before starting your painting project, you will need to collect all the necessary equipment. Depending on whether you are painting indoors or outdoors, you may need some or all of the following equipment:

Brushes

You will need a variety of types and sizes of brushes — small brushes for trim and touch-up and 4" brushes for covering large areas. Natural bristle or "china bristle" brushes are the highest quality brushes for oil paints. Do not use these for latex however, the water will cause the bristles to curl like bad hair on a humid day.

Synthetic bristle brushes made of nylon work well for both oil and latex paint. Look for the highest quality rating. The bristle ends should be frayed (similar to split ends) and the bristles soft but not limp. Flat wall brushes for painting large areas should be cut straight across. Smaller brushes for trim work should have the bristles cut at an angle. These are called "sash" brushes because the angled cut allows you to work paint into corners and maintain a straight paint line around trim. Don't try to save money on cheap brushes — they will always do a bad job. Properly maintained quality brushes will produce a superb finish for years.

Rollers

Paint rollers come in several varieties for different jobs. The rollers are made up of the roller handle, the roll itself, and some type of pan to hold the paint. Most handles come with a threaded hole which will accept an extension handle, allowing you to reach tall areas without a ladder. The most common roller size is 9". The main difference between rollers are the size and composition of the roller covers. Rolls with a close "nap" look like velvet and apply the smoothest finish and should be used on the smoothest surfaces only. They don't hold much paint however, so frequent trips to the paint are necessary. The largest "nap" looks like wool fleece and will hold a great deal of paint. Their thick nap allows the roller to deliver a lot of paint to rough surfaces such as stucco, masonry, and rough siding. These "wooly mammoth" style rollers are great for outside jobs — but they create a lot of overspray. Make sure you have an ample covering of drop cloths. The most popular rollers fall somewhere between these two extremes.

Power rollers are available that force paint up a hose to the roller, eliminating the need to go back to the paint can. You can paint continuously. This makes the smaller nap rollers more practical, so your paint job will usually be smoother with less overspray. Power rollers are available at most hardware stores and are well worth the investment for large paint jobs.

A large flat pan is available to apply paint to the roller and allows you to "roll out" the excess paint on a ramp built in to the pan. If you are starting a large paint job, you can forgo the pan and use a roller "screen" instead. This is a piece of wire mesh designed to fit into 5 gallon professional paint cans. You can dip your paint roller right into the 5 gallon can, which is wide enough to accept a standard roller. The wire screen allows you to roll out the excess paint from the roller. If you expect to use more than 3-4 gallons of paint for the job, these large paints cans are more economical and convenient.

Sprayers

Paint sprayers come in four types: the hand held airless sprayer, the high pressure compressed air sprayer, the professional airless pump sprayer, and the new low pressure air sprayers. The hand held airless sprayers are the common sprayers many consumers buy for small paint jobs. They don't hold much paint, they are loud, and they emit paint in a diffuse cloud of spray that is difficult to direct to the painted surface. You are likely to get as much paint on yourself as the target area. These are not appropriate for large jobs.

Traditional air compressor sprayers will work for larger jobs but they require that the paint be thinned out so that the paint will "atomize" properly. The sprayer has several adjustments to change the paint pattern, volume, and pressure of the paint. Thick paint will "splatter" out

Wood	Ease of keeping well painted (I=easiest, V=most exacting[a])	Weathering		Appearance	
		Resistance to cupping (1=best, 4=worst)	Conspicuousness of checking (1=best, 4=worst)	Color of heartwood[b]	Degree of figure on flat-grained surface
Softwoods					
Ceders					
Alaska	I	1	1	Yellow	Faint
(California) incense-ceder	I	—	—	Brown	Faint
Port-Oxford-cedar	I	—	—	Cream	Faint
Western red cedar	I	1	1	Brown	Distinct
White cedar	I	—	—	Light brown	Distinct
Cypress	I	1	1	Light brown	Strong
Redwood	I	1	1	Dark brown	Distinct
Products[c] overlaid with resin-treated paper	I	—	1	—	—
Pine					
Eastern white	II	2	2	Cream	Faint
Sugar	II	2	2	Cream	Faint
Western white	II	2	2	Cream	Faint
Ponderosa	III	2	2	Cream	Distinct
Fir, commercial white	III	2	2	White	Faint
Hemlock	III	2	2	Pale brown	Faint
Spruce	III	2	2	White	Faint
Douglas fir (lumber and plywood)	IV	2	2	Pale red	Strong
Larch	IV	2	2	Brown	Strong
Lauan (plywood)	IV	2	2	Brown	Faint
Pine					
Norway	IV	2	2	Light brown	Distinct
Southern (lumber and plywood)	IV	2	2	Light brown	Strong
Tamarack	IV	2	2	Brown	Strong
Hardwoods					
Alder	III	—	—	Pale brown	Faint
Aspen	III	2	2	Pale brown	Faint
Basswood	III	2	2	Cream	Faint
Cottonwood	III	4	2	White	Faint
Magnolia	III	2	—	Pale brown	Faint
Yellow poplar	III	2	1	Pale brown	Faint
Beech	IV	4	2	Pale brown	Faint
Birch	IV	4	2	Light brown	Faint
Cherry	IV	—	—	Brown	Faint
Gum	IV	4	2	Brown	Faint
Maple	IV	4	2	Light brown	Faint
Sycamore	IV	—	—	Pale brown	Faint
Ash	V/III	4	2	Pale brown	Distinct
Butternut	V/III	—	—	Light brown	Faint
Chestnut	V/III	3	2	Light brown	Distinct
Walnut	V/III	3	2	Dark brown	Distinct
Elm	V/IV	3	2	Brown	Distinct
Hickory	V/IV	4	2	Light Brown	Distinct
Oak, white	V/IV	4	2	Brown	Distinct
Oak, red	V/IV	4	2	Brown	Distinct

[a]Woods ranked in group V for ease of keeping well painted are hardwoods with large pores that must be filled with wood filler for durable painting. when so filled before painting, the second classification in the table applies.

[b]Sapwood is always light.

[c]Plywood, lumber and fiberboard with overlay or low-density surface.

FINISH	INITIAL TREATMENT	APPEARANCE OF WOOD	COST OF INITIAL TREATMENT	MAINTENANCE PROCEDURE	MAINTENANCE PERIOD OF SURFACE FINISH	MAINTENANCE COST
Preservative oils (creosotes)	Pressure, hot and cold tank steeping	Grain visible. Brown to black in color, fading sightly with age	Medium	Brush down to remove surface dirt	5-10 year only if original color is to be renewed, otherwise no maintenance is required	Nil to low
	Brushing	Brown to black in color, fading slightly with age	Low	Brush down to remove surface dirt	3-5 year	Low
Waterborne preservatives	Pressure	Grain visible, greenish in color, fading with age	Medium	Brush down to remove surface dirt	None, unless stained, pointed or varnished as below	Nil, unless stains, varnishes or paints are used. See below.
	Diffusion plus paint	Grain and natural color obscured	Low to medium	Clean and repaint	7-10 year	Medium
Organic solvents preservatives*	Pressure, steeping, dipping, brushing	Grain visible. Colored as desired	Low to medium	Brush down and reapply	2-3 year or when preferred	Medium
Water repellent**	One or two brush coats or clear material or, preferably, dip applied	Grain and natural color visible, becoming darker and rougher textured.	Low	Clean and apply sufficient material	1-3 year or when preferred	Low to medium
Stains	One or two brush coats	Grain visible. Color as desired	Low to medium	Clean and apply sufficient material	3-6 year or when preferred	Low to medium
Clear varnish	Four coats (minimum)	Grain and natural color unchanged if adequately maintained.	High	Clean and stain bleached areas, then apply two more coats	2 year or when breakdown begins	High
Paint	Water repellent, prime and two topcoats	Grain and natural color obscured	Medium to high	Clean and apply topcoat, or remove and repeat initial treatment if damaged	7-10 year***	Medium to high

Source: This table is a compilation of data from the observations of many researchers.
*Pentachlorophenol, bis (tri-n-butyltin, oxide), copper naphthenate, copper-8-quinolinolate and similar materials.
**With or without added preservatives. Addition of preservative helps control mildew growth and gives better performance.
***Using top-quality acrylic latex topcoats.

the nozzle in lumps. This type of spray gun works best for thinner stains and varnishes. It also requires an air compressor. Most paint stores rent these machines.

A new type of sprayer has taken the market by storm in recent years — the high volume, low pressure sprayer (HPLV). These devices come with their own high volume turbine compressors that look and act like vacuum cleaners in reverse. The high volume of compressed air "drives" the paint toward the target, so more paint reaches its destination with less overspray. The compressed air is hot and helps to speed adhesion and drying of the paint. The adjustments on the sprayer provide a variety of paint patterns and volume. This sprayer can produce a very professional job and can be adjusted to a fine spray pattern for working around trim areas. Like the traditional air gun, this type works best with thinner paints.

The airless diaphragm sprayers are the type you will most likely see professional painters using. They contain a high volume airless compressor the pumps paint through a long hose to the spray nozzle. Most are designed to fit right on the edge of a five-gallon paint can. They deliver a high-volume stream of paint with low overspray and can be used for inside and outside painting. It is amazing to watch a professional painter use one of these machines. They work well with the thicker latex paints and can paint an entire house in one day. You can rent these machines from paint stores as well.

Choosing the Painting Method

If you decide to paint the project yourself, you will need to determine the most effective painting method to use — brushing, rolling, or spraying. Each has its advantages and disadvantages. Generally, clean up and preparation time is inversely proportional to the speed of painting. Brushing is the slowest method but requires the least masking and preparation. Spraying is extremely fast but requires careful preparation to protect surrounding areas from overspray. Paint rollers fall somewhere in between.

TYPE OF EXTERIOR WOOD SURFACES	WATER-REPELLENT PRESERVATIVE		STAINS		PAINTS	
	SUITABILITY	EXPECTED LIFE* (YR)	SUITABILITY	EXPECTED LIFE** (YR)	SUITABILITY	EXPECTED LIFE*** (YR)
Siding						
Cedar and redwood						
Smooth (vertical grain)	High	1-2	Moderate	2-4	High	4-6
Rough sawn or weathered	High	2-3	Excellent	5-8	Moderate	3-5
Pine, fir, spruce, etc.						
Smooth (flat grain)	High	1-2	Low	2-3	Moderate	3-5
Rough (flat grain)	High	2-3	High	4-7	Moderate	3-5
Shingles						
Sawn	High	2-3	Excellent	4-8	Moderate	3-5
Split	High	1-2	Excellent	4-8	–	–
Plywood (Douglas-fir and Southern pine)						
Sanded	Low	1-2	Moderate	2-4	Moderate	3-5
Rough sawn	Low	2-3	High	4-8	Moderate	3-5
Medium-density overlay†	–	–	–	–	Excellent	6-8
Plywood (cedar and redwood)						
Sanded	Low	1-2	Moderate	2-4	Moderate	3-5
Rough sawn	Low	2-3	Excellent	5-8	Moderate	3-5
Hardboard, medium density§						
Smooth						
Unfinished	–	–	–	–	High	4-6
Preprimed	–	–	–	–	High	4-6
Textured						
Unfinished	–	–	–	–	High	4-6
Preprimed	–	–	–	–	High	4-6
Millwork (usually pine)						
Windows, shutters, doors, exterior trim	High	–	Moderate	2-3	High	3-6
Decking						
New (smooth)	High	1-2	Moderate	2-3	Low	2-3
Weathered (rough)	High	2-3	High	3-6	Low	2-3
Glued-laminated members						
Smooth	High	1-2	Moderate	3-4	Moderate	3-4
Rough	High	2-3	High	6-8	Moderate	3-4
Waferboard	–	–	Low	1-3	Moderate	2-4

Source: This table is a compilation of data from the observations of many researchers. Expected life predictions are for an average continental U. S. location; expected life will vary in extreme climates or exposure (desert, seashore, deep wood, etc.).

*Development of mildew on the surface indicated a need for refinishing.

**Smooth, unweathered surfaces are generally finished with only one coat of stain, but rough-sawn or weathered surfaces, being more absorptive, can be finished with two coats, with the second coat applied while the first coat is still wet.

***Expected life of two coats, one primer and one topcoat. Applying a second topcoat (three-coat job) will approximately double the life. Top-quality acrylic latex paints will have best durability.

§Medium-density overlay is generally painted.

Exterior millwork, such as windows, should be factory treated according to industry Standard IS4-81. Other trim should be liberally treated by brushing before painting.

If your project is extensive, such as exterior painting, and everything needs to be painted, spraying is probably worth the extra preparation time. Once you prepare the site, you can often finish the job in 1-2 days. This helps to avoid interruptions for bad weather. Proper spraying requires a certain level of skill however, so make sure you are up to the task.

Saving Time

You can save a significant amount of time if you use a power roller. These units keep adequate paint on the roller with a push of a button. Avoid using paint sprayers for interior walls except for the primer coat. Sprayers do not hide minor blemishes and drywall sanding marks like paint rollers do and will give disappointing results.

You can save a tremendous amount of time painting if you coordinate the paint, drywall and trim process between subs. Apply both coats of wall paint and the first coat of trim paint before the doors and trim are installed. This will allow you to roll paint the walls right up to the door openings without worrying about getting paint on the trim. Put one coat on the doors and trim before installation. These two steps will cut interior painting time in half. Make sure to coordinate this with your trim carpenter first. Many trim subs do not like to work with pre-painted trim. It creates an irritating dust when sawed.

Test masking tape to be sure that it does not pull off paint when removed.

Painting Tips

- If you plan to do all or most of this yourself, don't forget to wear your worst clothing and bring your portable radio.
- Punch holes in the inside rim of opened paint cans with a nail or ice pick. This will allow the extra paint to drip back in the can.
- Tapered, angle-cut china-bristle brushes are recommended for a smoother finish. Avoid cheap brushes that shed bristles into your freshly painted surfaces.
- It is not necessary to use a water-based primer for a water-based final coat and an oil-based primer for an oil-based final coat. But the base coat must be dry and stable. Consult with your local paint supplier.
- Don't worry about what the surface looks like. If it feels smooth, it's ready to paint. When in doubt, sand more.
- Always use nose protection or respirator when sanding.
- Wrap oil-based rollers and brushes in plastic wrap (for short periods of time) or place them in a jar of paint thinner.
- Rinse water based rollers and brushes in water when stopping for several hours.
- Try to paint in daylight. If you must paint at night, use powerful halogen lamps.
- Allow coats of paint to dry before recoating.
- Apply paint evenly. Applying too heavily will leave trails of paint on the wall that must be sanded later.
- Be careful around flammable paints and solvents. No smoking, no open flames and no matches.
- Be careful of harmful vapors in paints and solvents. Obey all cautions printed by manufacturers.
- Stain grade moldings and handrails should be sanded with a very fine grit sandpaper before staining. Fill holes with a wood filler that accepts stain or color match filler to stain used. Use a good quality wiping stain and don't leave the stain on too long.

Wallcoverings

Wallpaper is an excellent do-it-yourself project. Most wallpaper outlet stores will be glad to show you the ins and outs of quality wallpaper selection and installation. Most wallpaper comes in double rolls of 32 feet per roll. Single rolls can be purchased at a premium. Wallpaper has a standard width of 24".

When purchasing wallpaper, make sure all rolls have the same run number. The same number tells you that wallpaper of a certain pattern was all made at the same time, and the colors will be exactly the same from roll to roll. Repeat pattern is a big factor in estimating the number of rolls that you will need. Repeat pattern is the width of the pattern before it starts over. On large repeat patterns, you will waste a lot of paper getting patterns to match up.

Wallpaper comes in unpasted and prepasted forms. Many wallpaper hangers still use a paint roller to apply a wallpaper glue even on prepasted wallpaper. Wallpaper that will be used for baths should be either foil or vinyl due to the humid environment.

PT1   SELECT paint scheme. Minimize number of colors. Try not to rely only on paint samples since final colors may vary. Try a few test samples on your wall. Check them in daylight, and with a portable lamp at night. Colors often have different "looks" between day and night. If you need to mix custom colors, one quart is normally the minimum. Some custom colors can only be mixed in gallon sizes due to the minute amounts of some pigments required. Custom colors are not returnable.

PT2   PERFORM standard bidding process. You may want to get this bid broken out between materials and labor. You may wish to use the painter as labor and furnish your own paint. If you do, have the painter help you select your paint. The painter might be able to get you a better price. It is advisable to buy the best paint you can afford. More expensive paint has more pigment and its durability can withstand more scrubbing. Consumer's Reports can help you identify which brands of paint have been tested for best quality, durability and value. If you are planning or considering doing the painting yourself, this will help you to determine how much you will save by doing the job yourself. If you are not up to doing the entire job, consider hiring the painter for the exterior work only and doing the interior work yourself.

PT3   PURCHASE all painting materials if you plan to paint yourself. Normally you'll want flat water-based latex for walls and ceilings. Semi-gloss is normally recommended for baths and children's rooms. Materials needed will include:

- 1 gallon or quart plastic/cardboard buckets for carrying around small amounts of paint for small work.
- Bucket of drywall mud (for fixing walls and ceilings)
- Can opener
- Cans of interior and exterior wood filler (for fixing millwork)
- Caulk gun
- Disposable wooden paint stir sticks (normally free where you purchase your paint).
- Dropcloths. Rosin paper, which comes in large rolls, can be used as a substitute. Tape it down flat and it will stay out of your way. It is good at absorbing small spills and drips.
- Extendable sanding poles with a swivel head. Fiberglass composition is recommended.
- Extension cords for power sprayers, sanders, vacuum cleaners, lamps and portable radios.
- Halogen lamps for painting at night
- Liquid soaps and hand cleaners, sponges and buckets for cleanup.
- Medium and fine-grained sandpaper
- Paint brushes (1", 2" and 3" for painting millwork)
- Paint brushes (4" for doing inside wall corners and around trim)
- Paint rags
- Paint roller (fine nap) or power roller (self feeding). Consider getting a professional type roller that has a 20" coverage to save time.
- Paint sprayer
- Paint stir sticks
- Paint thinner (Penetrol for oil-based paints, Flotrol for water-based paints).
- Paint tray
- Paint tray and paint tray liners. Liners are inexpensive and eliminate the need to clean the pan out each day.
- Rolls of masking tape. 1" and 2". If tape stays down too long, it can damage the wall surface. You may wish to tape windows to minimize clean-up later.
- Single-edged razor blades and handle
- Six-foot step ladder
- Trim guard
- Tubes of trim caulk
- Tubes of trim caulk. A large home may require several dozen tubes.
- Vacuum cleaner to suck up paint sanding dust

Exterior

PT4   PRIME and caulk all exterior surfaces: related trimwork, windows, doors, exterior corners and cornice. Exterior millwork should be primed immediately after installation. An all-weather exterior primer should be used.. Priming protects this raw millwork from moisture damage. In this step, all finish nails should be set below the surface with a nail punch. Holes should be filled with exterior water-based wood filler or oil-based wood dough.

PT5   PAINT exterior siding, trim, shutters and wrought iron railing. Iron railing should be painted with a rust-retarding paint. Painting two story fixtures can be simplified by coordinating with stucco, siding, or masonry subs. They will have scaffolding on site that can be used for painting.

PT6   PAINT all cornice work. For two-story homes, this usually requires very long, adjustable ladders. A professional is recommended.

PT7	PAINT gutters if needed. It is often handy to paint metal gutter sections on the ground prior to installation. If you do so, don't paint the joints. These must be kept clean so that they can be jointed properly. Gutters may still get scratched up during the installation process, but this gives you a head start on a difficult painting process. You should use either a latex or alkyd-based metal primer for metal gutters. NOTE: If the gutter seams require soldering, don't paint them until the soldering is finished.
Interior	
PT8	PREPARE painting surfaces. This is one of the most important steps in this book. Nothing is more important in the final appearance of your home than good paint preparation. Sand all trim and repair as needed with wood filler. Ceilings and walls should be sanded smooth with a pole sander. If you intend to use an oil-based enamel finish coat, your millwork should be a smooth as a baby's behind; imperfections are magnified greatly once a gloss or hi-gloss finish coat is applied. Apply trim caulk between base moldings and wall, between crown moldings and wall, between crown moldings and ceiling and between adjoining pieces of crown molding, base molding, wainscotting. Water-based caulk is normally recommended. This involves performing the following tasks: • DUST off all drywall with a dry rag. • APPLY trim caulk to joint between trim and wall. Smooth it with your finger. Wipe off excess. When dry, sand to a smooth joint. This is an important step which will help to yield professional quality results. • REPAIR any dents in drywall or moldings with spackle or wood filler. • Sand all repaired areas to a smooth finish. • NOTE: If ceiling is to be stippled, you don't need to do a thing to it. • NOTE: If you plan to paint stair spindles, do them before installing them. The painting process will be much easier.
PT9	PAINT prime coat on ceilings, walls and trim. Water-based primer takes about four hours to dry. Less in hot, dry weather. The prime coat provides a good bonding surface for the finish coat. Just about any good painting job involves two coats of paint. Before you even start, make sure you have excellent lighting available in every room to be painted. Even if you intend to wallpaper, paint a primer coat so that it will be easier to remove the wallpaper if you ever want to later. If you have Masonite® interior doors, prime them also. Ceiling white makes a good neutral primer. Since there is nothing in the house that you can harm, you need not worry about drop cloths. You can either rent a commercial paint sprayer, purchase your own or use a roller. For a professional job, lightly sand the walls before the final coat after the primer has dried. Sand trim also.
PT10	PAINT or stipple ceilings. A paint roller with an extension arm makes this job easier. Wear goggles to keep paint splatters out of your eyes. If you have stippled ceilings, you won't be painting them.
PT11	PAINT walls. Even if you want switch and outlet plates to match your paint. Paint them separately. Don't install them first. Start with a small brush and paint around the ceiling, all windows and doors. Then cover the large areas with a roller. Do one wall at a time.
PT12	PAINT OR STAIN trim. This also includes kitchen cabinets, window frames, doors and door jambs, handrails, treads and risers as needed. If you plan to stain these, a wiping stain is recommended. Test all materials on a sample piece of like material before applying to finished product. All areas to be stained should be prepared using stain grade putty. If you will have exposed, stained stair treads and painted stair rail pickets, it is easier to paint stair pickets prior to installing them to avoid getting paint on the rail and treads. Once dry, stained surfaces should be coated with BIN — an alcohol-shellac-based primer. Start with the highest trim and work down. Allow the walls to dry first. It will be easier if you use a 1½" sash brush for doing windows and a 2" brush for all other trim. Some painters don't bother to cover all the little trim in multi-paned windows; they just scrape all the dried paint off with a razor blade. Stained trim should be coated with a varnish or other sealer. Hardwood floors are stained and sealed after sanding. This should normally be done by a professional.
PT13	REMOVE paint from windows with one-sided utility razor blades. Do this only when the paint has completely dried or the soft paint shavings will stick to the surface and make a mess.
PT14	INSPECT paint job for spots requiring touch-up. Make sure you are looking at the paint job either in good natural light or under a bright light.
PT15	TOUCH UP paint job.
PT16	CLEAN UP. Latex paints will come off with soap and warm water. Enamel, which you may have used on your trim, will require mineral spirits or Varsol for removal. Dry out the brushes and store them wrapped up in aluminum foil with the bristles in a smooth position. Place plastic wrap on the paint cans prior to storing them.
PT17	PAY painter (if applicable) and have him sign an affidavit.
PT18	PAY retainage after final inspection.

PAINTING — SPECIFICATIONS

Interior Painting
- Bid is to include all material, labor and tools.
- Contractor to ensure that all cans of custom mixed paint match. (This is important!).
- Primer coat and one finish coat to be applied to all walls including closet interiors.
- Walls to be touch sanded after primer has dried.
- Flat latex to be used on drywall; gloss enamel to be used on all trim work.
- Ceiling white to be used on all ceilings.
- All trim joints to be caulked and sanded before painting.
- All paint to be applied evenly on all areas.
- Window panes to be cleaned by painter.
- Painting contractor to clean up after job.
- Excess paint to be labeled and to remain on site when job is completed.

Exterior Painting
- Five-year warranty exterior grade latex paint to be used on exterior walls and shutters.
- All exterior trim and shutters to be primed with an exterior grade primer.
- All trim area around windows, doors and corners to be caulked with exterior grade caulk before final coat. Caulk color to match paint color.

PAINTING — INSPECTION

- ☐ Proper paint colors used.
- ☐ Paint application appears even with no variation between cans of custom-mixed paint.
- ☐ All ceilings and walls appear uniform in color with no visible brush strokes.
- ☐ Trim painted with a smooth appearance. Gloss or semi-gloss enamel used as specified.
- ☐ All intersections (ceiling-wall, trim-wall, wall-floor) are sharp and clean. Clean, straight line between wall and trim colors.
- ☐ Window panes free of paint inside and out.
- ☐ Extra touch-up paint left at site.
- ☐ All exterior areas painted smoothly and evenly.
- ☐ No dried paint drops.

CABINETRY AND COUNTERTOPS

Cabinets

Cabinetry and countertops contribute significantly to the typical high cost of kitchens and baths. While they can be expensive, cabinets are some of the most frequently used accessories. Money spent here can really add value to your home. Cabinets are also perfect for other built-in uses in lieu of furniture: as media centers, laundry room units, bookshelves in hallways, and as bookcases. Make sure you hire an experienced and qualified cabinet installer. You will be looking at these units every day.

Cabinets come in three quality grades as defined by the Architectural Woodwork Institute (AWI): economy, custom, and premium. Ask your cabinet sub for custom grade or better. The AWI quality standards will specify the finishing details necessary for each grade. Premium quality cabinets are extremely expensive and are usually not worth the extra price. Economy grade is not suitable for most residential installations, so be wary of bids based on this classification. Manufactured cabinets have their own classification known as modular grade. It is generally comparable to custom grade cabinetry.

Factory assembled cabinetry is popular and cost effective. They are typically sold in standard widths and can be mixed and matched to fit the dimensions of your kitchen or bath. Pre-fabricated cabinets can either be purchased from kitchen supply companies or custom cabinetry sub-contractors. In either case, a salesman or kitchen designer will help to choose the proper style and component set. Most kitchen designers have software that will aid in designing and costing cabinetry.

The other option is to go with custom built cabinetry. In this case, the cabinet sub will measure the available space and fabricate a custom design that matches the blueprints. Custom cabinetry is significantly more expensive but offers more flexibility in design. This is a good choice if you are using recycled lumber or special wood. Pre-fabricated cabinets limit design options, but the sky's the limit with custom cabinetry. Consider building full length cabinets that reach the ceiling. This eliminates the need to box in a soffit above the cabinets and offers more storage space for a small additional cost. A small stool will be needed to access this upper area. If you go with custom cabinetry, ask your cabinet sub if he can add built in stools. These are commonly constructed as pull-out drawers that occupy the toe kick area at the bottom of the cabinet. The drawers can be pulled out and unfolded to reveal a stool perfect for reaching the high shelves.

When shopping for cabinets, pay attention to space-saving devices such as revolving corner units, drop down shelves and slide-out shelves. These options can significantly increase usable space. Slide out pantry drawers are popular, but you may find it more cost effective to build shallow shelves on an unused wall or to place the pantry in a nearby laundry room.

Cabinet Types

American Style cabinets commonly have a face frame attached to the front of the bulkhead (box carcass of the cabinet). Cabinet doors in any number of styles are then attached directly to the face frame, either overlapping or flush. They produce attractive, practical designs with a minimum of hardware. European Style cabinets are gaining in popularity, mainly because the cabinet bulkhead has no face frame. The door and drawers cover the entire face of the cabinet. This allows for more flexible and inexpensive construction and will make it far easier to upgrade cabinets in the future. If you decide later to change the cabinet style, you need only replace the doors and drawer fronts, which will be far less expensive than replacing the entire cabinet set. European cabinets depend on special hinge hardware that is invisible from the front and makes for a very clean and sophisticated design. Their flexible design makes European style cabinets perfect for custom installations.

Cabinetry selections have expanded to include many special use prefab units.

Pull out drawers can maximize storage space for oddly shaped objects.

Cabinet Styling

Visit a few cabinet stores or kitchen design stores to become acquainted with the many functional types of cabinets that are available. Look at sample kitchen layouts to get ideas. Sample kitchens allow you to visualize how the flooring, cabinetry, and countertops coordinate. Choose your material (wood, Formica, etc.), hinge and knob styles and finish. Determine how much cabinet space you need. The amount of cabinet space is not as important as the function of the space. You can offload some storage to the pantry or laundry room to save money on expensive cabinetry. The cabinetry should be an integral part of your kitchen design, complementing appliances, lighting and space. Many appliances are now designed to accept cabinet fronts to better blend in with other kitchen elements.

Bathroom Vanities

Purchase all your cabinets from the same place if you can. Not only does this make your cabinet selection more uniform, but you can probably get a better price from your cabinet sub. Always look for bathroom vanities with drawers for storing toiletries. If you choose instead to go with freestanding sinks, you will need to install a larger medicine cabinet area to store toiletries and bathroom appliances.

Counter Tops

Counter tops can be made from many different materials. The selection of material largely depends on your budget and the look you want. The cost of countertops can vary widely, so shop carefully. Counter tops typically should have a standard depth of 24" and may have a 3" or 4" splash block along the wall. If you desire, add a few inches of additional depth to the countertop. This will leave more room to clean behind the sink. Countertops are priced and installed by the foot, while special charges may be incurred to cut openings for stoves, sinks and fancy edgework.

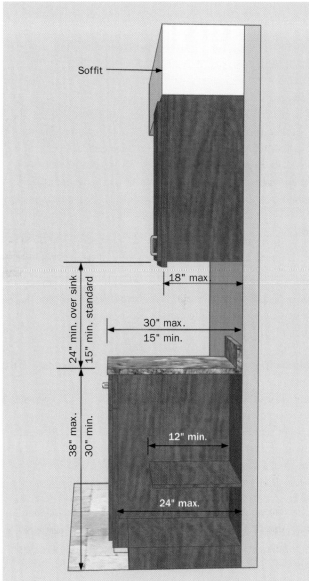

Kitchen cabinet dimensions.

Formica

Formica is by far the most cost effective and popular counter surface. It's made from a high density plastic that is durable and can be formed into countertops with integral curved fronts and backsplashes. The latest designs have improved in style and sophistication and can often be indistinguishable from some of the far more expensive surfaces. Formica is not indestructible however. The surface can be chipped and the material can delaminate from its base if hot pots and pans are placed directly on it.

Granite

Granite is the premium choice for countertops and the most expensive. It is dense, scratch and stain resistant, heat resistant and simple beautiful. Avoid using marble, which is a softer and more porous alternative. Food acids can etch the surface of marble.

Corian

Corian is an artificial surface that simulates granite in many respects. It also has the unique ability to be resurfaced if stains or scratches occur. The texture and color runs through the entire substrate. If a large scratch occurs, simply sand down the entire surface. Corner pieces can be joined in a virtually invisible bond by skilled technicians.

Tile

Tile looks great and is very durable, but makes a dubious countertop surface. The grout between tiles is very porous and will easily pick up stains if not thoroughly sealed. Most people tire of cleaning the grout lines. It does work well as a backsplash and can add texture and color to a kitchen design.

Concrete

Concrete is a unique and versatile alternative for countertops. It can be colored to look like granite, slate, or other stone surfaces and is typically cast in place, resulting in a strong, seamless surface. It's extremely durable and stain resistant and can be surprisingly inexpensive. If you have granite tastes and a Formica budget, consider concrete. Decorative or sentimental items can be embedded in the surface for a custom look. The surface will need extensive polishing however, so plan for a few extra days for this step. Make sure your cabinet or countertop sub is thoroughly familiar with its properties if you choose this option.

Pull-out pantry.

Cabinet Sub

The cabinet sub will normally produce or supply all cabinetry and counter tops. This includes making all the cuts for sinks, stoves, etc. Make sure you have these units on site when the cabinets are being installed so the cabinet sub can cut openings to the correct size. The plumber and electrician will come in later to finish the connections. Cabinet subs will normally bid the job at one price for material and labor.

The cabinets should be installed before flooring and wallcovering is applied. Make sure that all electrical, plumbing, and ventilation units are in place before cabinet installation. Walls should be primed before installation. This will provide a solid base for wallcovering and seal the walls behind the cabinets. Ask the cabinet sub to cover countertops with a protective covering or plywood after installation to protect them from potential scratches and dings made by other sub-contractors.

Your cabinetry sub-contractor will also bid and install bathroom vanities and other cabinetry.

Built-in cabinetry can be used in more rooms than just kitchen and bath, eliminating the need for separate furniture.

CABINETRY — STEPS

CB1	CONFIRM kitchen design. Make sure you have enough cabinets to store all pots, pans, utensils and food items.
CB2	SELECT cabinetry/counter top styles/colors.
CB3	COMPLETE diagram of cabinet layout for kitchen and baths. This is a critical step to perform prior to electrical and plumbing installation. Show locations of all plumbing connections on diagram: Ice maker, dishwasher, and sink.
CB4	PERFORM standard bidding process. There are lots of independent cabinet makers who can copy expensive European cabinet designs.
CB5	PURCHASE or have constructed all cabinetry and counter top surfaces. If your counter tops come from a separate sub, coordinate the cabinetry and counter top sub. You can't install tops until vanities are in. Unless you know what you are doing, wait for the drywall to be hung before having the cabinet sub measure. This gives him more exact figures to work from and a place to mark on the wall. But do order your cabinets as soon as possible because they can be a source of big delays later.
	Give your cabinet sub specs for all appliance dimensions and locations of all doors and windows. This will insure that the cabinets will install properly under the window sill. At a minimum, wait for framing to be finished. Walk through the structure with the cabinet sub, covering kitchen, baths and utility room. Countertops are typically 36" above the floor, kitchen desks 32" high with a minimum 24" knee space. When measuring, remember to account for ½" drywall on walls and widths of adjacent trim. Mark dimensions on the floor with builder's chalk or crayon.
CB6	STAIN and seal, paint or otherwise finish all cabinet woodwork. Stains will stain the grout and other nearby items, so cabinets must be stained before they are installed.
CB7	INSTALL bathroom vanities. These must be in place before your plumber can connect the sinks, drain pipes and faucets.
CB8	INSTALL kitchen wall cabinets. If you install base cabinets first, they get in your way while installing wall units.
CB9	INSTALL kitchen base cabinets. Also make necessary cutouts for sinks and cooktops. Show the cabinet sub where cutouts are to be made. Don't make any cutouts until the units have arrived.
CB10	INSTALL kitchen and bath counter tops, aprons and backsplashes.
CB11	INSTALL and adjust cabinetry hardware: pulls, hinges, etc.
CB12	INSTALL utility room cabinetry.
CB13	CAULK all wall/cabinetry joints as needed.
CB14	INSPECT all cabinetry and counter tops.
CB15	TOUCH up and repair any scratches and other marks on cabinetry and counter tops.
CB16	PAY cabinet sub. Have him sign an affidavit.

CABINETRY — SPECIFICATIONS

- Bid is to provide and complete cabinetry and counter top installation as indicated on attached drawings.
- All cabinets are to be installed by qualified craftsman.
- All cabinets are to be installed plumb and square and as indicated on attached drawings.
- All hardware such as hinges, pulls, bracing and supports to be included in bid.
- All necessary counter top cutouts and edging to be included in bid.
- All exposed corners on counter tops to be rounded.
- Install one double stainless steel sink (model/size).
- Special interior cabinet hardware includes:
  - Lazy Susan
  - Broom rack
  - Pot racks
  - Sliding lid racks

Vanity Tops — Specifications

- Bid is to manufacture and install the following cultured marble tops in ¾" solid color material. Sink bowls to be an integral part of top.
- Vanity A to be 24" deep, 55" wide, finished on left with backsplash; 19" oval sink cut with rounded edge with center of sink 22" from right side. Color is solid white. Spread faucet to be used.
- Vanity B to be 24" deep, 45" wide, finished on right with backsplash; 19" oval sink cut with rounded edge with center of sink 18" from left side. Color is solid almond. Standard faucet to be used.

CABINETRY — INSPECTION

- ☐ All cabinet doors open and close properly with no binding or squeaking. Doors open completely and remain in place half opened.
- ☐ All pulls and other hardware are securely fastened. Check for excessive play in hinges.
- ☐ All specified shelving is in place and level.
- ☐ All exposed cabinetry has an even and smooth finish.
- ☐ No nicks, scratches, scars or other damage/irregularity on any cabinetry and counter tops.
- ☐ Counter tops are level by placing water or marbles on surface. All joints are securely glued with no buckling or delamination.
- ☐ All drawers line up properly
- ☐ All mitered and flush joints tight.
- ☐ Cutouts for sinks and cooktops done properly and units fitted into place.

FLOORING, TILE AND GLAZING

The most widely used floorings include:
- Carpet
- Hardwood (oak strip, plank and parquet)
- Vinyl floor covering
- Ceramic tile

Carpet

Carpet is measured, purchased and installed by the square yard and comes in rolls 12' wide. Do not try to install carpet yourself unless you have experience; it is not as easy as it appears. Carpet should be installed with a foam pad, which adds to the life of the carpet and gives carpet that plush feel. A stretcher strip will be nailed around the perimeter of the room and the carpet will be stretched flush with the wall. Stretching the carpet prevents it from developing wrinkles. After two to three months, the carpet may need to be stretched again. Make sure your carpet company will agree to do this at no cost.

When pricing carpet, always get the installed price with the cost of the foam pad included. A thicker foam pad will make your carpet feel more luxurious but may cost extra. Do not install carpet until all worker traffic has ceased. Make sure that all bumps and trash have been cleaned from the floor before installation. Even the smallest bump will be noticeable through the carpet after installation.

Hardwood

Hardwood flooring is measured by the board foot (oak or other plank flooring) or by the square yard (parquet). Oak is laid in tongue and groove strips nailed to the subfloor, stained and sealed. Parquet is normally installed as 6" squares with an adhesive. Parquet squares are available in stained and unstained styles. Avoid laying hardwood floors in humid weather because when the air dries out, gaps will appear between the boards. Buy the flooring a week or two before it is needed and store it in the rooms where it will be laid to let the moisture content stabilize. "Select and better" is the normal grade used in homes. It has a few knots in it. "Character grade" has more knots but is less expensive.

After installation you may want to sand the floor for an extra smooth finish. Pine flooring normally isn't sanded after installation. On all other wood floors, sanding should only be done with the grain, or sanding marks will show up. Use a professional-sized drum sander. An "edger" should be used to sand the edges. If you lay hardwood floors on a concrete slab, you need to seal the concrete and use a concrete adhesive. A better method will be to lay down runner strips and nail the floor to the strips. Wood strip floor installers usually work as a pair: one nailing strips in place while the other saws pieces to the proper length and sets them in place. If you can avoid it, do not lay wood floors on slabs.

Classic hardwood flooring adds warmth and versatility to any room.

Vinyl

Vinyl floor covering is measured by the square yard and comes in a variety of styles and types. Padded vinyl rolls have become very popular because of their ease of installation and maintenance. Installation procedures vary depending on the material and manufacturer, so take care to follow the manufacturer's recommendations when installing. Warranties can be voided if the material is not installed properly.

High quality vinyl can easily pass for more expensive tile.

Tile is a very durable floor covering but can be hard on the feet when working in the kitchen. Mortar joints should be sealed to protect against the inevitable food spills.

Custom mirrors and cabinetry lend an old world charm.

When installing vinyl floors, consider laying an additional sub-floor of plywood over the area to be covered. This additional layer provides a smoother surface for the floor and brings the level of the floor up to the same level as any surrounding carpet.

Tile

Tile is measured and sold by the square foot although tile units can come in any size. Tile comes either glazed or unglazed and can be installed over a wood sub-floor, although a special adhesive cement and latex grout must be used. Consider using a colored grout which can hide dirt. If you coat the tile grout with a protective coating, you may void the tile warranty.

Tile installation is a tedious and critical procedure. Make sure your tile sub is experienced and quality conscious. Improper installation can result in buckling, cracking and delamination. To insure that tiles adhere properly, avoid installation in cold weather or make sure that the floor is heated for 24 hours after installation.

Colored grouts add a special touch without adding to costs. Don't bother with a colored grout unless the joint is at least ⅛" wide. You may wish to use a grout additive instead of water. This will strengthen the grout and bring out the color. To save time, also consider using a special protective coating applied to tile prior to grouting. This will allow you to wipe the grout film off the tile easily when the job is done.

If you plan to install tile in kitchens or other large areas, check the height of the installed floor (tile and mortar bed) to ensure that you aren't higher than the adjoining carpet or hardwood floors.

Remember that you have a ½" fudge factor around floor perimeters if you use shoe molding. If you tile yourself, ask your supplier to loan or rent you a tile cutter with a new blade or buy a new blade — this will make a world of difference.

Tiling shower and bath stalls is trickier than tiling floors. You may want to do the floors, but hire a sub for the rest. Use spacers when tiling the floor to keep the tiles straight.

Glazing

Glazing, while not part of flooring, normally occurs during the flooring phase. Glazing refers to any mirrors and plate glass windows that must be custom cut from sheets of plate glass. No special specifications are needed. Glazing will be done by a glass contractor.

GLAZING — STEPS

GL1	DETERMINE mirror and special glazing requirements. This means all of the following items: • Mirrors • Closet door mirrors • Shower doors • Fixed picture windows • Medicine cabinets
GL2	SELECT mirrors and other items. Here you must decide such trivia as beveled edge or non-beveled edge mirror glass, chrome or gold frame for shower doors, etc. Glazing subs will have catalogs for you to look through.
GL3	CONDUCT standard bidding process for glazing sub.
GL4	INSTALL fixed pane picture windows. Check that window has no scratches and is adequately sealed.
GL5	INSTALL shower doors and medicine cabinets. If you can, turn on shower head and point toward the shower door to check for leaks. Door should glide freely along track.
GL6	INSTALL mirrors and closet door mirrors. Mirrors should be installed with a non-abrasive mastic. Check for hairline scratches in the silver coating; this cannot be fixed aside from replacing the mirror.
GL7	INSPECT all glazing installation.
GL8	CORRECT all glazing problems.
GL9	PAY glazier and have him sign an affidavit.

FLOORING AND TILE — STEPS

FL1	SELECT carpet brands, styles, colors and coverage. Also select your padding. Keep styles and colors to a minimum. This helps your buying power, minimizes scrap and provides for easier resale.
FL2	SELECT hardwood floor type, brand, style, color and coverage.
FL3	SELECT vinyl floor covering brand, style, color and coverage.
FL4	SELECT tile brands, styles, colors and coverage. This includes tile for bath walls, bath floors, shower stalls, patios and kitchen counter tops. Once your sub-floor is in, borrow about a square yard of your favorite tile from your supplier. Lay it down where it will be used and see how it looks.
FL5	CONDUCT standard bidding process For best prices on carpet, go to a carpet specialist or carpet mill. When possible, have the supplier bid for material and labor. If your supplier does not install, ask for installer references.
Tile	
FL6	ORDER tile and grout. If you do the job yourself, borrow or rent a tile cutter from the tile supplier. Look for good bargains on "seconds" — tile with slight imperfections — to use as cut pieces. Don't use sanded grout with marble tiles. The sand will scratch the marble surface. Make sure that all tile comes from the same batch. Tile colors can vary from batch to batch.
FL7	PREPARE area to be tiled. Sweep floor clean and nail down all squeaky floor areas with ridged nails.
FL8	INSTALL tile base in shower stalls. This normally involves applying concrete over a wire mesh. In shower stalls, this is done on top of the shower pan. If you have tile in the kitchen do not tile under the cabinets or the island. Prior to tiling around a whirlpool tub, make sure the tub is secure and has been fully wired, grounded and plumbed.
FL9	APPLY tile adhesive. Use a trowel with grooves suitable for your tile and grout. Use large curved, sweeping motions.
FL10	INSTALL tile and marble thresholds. Don't forget to use tile spacers to assure even tile spacing. Make sure tubs are secure in place before tiling. Make sure to save a few extra tiles in case of subsequent damage to floor tiles. This will ensure a perfect color match if repairs are needed.
FL11	APPLY grout over tile. Apply silicone sealant between the tub and tile.
FL12	INSPECT tile. Refer to specifications and inspection criteria.
FL13	CORRECT any problems that need fixing.
FL14	SEAL grout. This should only be done after grout has been in place for about three weeks. Use a penetrating sealer to protect the grout from stains.
FL15	PAY tile sub. Get him to sign an affidavit.
FL16	PAY tile sub retainage after floor has proven to be stable and well glued.

HARDWOOD — STEPS

FL17   ORDER hardwood flooring. Hardwood flooring should remain in the room in which it will be installed for several weeks in order for it to expand and contract based on local humidity conditions. You may want to order a few extra bundles of shorter, cheaper lengths to do closets and other small areas.

FL18   PREPARE sub-floor for hardwood flooring. If you have a wood sub-floor, you should put down at least two layers of heavy-duty building paper. Make sure that all dirt and drywall compound are thoroughly cleaned off the floor. The surface must be totally clean and level. If you are laying hardwood on a slab, you have two choices:

- Seal the slab with a liquid sealer and install flooring with an adhesive
- Prepare a sub-floor:
- Sweep slab
- Apply 1×2" treated wood strips called bottom sleepers with adhesive and 1½" concrete nails, 24" apart, perpendicular to the oak strips
- Lay 6 mil poly vapor barrier over strips
- Lay second layer of 1×2" wood strips over vapor barrier.
- Nail wood flooring strips to the sleepers.

FL19   INSTALL hardwood flooring. You may want to wait until all drywall and painting work is complete before installing hardwood to protect the finish. Ask installers to leave some scraps. Store them away for repair work or other use.

FL20   SAND hardwood flooring. Although most floors are pre-sanded, you may still need to have them sanded for the smoothest possible finish. Sand only with the grain. Cross-grain sanding will not look bad until the stain is applied. Once floors are sanded, there should be no traffic until after the sealer is dry.

FL21   INSPECT hardwood flooring. Refer to specifications and inspection guidelines.

FL22   CORRECT any problems with hardwood flooring. Common problems are split strips, hammer dents in strip edges, uneven spacing, gaps, squeaks, non-staggered joints and last strip not parallel with wall.

FL23   STAIN hardwood flooring. The floors must be swept very well before this step is performed. Close windows if it is dusty outside.

FL24   SEAL hardwood flooring. Polyurethane is the most popular sealant. You normally have an option of either gloss or satin finish. Place construction paper over dry floors to protect them from construction traffic.

FL25   INSPECT hardwood flooring again, this time primarily for finish work.

FL26   PAY hardwood flooring sub. Have him sign an affidavit.

FL27   PAY hardwood flooring sub retainage.

VINYL FLOOR COVERING — STEPS

FL28   PREPARE sub-floor for vinyl floor covering. Sweep floor clean and nail down any last squeaks with ringed nails. Plane down any high spots.

FL29   INSTALL vinyl floor covering and thresholds. Ask installers to leave behind any large scraps that can be used for repair work.

FL30   INSPECT vinyl floor covering. Refer to specifications and inspection guidelines.

FL31   CORRECT any problems with vinyl floor.

FL32   PAY vinyl floor covering sub. Have him sign an affidavit.

FL33   PAY vinyl floor covering sub retainage.

CARPET — STEPS

FL34   PREPARE sub-floor for carpeting. Sweep floor clean and nail or plane down any high spots. This is your last chance to fix squeaky floors easily and install hidden speaker wires.

FL35   INSTALL carpet stretcher strips.

FL36   INSTALL carpet padding. You should use an upgraded pad if you use a tongue and groove sub-floor. If you don't use a carpet stretcher on this step, you are doing the job wrong.

FL37   INSTALL carpet. This should be done with a carpet stretcher to get a professional job. Ask installer to leave larger scraps for future repair work.

FL38   INSPECT carpet. Refer to specifications and inspection guidelines.

FL39   CORRECT any problems with carpeting installation.

FL40   PAY carpet sub. Have him sign an affidavit.

FL41   PAY carpet retainage.

FLOORING AND TILE — SPECIFICATIONS

Carpet

- Bid is to furnish and install wall-to-wall carpet in the following rooms:
  - Living room
  - Dining room
  - Master bedroom and closets
  - Second bedroom and closets
  - Third bedroom and closets
  - Second floor stairway
  - Second floor hallway
- 8" accent border to be installed in master bathroom. Color to be of builder's choice.
- Full-size upgraded pad is to be provided and installed under all carpet.
- Stretcher strips are to be installed around carpet perimeters.
- All seams to be invisible. (This is critical to a good carpet job.) Seams should be located in inconspicuous and low traffic areas.
- All major carpet and padding scraps to remain on site.
- All necessary thresholds to be installed.
- All work to be performed in a professional manner.

Oak Flooring

- Bid is to provide and install "select and better" grade flooring in the following areas:
  - Foyer and first floor hallway
  - Den
- Oak strips to be single width, 3¼" strips.
- Two perpendicular layers of heavy gauge paper to be installed between sub-floor and oak flooring.
- All oak strips to be nailed in place with flooring nails at a 50-degree angle. Nails to be installed only with a nailing machine to prevent hammer head damage to floor.
- Shorter lengths to be used in closets.
- All oak strips to be staggered randomly so that no two adjacent joint ends are within 6" of each other.
- Oak flooring and oak staircase treads to be sanded to a smooth surface.
- Oak flooring and oak staircase treads to be swept, stained, sealed and coated with satin finish polyurethane.
- Stain to be a medium brown.
- Ridged or spiral flooring nails to be used exclusively.
- Only first and last strip to be surface-nailed, concealable by shoe mold.

Parquet Flooring

- Bid is to provide and install parquet flooring in the following areas:
  - Sunroom
  - Sunroom closet
- Parquet flooring to be stained and waxed.
- Install marble thresholds in the following areas:
  - Between hallway and living room
  - Between hallway and dining room
  - Between hallway and den
  - Between kitchen and dining room

Vinyl Floor Covering

- Bid is to provide and install vinyl floor covering in the following areas:
  - Kitchen and kitchen closet
  - Utility room
- All seams to be invisible.
- All major scraps to remain on site.
- All necessary thresholds to be installed.

Tile

- Bid is to provide and install specified tile flooring in the following areas:
  - Master bath shower stall walls. One course of curved base tile around perimeter. Tile to go all the way to ceiling.
  - Master bath shower floor.
  - Surround and one step for whirlpool bath.
  - Second bath shower stall walls all the way to ceiling.
  - Second bath floor. Per attached diagram.
  - Half bath floor. Per attached diagram. One course of base tile around perimeter.
  - Kitchen floor (per diagram). Tan grout sealed. One course of base tile around perimeter.
  - Kitchen wall backsplash. See drawings attached.
  - Kitchen island top.
  - Hobby room and closet.
- Grout colors to be as specified by builder.
- All necessary soap dishes, tissue holders, towel racks and toothbrush holders to be supplied and installed in white unless otherwise specified by builder in writing.
- Matching formed base and cap tiles to be used as needed to finish edging and corners.
- ¼" per foot slope toward drain in shower stalls.
- All necessary tub and shower drains installed.
- All grout to be sealed three weeks after grout is installed.
- Extra tile and grout to remain on site.

FLOORING AND TILE — INSPECTION

Carpet
- [ ] Proper carpet is installed.
- [ ] No visible seams.
- [ ] Carpet is stretched tightly and secured by carpet strips.
- [ ] Colors match as intended. Watch for color variations. Carpet installed on stairs fits tight on each riser.

Oak Flooring and Parquet
- [ ] Proper oak strips and/or parquet installed. (Make, color, etc.)
- [ ] No flooring shifts, creaks or squeaks under pressure.
- [ ] Even staining and smooth surface.
- [ ] No scratches, cracks or uneven surfaces. Scratches normally indicate sanding across the grain or using too coarse sandpaper for final pass.
- [ ] Smaller lengths used in closets to reduce scrap and waste.
- [ ] All necessary thresholds are installed and are parallel to floor joints.
- [ ] Only edge boards have visible nails. No other boards have visible nails.
- [ ] Edge boards are parallel to walls.
- [ ] No irregular or excessive spacing between boards.
- [ ] No floor registers covered up by flooring.

Vinyl Floor Covering
- [ ] Proper vinyl floor covering is installed. (Make,
- [ ] color, pattern, etc.)
- [ ] Vinyl floor covering has smooth adhesion to floor. No bubbles.
- [ ] No movement of flooring when placed under pressure.
- [ ] No visible or excessive seams.
- [ ] No scratches or irregularities.
- [ ] All necessary thresholds have been installed and are parallel to floor joints.
- [ ] All necessary thresholds are of proper height for effective door operation. No more than 1/8" gap under interior doors.
- [ ] Pattern is parallel or perpendicular to walls as specified.

Tile
- [ ] Proper tiles and grouts are installed. (Make, style, dimensions, color, pattern, etc.)
- [ ] Tile joints are smooth, evenly spaced, parallel and perpendicular to each other and walls where applicable.
- [ ] Tiles are secure and do not move when placed under pressure.
- [ ] No scratches, cracks, chips or other irregularities.
- [ ] Grout is sealed as specified.
- [ ] Grout lines parallel and perpendicular with walls.
- [ ] Tile pattern matches artistically.
- [ ] Tiles are laid symmetrically from the center of the room out. Cut pieces on opposite sides of the room are of the same size.
- [ ] All tile cuts are smooth and even around perimeter and floor register openings.
- [ ] Marble thresholds are installed level with no cracks or chips.
- [ ] Marble thresholds parallel with door when closed.
- [ ] Metal thresholds installed between tile and wood joints.
- [ ] Marble thresholds are at proper height.
- [ ] Matching formed tiles used at base of tile walls as specified.
- [ ] Cap tiles used at top of tile walls.
- [ ] Proper tile edging for tubs, sinks and shower stalls.
- [ ] No cracked or damaged tiles.
- [ ] Tile floor is level.
- [ ] Extra tile and grout available on site.

MISCELLANEOUS FINISH

Weatherstripping
- [ ] All exterior doors have weatherstripping and threshold installed.
- [ ] Doors open and close securely with a tight seal.
- [ ] No leaks after a heavy rain.

Mirrors and Shower Doors
- [ ] Door fixtures match faucets.
- [ ] Doors slide properly with no friction or binding.
- [ ] No scratches on mirror.

LANDSCAPING

When you consider how much resale value and curb appeal a good landscaping job can add to a home, it is surprising that so many builders scrimp on this important effort. Many builders just throw up a few bushes and then seed or sod. Hopefully you will be willing to spend just a bit more time and effort to do the job right. If you don't want to spend all the money at once, you can devise a two- or three-year implementation plan, beginning with the critical items such as ground coverage.

Important items to keep in mind are:

- Plan for curb appeal. Your lawn should be a showplace.
- Plan for low maintenance. If you don't want to trim hedges and rake leaves forever, be smart about the trees and bushes you install or keep.
- Work with the sun, not against it. Maximize sunlight into the home by careful planning of trees.
- Plan for proper drainage. Water must drain away from dwelling and not collect in any low spots.
- Consider doing much of this work yourself.
- Consider assistance from a landscape architect.

Landscape Architect

Every lot has one best use and many poor ones. A good landscape architect can help you do more than select grass. Landscape architects can be helpful in:

- Determining best position of the home on the lot.
- Suggesting attractive driveway patterns.
- Choosing which trees should remain and additional trees and/or bushes to plant for privacy and lowest maintenance.
- Analyzing soil condition and recommending soil additives.
- Determining proper drainage and ground coverings.

You can normally pay an independent landscape architect by the hour, but if you do a lot of work with him, work up a fixed price. Meet him at your lot with the lot survey. Walk the lot with him, getting his initial impressions and ideas. From there, you may request a site plan. The site plan will normally be a large drawing showing exact home position, driveway layout, existing trees and bushes, sculptured islands and other topographical features.

Site Plan and Design

Before you even lay out the foundation of your house, you should examine the site and plan the placement with landscaping and sun orientation in mind. A well-oriented, well-designed house will maximize energy savings. When purchasing the lot, consider its orientation and the exact site where you may want to situate your home. Look at the lay of the land and the way the lot drains. Examine your home site's exposure to sun, wind, shade and water. Also note the location and proximity of any nearby buildings,

fences, water bodies, trees, and pavement that might have climatic effects. Nearby buildings provide shade and a windbreak. Nearby fences and walls can block or channel wind. Bodies of water can moderate temperature but can increase humidity and produce glare. Trees will provide shade, windbreaks, or wind channels. Dark pavement absorbs heat; light pavement reflects heat. Take a soil sample and have the local county extension office do a complete soil analysis. This will tell you if soil supplements are needed.

Before you start landscaping, you need to draw a landscape site plan. The landscaping plan should include all the trees, shrubs, flowers, berms, walls, fences, trellises, sheds, and garages, drainage paths. This will help you refine the landscape plan before you begin construction. To develop a landscape plan:

- Use plan paper with 1/4" or 1/8" grids.
- Sketch a simple, scaled drawing of your yard's basic features, showing buildings, walks, driveways, and utilities.
- Show the location of streets, driveways, patios, and sidewalks.
- Identify potential uses for different zones of the yard: vegetable gardens, flower beds, patios, and play areas. Use arrows to represent sun angles and prevailing winds for both summer and winter.
- Circle the areas of your yard needing shade or wind protection.
- Mark areas that may have poor drainage or standing water. You will need to plant moisture tolerant plants here, or even install a drain pipe.
- Pencil in the name and location of each plant. Draw the plant to scale.

Landscaping Isn't Just for Looks

A well designed landscaping job serves many valuable purposes in addition to improving the appearance of the yard. These other advantages can actually add up to more benefits than the cost of the landscaping.

Landscaping Can Save Energy

By placing trees in strategic locations, you can reduce heating and cooling costs by up to 25%. How? In the summer, deciduous trees can shade the house from the sun, reducing cooling costs. In the winter, that same tree will drop its leaves, allowing the low winter sun to enter window and warm the house. The U.S. Department of Energy estimates that three properly placed trees can save $100 to $250 in annual energy costs enough savings to return your initial investment in less than 8 years. Shade from trees can also reduce the temperature in the yard by 5-10%.

The sun's heat, passing through windows and being absorbed through the roof, is the main cause of heat buildup in the summer. Shading rooftops will reduce solar heat gain and air-conditioning costs. However, using shade effectively requires knowledge of the position of the sun and the size, shape, and location of the object casting the shadow.

The Sun's Changing Angle is the Key

As the seasons change, the sun's angle in the sky changes, affecting the placement of plants, trees, and windows. At the Winter solstice, the sun rises 30% South of East and stays low in the sky even at noon. The sun then begins to rise in the sky, tracing a more northerly path, as the season progresses and approaches Summer. At the summer solstice, the noon sun is almost straight overhead and the sun sets 30% North of West. The exact angle of the sun will vary, depending on whether you live in Northern or Southern latitudes. The sun's angle is higher in Southern latitudes, but the general effect is the same. These changing angles will affect the sun's ability to cast shadows. When the sun is low in the sky, objects in its path cast long shadows horizontally. Horizontal overhangs or tall trees cast short vertical shadows because the sun shines underneath them. However, short shrubbery can block the low sun from windows on the East and West sides of the house. In the summer, on the other hand, overhangs and tall trees can block much of the high, direct sunlight from entering South facing windows, keeping the house much cooler. Knowing these angles will help you to place plants properly (see The Lot chapter for a sun analysis done with SketchUp).

Trees should be selected for appropriate size, density, and shape for the specific shading application. To block solar heat in the summer and allow it in the winter, use hardwoods (deciduous trees). To provide continuous shade or to block heavy winds, use evergreen trees or shrubs that don't drop leaves in the Winter. Tall, deciduous trees should be planted South of the house to provide maximum Summertime roof shading. Trees with lower branches are more appropriate to the West, where shade is needed from lower afternoon sun angles. Plant trees or shrubs so that they shade the air conditioning unit. This can increase its efficiency by as much as 10%. Trees, shrubs, and groundcover can also shade the ground and pavement around the home, reducing heat radiation and cooling the air before it reaches your home's walls and windows.

Plants Also Provide Wind Protection

Properly placed landscaping can provide excellent wind protection, reducing winter energy costs. You should place the windbreaks on the North and West side of the house to provide shelter from cold Winter winds. Plant trees and shrubs together, to block wind from the ground to the treetops.

Evergreen shrubs and small trees can be planted as a solid wall at least four to five feet away from the North side, providing a windbreak. However, it is better to have dense growth further away so that air movement can occur during the summer.

If snow tends to drift and collect in your area, plant low shrubs on the windward side of your windbreak in order to trap snow before it blows next to your home. In addition to more distant windbreaks, consider planting shrubs, bushes, and vines adjacent to the dwelling in order to create dead air spaces that insulate the dwelling in both winter and summer. Allow at least one foot of space between full-grown plants and the dwelling's wall.

Landscaping Can Improve the Environment

Well placed trees, shrubs, and flowers can have many other health benefits:

- Trees and ground cover can control erosion in areas where grass will not grow, such as on sloped or rocky areas.
- Large scale tree and shrubbery planting will clean the air by absorbing carbon dioxide and releasing oxygen.
- Flowers can provide a home for beneficial insects and birds that can control harmful pest infestations, such as Japanese beetles, mosquitoes, and leaf eating caterpillars.

Choosing Plants for Your Area

By choosing plants properly, you can reduce the amount of care and feeding needed to keep your yard in prime condition. Some species of trees, bushes, and grasses require less water than others. Other species are naturally more resistant to certain pests. Using these species can reduce energy, water, and time associated with lawn care, watering, and trimming.

Xeriscaping

Xeriscaping is a landscaping technique that uses drought-resistant plants able to survive with very little water. These are usually plants from arid regions and various types of cactus. When planted properly, these sites are attractive, grow slowly, and require virtually no watering. Xeriscaping saves energy and reduces water consumption. Make sure these plants will grow in your area. Excessively damp climates or poorly drained soil will encourage diseases.

Select Plants to Match Your Climate

Your goals and plant selection will differ depending on the climate in your area. The plants will serve different purposes in different climates. Climates fall into four basic categories:

Cool Climate

- Use evergreen shrubbery as yard borders to block cold winter wind.
- Leave south facing areas clear of trees to allow winter sun to enter southern windows.

- Use deciduous trees that will drop leaves in the winter.
- Place evergreen shrubbery to shelter north facing windows and reduce radiant heat loss.

Hot and Dry Climate
- Use dry-loving trees and bushes
- Plant tall evergreens to create maximum cooling shade for roof, walls, and windows.
- Funnel summer winds toward house to cool naturally.

Hot and Humid Climate
- Channel cool summer breezes toward the dwelling.
- Maximize summer shade with deciduous trees that allow penetration of low-angle sun in winter.
- Avoid placing plants in low lying areas that drain poorly.
- Choose disease resistant plant varieties.

Temperate Climate
- Use deciduous trees to maximize the cooling effects of sun in the summer and warming effects of bare trees in the winter.
- Deflect winter winds away from dwelling.
- Funnel summer breezes toward dwelling.

Growth Rate and Root Systems

Slow-growing trees require many more years of growth before shading a roof, but it will generally live longer than a fast-growing tree. Also, slow-growing trees often have deeper roots and stronger branches, making them less prone to breakage by windstorms or snow loads, and more drought resistant than fast-growing trees. Avoid placing fast growing trees with large root systems near the house or walkways. For example, never place a Weeping Willow tree near the house. Its massive root system can literally uproot foundation footings. Magnolia trees have roots that like to grow right on the surface, making lawn mowing difficult.

Grading and Drainage

Before your planting begins, the landscaper or grader will need to complete the final grade. This will be your last chance to make sure the lot drains well. Lay out a grid of stakes and strings that have been leveled with a string level. Criss-cross the lot and examine the lay of the land. It is almost impossible to judge drainage sloping if the lot is relatively flat. The string grid will help you to follow the lay of the land and calculate the path of water runoff. This is your final opportunity to get the drainage just right.

If the grading contractor has to move a lot of dirt or install a retaining wall, he will probably use a loader; switching to a tractor for the finish grading. He should rake the yard to remove large and medium size rocks and to mix in any topsoil that you have saved for this purpose. The string grid will have to be removed during grading, so check the drainage again with the string once the grader

is finished, but before he leaves. Have him make any final adjustments. Make sure the drainage is perfect now. It is much harder to correct drainage problems later. If your yard has a low lying area that is impossible to build up, dig a trench and install drainage pipe to provide an underground drainage path.

Retaining Walls

Retaining walls are used to alter topography or to provide improved drainage. In some local jurisdictions a special permit is required to erect a retaining wall in excess of a given height, such as 36". Materials used for constructing retaining walls include pressure treated wood, masonry, and poured concrete. If your site requires a large concrete retaining wall, chances are you will construct it while the concrete subcontractor is pouring the foundation. This will save money and help prevent mudslides and flooding during construction. Refer to the foundation section for more information on poured concrete walls. If you are installing wood or masonry retaining walls, you might consider doing this yourself.

Wood Retaining Walls

Pressure treated rectangular wood timbers or railroad ties may be used to construct retaining walls. The timbers are stacked so that the butted ends of the members in one course are offset from the butted ends of the members in the courses above and below. The bottom course should be placed at the base of a level trench. In well drained, sandy soil, the footing is unnecessary. In less well drained soils, apply 12" to 24" of gravel backfill behind the wall and a 6"-deep gravel footing below the bottom course. Nail the courses together with galvanized spikes. Every second course of timbers should include members inserted into the face of the wall that extend horizontally into the soil behind the wall. This "tieback" timber will anchor the wall to the soil and prevent the weight of the soil from collapsing the retaining wall. The tieback timbers should extend into the soil for the same distance as their distance above the base of the wall. The end of this tieback member should have a "deadman" timber 24" long, attached to it in the shape of a "T." This works like the anchor of a boat. These tiebacks and deadmen should be installed every 4' to 6' feet along the retaining wall.

An alternative retaining wall design consists of pressure treated rectangular timbers or railroad ties set in holes spaced 4' apart. Rough sawn, pressure treated 2" lumber is then placed behind the vertical members. The 2" cross pieces are held in place by backfilling as they are inserted. In poorly drained soils, the backfill should consist of 12" to 24" of gravel. In this design, the vertical members should be set in post holes to a depth of 4' or to frost line depth, which ever is greater, in order to resist tipping from the pressure of the retained soil.

Block Retaining Walls

A reinforced concrete block retaining wall can be constructed from several different types and styles of block specially designed for this purpose. Check with your local landscaping supplier. Several new types of block have been developed that don't require mortar or footings, as long as the wall is not more than 36" tall. Some blocks are designed with a cavity for planting cover plants or vines.

For taller walls, use a mortared concrete block wall. Pour an extra wide footing to a depth below the frost line. After the footing concrete has hardened, the retaining wall blocks are laid. After the wall has set, install 12" to 24" of gravel behind the wall as backfill to provide drainage and to minimize the pressure from the soil freezing behind the wall.

Poured Concrete Wall

The retaining wall can be constructed solely of poured concrete instead of concrete block. This wall will be identical to poured walls used for the foundation. Because of the cost of setup, fabrication and pouring, try to install these retaining walls at the same time as poured foundation walls. The same subcontractor can do both. The form for the face of the wall should be vertical but the back of the wall should be built at an angle to provide a wall that is thicker at the base. Reinforcing rods should be placed in the form and wired together. The concrete should be poured in the form to the depth of the footing and allowed to partially set before the remaining concrete is poured. Backfilling the wall with 12" to 24" of gravel is recommended.

Lawns

It is hard to imagine a suburban landscape without its well manicured lawns. Regardless of your landscaping design, most of the square footage of your yard will be covered with grass, unless you plan an "all natural" landscape. The trick is to get this lawn established before rain and wind carries off the valuable topsoil.

Installing Sod vs. Seed

Sodding a lawn provides instant beauty and maturity. Turf is attractive and, in most cases, relatively easy to maintain. It's the perfect way to quickly stop erosion and runoff. Sod is also good for yards will little or not topsoil, since the sod brings its own topsoil along. Sod may be too expensive to use on the entire yard. You can compromise by using it in the front yard and using seed in the back yard. Sod can be successfully installed in every month of the year. It can be harvested and laid almost anytime during the winter provided the ground is not frozen. There is not a day in the summer that is too hot for sodding provided it is watered immediately after laying. Freshly laid sod requires frequent watering until the roots have grown into the soil.

Seeding the Yard

Sowing grass is not simply a matter of scattering seed. Soil preparation and composting can really make a difference in the amount of water your lawn needs and uses. This is especially important in yards with little or no topsoil. Properly prepared soil allows moisture to be absorbed and retained. The end result is less surface run-off, and less overall water consumption.

Make sure to choosing the right type of grass for your climate and maintenance requirements. Do you need heat tolerance or cold tolerance? Does the area receive a lot of traffic? Is it in full sun or shade? Will it be irrigated regularly or is it subject to drought conditions? Here is a brief summary of the major grass types:

- Bluegrass cool weather grass with fine texture, ability to withstand wear, and rich green color. Shade tolerant. Needs good soil, fertilizer, and regular watering.
- Fescue moderate weather grass with tough, course texture. Tolerates wear, drought, disease, and hot temperatures. Grows easily from seed.
- Ryegrass cold weather grass with fine blades. Comes in annual and perennial varieties. Annual variety used as a cover crop to allow other grasses to establish themselves. Will die in warm weather. Grows easily from seed.
- Bermudagrass hot weather grass with thick, fine blades. Tolerates heat, drought, poor soil, salt, and heavy traffic. Makes an excellent turf. Turns white in Winter. Requires regular watering and fertilizer. Needs full sun and will spread into nearby planting beds. Best grown from turf or sprigs.
- Zoysiagrass Excellent smooth, even textured grass that grows low to the ground and creates a smooth thick turf. Tolerant of poor soil, drought, and heat. Turns white in the winter. Slow growing and takes one or two years to cover yard. Must be grown from sprigs.
- Centipedegrass Hot weather grass that grows low to the ground. Very low maintenance with tolerance for poor soil and heat. Tolerates shade and chokes out weeds. Disease and insect resistant. Can be grown from seed or sprigs.

To establish a seeded yard, make sure the soil has been smoothed completely and has a good soft texture. The challenge will be to get the grass established quickly. Otherwise, rain will wash away the seeds and soil. Use a mixture of your final grass selection and annual rye grass. Rye germinates and grows quickly, establishing a root system quickly. This will hold the soil in place while the final grass gets established. Rye has the additional advantage of fixing nitrogen into the soil, actually improving it for the other grass. The rye will eventually die off, allowing the final grass selection to take over.

Use a commercial seed spreader to spread an even layer of seeds. Cover the seeds with some type of mulch, such as wheat straw or commercial spray-on mulch. Don't apply it too heavily, or you will block out the sun and smother the newborn grass seedlings. Keep the ground moist until the grass is fully established. After about two weeks, you should have a pretty green carpet of rye. Go back and fill in any bare spots with additional seed.

Sprinkler Systems

Sprinkler systems are great for saving time on watering. Underground sprinkler systems should be installed after the final grading and before sodding or seeding. Sprinkler systems can actually save water, and money. Difficult soils absorb water very slowly, and sprinkler systems can actually help. The timer can be set to water during the night, for short intervals, or repeatedly. This method will give your lawn an adequate soaking, without leaving gallons of water running down the street.

A sprinkler system is comprised of:
- An electronic control system
- Anti-siphon valves, where required
- PVC pipe
- Valve stems and sprinkler heads.

Landscape supply companies stock all types of sprinkler systems. Because of the complexity of sprinkler systems, it is best to let the suppliers or subcontractors install them.

LANDSCAPING — STEPS

Before Construction

LD1	EVALUATE your lot in terms of trees to keep, drainage, privacy, slopes, high and low spots and exposure to morning and afternoon sun. A landscape architect can help you greatly with this step. He can also help you determine if you will have too little or too much soil after excavation and what to do about it. Excavated soil may be used to fill in low spots or otherwise assist in developing a drainage pattern or landscaping feature.
LD2	DETERMINE best possible position for home on lot. Considerations should include: • Good southern exposure for parts of the house with a lot of glass. • Best view for most important rooms when at all possible. Don't give your best view to a garage or windowless wall. • Best angle for curb appeal. Consider angling your home if on a corner lot.
LD3	DEVELOP site plan. This is a bird's-eye view of what you want the finished site to look like, including exact position of home, driveway and all other paved surfaces, such as walks and patios, major trees, islands, grass areas, flower beds, bushes and even an underground watering system if you intend to have one. The site plan should show all baselines and easements. If any portion of your home falls outside a baseline, you will need a land variance. When determining grass areas, consider shade. Most grasses do not grow well or do not grow at all in shade. No amount of fertilizer or lawn care can compensate for a lack of sun.
LD4	CONTACT a landscape architect to critique and/or improve on your design. They normally have more experience and can surprise you with new and interesting ideas. For example, instead of having a driveway go straight from the street to the garage, consider putting a lazy curve in the drive and a slight mound on the street side. This helps hide the expanse of concrete which is not normally a focal point. Landscape architects can recommend specific types of bushes to suit special privacy needs.
LD5	FINALIZE site plan. Make several copies in blueprint form. You will need at least one for yourself, one for a landscaper if you hire one and one for the county.
LD6	SUBMIT site plan to building officials and bank along with your blueprints.
LD7	DELIVER fill dirt, if needed, to site. Choose a spot to pile topsoil from the foundation grading. Do not mix this dirt with fill dirt unless both types are topsoil. Make sure fill dirt is clean and dry. Wet dirt will form lumps and clods as it dries.
LD8	PAY landscape architect (if you used one).

After Construction

LD9	CONDUCT soil tests. Soil samples should be made for every several thousand feet of yard. Soil to be tested should be taken from 6" down. Soil testing will help to determine pH (level of acidity) and the lack of essential nutrients. Most counties have an extension service which will test the samples for you at little or no charge. Good general purpose soil conditioners are peat moss, manure and lime.
LD10	TILL first 4" to 6" of soil, adding soil conditioners as needed. Cut any unwanted trees down and grind stumps before sod is laid down. Rake surface smooth first one way and then another.
LD11	APPLY soil treatments as needed. You may also wish to add a slow release, non-burning fertilizer (no stronger than 6-6-6) to help your young grass grow. Apply gypsum pellets to break down clay.
LD12	INSTALL underground sprinkling system as needed. This normally consists of a network of ½" PVC pipe purchased from a garden supply store.
LD13	PLANT flower bulbs. This is the easiest and best time to plant daffodils, crocus, hyacinths, gladioli, etc.

LD14	APPLY seed or sod, depending upon your choice. Seed is much cheaper, but sod means an instant thriving lawn and elimination of topsoil runoff. Seed is hard to start on a hill or in other adverse conditions. If you do choose seed, use a broadcast spreader or a special seed dispensing device. Cover the seed with hay or straw to protect seeds from washing away with rain. If you sod, this will be a tougher job, but doing it yourself could really save some money. Sod always has at least three prices: direct from a sod farm, delivered by a sod broker or delivered and installed by a sod broker. If you can, buy direct from a sod farm. Visit them if you have time to see before you buy. Have them drop the sod in shady areas all around the grounds to be covered to reduce carrying. Keep sod slightly moist until installed. Have a good crew to help you lay sod because it will only live on pallets for a few days.
LD15	SOAK lawn with water if sod is applied. Sprinkle lightly for long periods of time if a seeded lawn is used. Roll lawn with a heavy metal cylinder.
LD16	INSTALL bushes and trees. If you plan to install trees, please keep them at least 4' from the driveway area. Trees are notorious for breaking up a driveway when the roots begin to spread.
LD17	PREPARE landscaped islands. Place pine bark, pine straw, mulch or gravel in designated areas. Over time, pine straw will kill some grasses, particularly Bermuda. Keep pine straw in islands only.
LD18	INSTALL mailbox if you have the kind that sits on a post. If you have a masonry or large frame mailbox, it will be installed during the framing or masonry stages of construction.
LD19	INSPECT landscaping job. Refer to specifications and checklist.
LD20	CORRECT any problems noted.
LD21	PAY landscaping sub for job. Have him sign an affidavit.
LD22	PAY landscaping sub retainage after grass is fully established without bare areas.

LANDSCAPING — SPECIFICATIONS

- Bid is to landscape as described below:
  - Loosen soil in yard to a depth of 6" with no dirt clumps larger than 6".
  - Finely rake all area to be seeded or sodded.
  - Install the following soil treatments:
    Peat moss 21 cubic feet per 1,000 square feet
    Sand   500 lbs. per 1,000 square feet
    Lime   200 lbs. per 1,000 square feet
    6-6-6   as prescribed by mfr. for new lawns
  - All grass to be lightly fertilized with 6-6-6 and watered heavily immediately to prevent drying.
  - Sod to be Bermuda 419, laid within 48 hours of delivery.
  - Sod pieces to butt each with no gaps with joints staggered.
  - Islands to be naturally shaped by trimming sod.
  - Sod to be rolled after watering.
  - Prepare three pine straw islands as described in attached site plan.
  - Install trees and bushes according to attached planting schedule and site plan.
  - Cover all seeded areas with straw to prevent runoff.
  - Install splash blocks for downspouts.
  - Burlap to be removed from root balls prior to installation.
  - Trees and shrubs to be watered after installation.

LANDSCAPING — INSPECTION

- ☐ All areas sodded/seeded as specified.
- ☐ All trees and bushes installed as specified.
- ☐ Sod and/or seed is alive and growing.
- ☐ Sod is smooth and even with no gaps.
- ☐ Straw covers seed evenly with no bare spots.
- ☐ All specified bushes and trees are alive and in upright position.
- ☐ Splash blocks installed.
- ☐ Grass cut if specified.
- ☐ No weeds in yard.
- ☐ No exposed dirt within 16" of siding or brick veneer.
- ☐ Pine bark or pine straw islands installed as specified.

DECKS AND SUNROOMS

These structures may, or may not be considered part of the house's heated living space. Sunrooms usually require a full foundation and roof and should usually be constructed along with the rest of the house. An exception is the pre-fabricated sunroom offered by many service companies. Decks are considered non-integral structures and will be the last structures to be added to your home. Sunrooms should usually be built by an experienced framer since they require the same structural integrity as the rest of the house.

Porches are usually integral to the design of the house and will be framed along with the rest of the main structure. Its construction is the same as the house framing and is not covered in the steps here.

Sunrooms

A sunroom is very popular because it adds additional light and conditioned space to the house. The least expensive installation will consist of fixed panes of glass, but without ventilation, this can cause the room to heat up unbearably in the summer. Sliding glass doors or screened windows will provide additional ventilation, but will also drive up the cost. Try to settle on a combination of fixed panes and screened openings that are strategically placed to provide cross ventilation. Ceiling fans will also help keep the air moving. Since heat buildup is a strong consideration, consider using energy efficient double pane glass. This is commonly called Low-E glass because of its "low energy" requirement. This glass is usually tinted to keep out excessive heat in the summer and the double panes help hold heat in during the winter. The additional cost will usually justify itself by creating a much more comfortable environment.

Solar heating in the winter is usually sufficient to heat a sunroom without additional heating systems, depending on its orientation. If the sun-space is open to the rest of the house the heat can also be distributed to reduce conventional heating needs. The summer is another story however. In southern climates heat buildup can be considerable. Make sure to provide for adequate cross ventilation. Also, consider installing several deciduous trees around the sunroom (but not too close) to provide additional shade. They will drop their leaves in the winter and allow the solar energy in, just when it is needed.

Many contractors specialize in installing new and retrofit sunrooms. They have all the specialized tools and knowledge to complete the job effectively. Many of these new prefab sunrooms are well designed and cost efficient. They usually consist of pre-manufactured metal channels, special "sunroom" windows, and foam core roofs. This job, unlike a deck, is best left to the professionals. Since most of the cost of a sunroom lies in the materials, you will save little by attempting the installation yourself.

The trend in sunrooms is for them to become integral parts of the house and conditioned space.

Decks

An outdoor deck is a standard do-it-yourself project for beginning and advanced homeowners. It provides maximum value for the amount invested and requires a moderate level of expertise to construct. Modern fasteners, custom lumber sizes, installation manuals, and videos make the construction of a world class deck an easy weekend project. Many building supply stores provide a computerized deck design program to help in determining your design and material costs. If you purchase the materials there, the service is usually free. If you decide to let the professionals do it, talk with the framer or trim carpenter that did the work on the main structure.

Decks are not considered part of the living space and can be added or expanded after initial construction, however, the building inspector will require a deck with a handrail on any back entranceway taller than 30". If you plan to add a more substantial deck later, plan a small section to meet code that can be easily expanded into the full size deck at a later date. You should wait until all the grading and landscaping is complete before framing the deck. The landscaping grader will need access to the area close to the house for smoothing and backfilling.

Even though it is not part of the heated structure, decks can add considerably to the value and traffic flow of the house. Decks can also extend the perceived space of the house by providing alternate areas for relaxing and socializing. Decks are very inexpensive space additions, which provide an escape from the confined space of the house.

Plan for the future. Are there any new trees close to the deck area? These can grow large in a few years and extend limbs over the deck area. Are you planning to convert the deck into a sunroom at some future date? If so, build your deck with standard footings and foundation. This will allow you to incorporate the existing deck structure into the sunroom. Otherwise, you may have to rip up the entire deck and start over. Think about the placement of steps leading down from the deck. Will they work structurally if the sunroom is added later? Is there room and an attachment point for a roof over the deck? Have you planned for the flashing that will be required between the house and the sunroom's roof?

TYPES OF DECKS

Wood Deck

The most common decks are low or high-level decks that are attached to the back of the house. Detached low-level decks are also popular on one story houses and around pool or patio areas.

Most decks are constructed from treated lumber nailed down or screwed, with a space between each board for ventilation and drainage. Less common are solid wood decks. These decks are made of caulked planking or exterior plywood with a waterproof coating. This kind of deck can serve double duty as both an upper level deck and a roof for a carport or playroom. It is imperative that solid decks are fully sealed or painted and that a waterproof membrane is installed underneath. New technologies have evolved to tackle the drainage issue of solid decks. They consist of a drainage or "deck gutter" system that is placed under a traditionally framed deck which functions as a roof for the space below.

In the past, redwood was a popular deck material. It was very attractive and stood up well to wind and weather. Unfortunately, redwood has become extremely expensive, which may lower your enthusiasm for using this traditional material. Cypress is another very attractive but expensive material suitable for decks. Consider these materials if you are looking for a distinctive "cost-is-no-object" project.

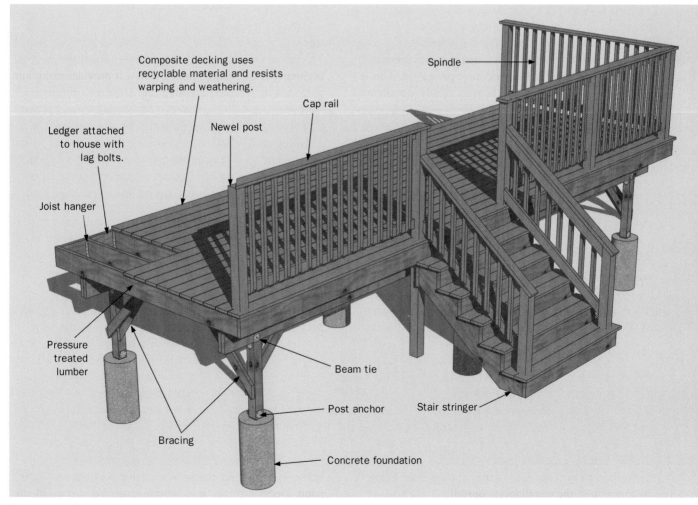

Components of a typical outdoor deck.

Currently, the most common deck material is pressure treated lumber. In the past, wood was pressure treated with a chromated copper arsenate solution, that provided a pest and rot proof material. However, toxicity concerns caused the Environmental Protection Agency to ban the use of CCA in 1994. The alternatives used today are two waterborne compounds: Alkaline copper quat and copper azole. These work just as well as the old CCA, but the strength and saturation of the chemicals had to increase. That resulted in greater corrosion on fasteners. Make sure to specify corrosion resistant fasteners specifically designed for this new product. Fasteners should be stainless steel or copper whenever possible. If coated fasteners are used, make sure to specify the highest rated hot dipped galvanized products.

Treated wood will require periodic maintenance and application of wood sealant every few years. The constant heat from the sun will eventually cause some warping and splintering on even the best laid decks. Using screws rather than nails will help to reduce the warping problem.

Composite Deck

Composite decking is usually manufactured from recycled plastic and wood fillers, making it a good sustainable product, despite the fact that it's made from a non-bio-degradable substance. It's better for plastics to substitute for other materials than to sit undisturbed in landfills for centuries. In fact, it is that permanence that gives composite decking its most desirable trait. Composite decking requires little or no maintenance, will not warp or crack, and will likely outlast the house itself. In most cases, composite decking can be recycled, extending its usefulness even further.

It is quite a bit more expensive than wood decking, so consider the long term durability and other factors when deciding whether to go with it. The dense, pliable nature of composites requires different installation procedures and hardware. Standard deck screws should not be used, because they will cause the surface of the decking around the screw to "bloom" or stick up. Specify composite deck screws that are designed to remove this ridge as the screw is tightened. Hidden fasteners are a better solution. If you've invested in expensive, durable decking, you might as well spend a bit more on fasteners that do not take away from the appearance of the decking. Hidden fasteners allow decking to "float" as it heats up and expands. They provide ease of installation, making the setting of the gap between boards automatic.

Not all composites are created equal. The less expensive brands may contain too much wood filler, sometimes as much as 70%, which can degrade over time, defeating the purpose of a durable, permanent deck product. When composite decking first came out, these inferior brands led to several class action lawsuits. If you can afford it, go with 100% recycled plastic or ask your supplier if the wood filler in the product has been pretreated with borate to retard decay.

Concrete Patio

This is the simplest and most cost effective surface to construct. Concrete patios are usually poured at the same time as the driveway. If the house foundation is more than 30" tall or the back yard area is uneven, concrete patios are impractical. Many people install a slab because it's inexpensive, planning to enhance it later by constructing a sunroom, porch, or deck on it at some later date. In this case, you must use the same standards as you would for a full foundation. Since concrete patios are not part of the structure, building codes don't usually require them to have reinforcement, edge footings, insect treatment, or waterproofing. But if you plan to build a structure on it later, this standard will be inadequate. The slab will crack

Like sunrooms, decks can become multi-functional additions to the home. With the popularity of grilling, decks have become outdoor entertainment centers.

Composite decking board comes in many different styles, compositions and colors. A natural wood grain surface is often cast into the material.

and settle under the weight of the new structure. Unless the concrete subcontractor knows your plans to build on the slab later, he will simply pour the least expensive slab possible. This may become the most expensive slab you ever poured, if you have to rip it up later. If you plan to build a structure on it, ask the contractor to pour a full foundation slab as described in the foundation section. It will save you headaches later.

Composite board can be applied with deck screws or with hidden fasteners that make the most of the board's appearance.

Hidden fasteners are designed to allow decking to slide freely during expansion and contraction.

Construction Hints

Decks constructed close to the ground are easy to layout and require a minimum foundation, sometimes consisting of nothing more than concrete blocks set in the ground on solid dirt. This type of deck can be attached to the house at the foundation with nails or lag bolts. Taller decks will require more substantial foundations and bracing. There have been many well publicized incidents of deck collapses in recent years. A deck full of robust Americans can add several tons of live load to a structure that was originally designed for a family of four or six. It's best to over-engineer the structure of your deck to protect the lives of your family and friends. Make sure tall decks have diagonal bracing on all posts and that the deck is securely fastened to the house following code. If you're not sure, beef the deck up some more.

Proper drainage on a deck is important, so avoid construction techniques that allow water to build up and sit

on parts of the deck. If the seam between the deck and the house is exposed to weather, cover it with flashing to prevent water from settling in the crack and rotting the siding of the house. On planked decks make sure to leave a drainage gap of at least ⅛" between each board. The easiest way to accomplish this is to use a 12d or 16d nail as a spacer when nailing down the boards. The surface of the deck should be slanted slightly away from the house to speed up drainage.

Hot Tip: The new wood treatments (alkaline copper quat and copper azole) are highly corrosive. Do not use aluminum flashing with treated wood as it will quickly accumulate pits and cracks, rendering the flashing useless. Go with copper flashing instead.

Galvanized joist hangers will greatly speed up the construction project and are much easier to use than "toe-nailing" the lumber into adjacent members. This is especially true when attaching the joists to a beam that is bolted to the house. Make sure the hangers are hot dipped galvanized and rated for the latest treated lumber.

Deck Foundation

The beams underneath the deck should rest securely on concrete foundation piers that have been poured. The piers should extend down to solid ground below the frost line for your area. Securely fasten the deck post to the foundation so that the post cannot slide laterally off the foundation. In older construction, the post was sometimes cast into the concrete. Avoid this technique if possible — the beam is exposed to a lot of moisture and will be difficult to replace if decayed. Galvanized strap anchors provide a convenient way to secure the deck beam to the foundation.

If the deck is close to the ground, you may have trouble with grass and weeds growing up between the boards, with no way to cut or trim them. You can prevent this by covering the ground with landscaping fabric and a layer of sand or fine gravel.

Composite decking provides superior longevity and appearance, with little or not warping, cracking, or discoloration over time. The splinter-free surface is much safer for bare feet.

Finishing

Deck surfaces are difficult to paint or stain because of the amount of abuse and foot traffic the surfaces will endure. Common paints and stains simply wear off the surface, leaving unattractive paths at points of heavy foot traffic. The best coating for a deck consists of transparent or semi-transparent water resistant compounds. This will seal the surface and reduce warping, splitting, and fading without changing the natural coloring of the wood.

Treated wood has been saturated with decay resistant compound and may be very moist when first installed. Make sure the wood is completely dry before applying penetrating water resistant compounds so that they can soak into the wood properly. This can take 6-8 weeks or longer in cold weather. If you purchase the wood ahead of time, find a dry area to stack the lumber, using spacers to allow ventilation. If the lumber has dried for at least 4 weeks, you can coat the underside of the lumber with penetrating finish, since this side will be difficult to reach after installation. Leave the other side uncoated so the lumber can continue to dry after installation. Apply the final coat to the exposed lumber after it has completely dried. See the painting chapter for more information. Water resistant compounds are not needed for composite decking.

Composite decking does not "breathe" like natural wood. Waterproofing membrane will help to protect floor joists from water damage.

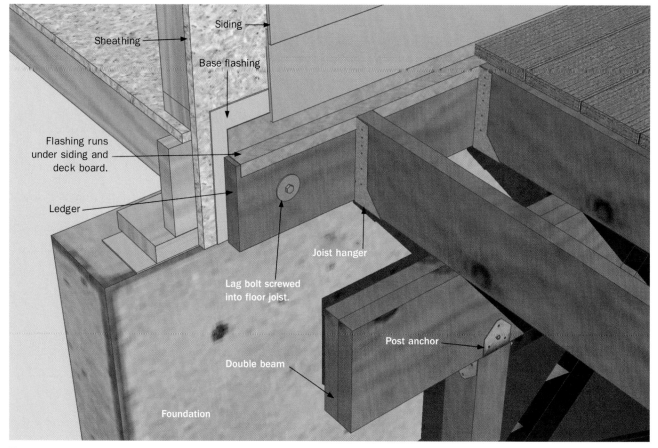

The attachment of the deck to the house is a frequent source of moisture issues. Make sure the ledger is securely fastened to the house and that flashing has been installed properly.

DECK — STEPS

DE1	EXAMINE the building site carefully, taking into consideration the height of deck in relation to the house and surrounding lot. Decks high above the ground will require larger supporting posts and perhaps cross bracing. Take this into consideration when ordering material.
DE2	DRAW a complete plan for the deck.
DE3	PURCHASE materials for deck, including nails, hangers, lumber, and concrete for footings. Lay out lumber so it can balance its moisture content with the local humidity. Store ready mix concrete inside or under a protective plastic cover to keep it from getting wet.
DE4	FINISH all grading and backfilling before starting deck construction.
DE5	LAYOUT the deck design with string and stakes to locate the position of posts and concrete piers. Measure the diagonal distances from corner to corner of the layout to square up the design. The diagonal distances should be the same from corner to corner.
DE6	DIG footings and pour concrete piers. Place the tops of the piers slightly above the height of the surrounding terrain to prevent water from collecting around the piers. The hole itself can function as the form for the pier. Use 1×4s at the top of the hole as forms to extend the concrete pier above ground level and to provide a smoother surface on the exposed part of the pier. Or use cardboard tubing made for this purpose. Insert any imbedded type post anchors into the concrete while it is still wet.
DE7	PULL a leveling line from corner to corner at the finish height of the deck and mark each corner post. The leveling line can be made with string and a string level or by using a garden hose. When using the garden hose, brace the posts in place and strap the ends of the garden hose to the posts. Fill the hose with water and adjust each end of the hose until each end is full of water all the way to the top. Keep adding water until this is accomplished. Since water always seeks its own level, this method produces very accurate results. Then mark the posts at each end of the hose. When marking the posts, make sure to allow for the height of the anchor plate that will attach the post to the concrete pier.
DE8	INSTALL posts on pier anchors and make sure they align with each other and are plumb. Use a carpenter's level to adjust the posts. Attach temporary braces to the posts to hold them in position.
DE9	ATTACH deck beams to the posts with lag bolts or galvanized beam hangers. This operation will take at least two people to position and attach the beams. Make sure the structure remains square as you attach each corner of the deck together.
DE10	INSTALL flashing on the tops of the beams where the beam and post are joined. This will keep rain from soaking the end grain of the post.
DE11	MOUNT the joist header to the house by running lag or stove bolts through the header, siding, and the joists of the house. Make sure to install a drip cap or flashing on top of the header to prevent standing water from accumulating between the header and the siding. The header should be placed so that the deck will slope away from the house slightly for better drainage.
DE12	PLACE the remaining headers on top of the beams and tie them together with bolts or angle irons.
DE13	INSTALL joists between the resulting frame with joist hangers or by resting them on ledgers nailed to the joist headers.
DE14	MOUNT cross bracing between the posts if the height of the deck is greater than 5 feet. This will strengthen the deck against side-to-side movement. Use 2×4s for the bracing and attach them to the posts and joist header with galvanized stove bolts.
DE15	APPLY the deck boards to the joists with galvanized deck nails, screws, or hidden fasteners. The best nails for the job are ring shank or spiral groove. Decking screws are the fasteners of choice. They have greater gripping power and can pull a warped board into alignment during installation. They have the added benefit of being removed easily. If one of the deck boards warps or splits, simply unscrew and replace it with a new board. As the deck boards dry out, you can easily go back and re-tighten the deck screws to maintain a snug fit. Leave a gap between each board for better drainage. The easiest way to do this is to use a 12d or 16d nail as a spacer. If the moisture content of the wood is high, leave a smaller gap — it will enlarge as the board dries and shrinks. Don't worry about cutting the exact length of each deck board. Leave a little extra hanging over the header joist. If composite decking is used, apply with the manufacturer's recommended fasteners.
DE16	SAW the deck boards flush with the header joist with a power saw after all boards are installed. This is much quicker, especially if you installed the deck boards diagonally. Pull a chalk line parallel to the header joist and follow the line with the power saw. This will give you a straight and accurate line.
DE17	ATTACH stairway stringers to header joist with galvanized strapping and mount the other ends of the stringers to the concrete piers with post anchors.
DE18	INSTALL the treads between the stringers.
DE19	INSTALL post rails around the perimeter of the deck and down the stair stringers. Attach the cap rails to the top of the posts.
DE20	TREAT wood decks with two coats of water resistant wood preservative. Apply the second coat after the first coat has soaked in but before it has dried. This will insure that the second coat can soak into the wood completely. Composite decks need no treatment.

Associations

APA - The Engineered Wood
  Association
7011 S. 19th St.
Tacoma, WA 98466-5399
(206) 565-6600
(www.apawood.org)

Air Conditioning & Refrigeration
  Institute
4301 N. Fairfax Dr., Suite 425
Arlington, VA 22203
(703) 524-8800
(www.ahrinet.org)

Air Conditioning Contractors of
  America
2800 Shirlington Road, Suite 300
Arlington, VA 22206
(703) 575-4477
(www.acca.org)

Air-Conditioning, Heating and
  Refrigeration Institute
2111 Wilson Blvd, Suite 500
Arlington, VA 22201
(703) 524.8800
(www.ahrinet.org)

Aluminum Association (AA)
818 Connecticut Avenue, N.W.
Washington, DC 20006
(www.aluminum.org)

American Institute of Timber
  Construction
7012 S. Revere Pkwy., Suite 140
Englewood, CO 80112
(303) 792-9559
(www.aitc-glulam.org)

American Architectural
  Manufacturers Association
1827 Walden Office Square, Suite 550
Schaumburg, IL 60173 4268
(847) 393-5664
(www.aamanet.org)

American Concrete Institute (ACI)
PO Box 9094
Farmington Hills, MI 48333-9094
(248) 848-3700
(www.concrete.org)

American Forest and Paper Association
1111 Nineteenth Street, NW, Suite 800
Washington, DC 20036
(800) 878-8878
(www.afandpa.org)

American Gas Association
400 North Capitol Street, NW Suite
  450
Washington, DC 20001
(202) 824-7000
(www.aga.org)

American Hardboard Association
1210 W. Northwest Hwy.
Palatine, IL 60067-3607
(708) 934-8800
(domensino.com/AHA)

American Hardware Manufacturers
  Association (AHMA)
801 North Plaza Dr.
Schaumburg, IL 60173-4977
(847) 605-1025
(www.ahma.org)

American Institute of Architects
1735 New York Avenue NW
Washington, DC 20006
(202) 626-7300
(www.aia.org)

American Institute of Steel
  Construction (AISC)
One East Wacker Drive Suite 700
Chicago, IL 60601-1802
(312) 670-2400
(www.aisc.org)

American Institute of Timber
  Construction (AITC)
7012 S. Revere Parkway Suite 140
Centennial, CO 80112
(303) 792-9559
(www.aitc-glulam.org)

American Insurance Association (AIA)
2101 L Street, NW, Suite 400
Washington, DC 20037
(202) 828-7100
(www.aiadc.org)

American Iron and Steel Institute (AISI)
1140 Connecticut Ave., NW
Suite 705
Washington, D.C. 20036
(202) 452-7100
(www.steel.org)

American Lighting Association
2050 Stemmons Fwy., Suite 10046
Dallas, TX 75258
(214) 698-9898 (800) 605-4448
(www.americanlightingassoc.com)

American National Standards
  Institute (ANSI)
1819 L Street, NW, 6th floor
Washington, DC 20036
(202) 293-8020
(www.ansi.org)

American Society for Testing
  Materials (ASTM)
100 Barr Harbor Drive
West Conshohocken, Pennsylvania
(610) 832-9500
(www.astm.org)

American Society of Heating,
  Refrigeration and Air Conditioning
  Engineers (ASHRAE)
1791 Tullie Circle, N.E.
Atlanta, GA 30329
(404) 636-8400
(www.ashrae.org)

American Society of Interior
  Designers (ASID)
608 Massachusetts Ave., NE
Washington, DC 20002-6006
(202) 546-3480
(www.asid.org)

American Welding Society, Inc. (AWS)
550 N.W. LeJeune Road
Miami, Florida 33126
(800) 443-9353
(www.aws.org)

American Wood Council
1111 19th St., NW, Suite 800
Washington, DC 20036
(202) 463-2766
(www.awc.org)

American Wood Preservers
   Association (AWPA)
P.O. Box 361784
Birmingham, AL 35236 1784
(205) 733-4077
(www.awpa.com)

American Wood Preservers Institute
   (AWPI)
2750 Prosperity Ave., Suite 550
Fairfax, VA 22031-4312
(703) 204-0500
(www.arcat.com)

Association of Home Appliance
   Manufacturers (AHAM).
1111 19th Street NW, Suite 402
Washington DC 20036
(202) 872-5955
(www.aham.org)

Association of the Wall and Ceiling
   Industries (AWCI) International
513 West Broad Street, Suite 210
Falls Church, VA 22046
(703) 538-1600
(www.awci.org)

Brick Industry Association - BIA
1850 Centennial Park Drive, Suite 301
Reston, VA 20191
(703) 620-0010
(www.brickinfo.org)

Brick Institute of America
1850 Centennial Park Drive, Suite 301
Reston, VA 20191
(703) 620.0010
(www.brickinfo.org)

Canadian Wood Council
99 Bank Street, Suite 400
Ottawa, Ontario K1P 6B9
(613) 747-5544
(ww.cwc.ca)

Carpet and Rug Institute
PO Box 2048
Dalton, GA 30722-2048
(706) 278-3176
(www.carpet-rug.org)

Cedar Shake & Shingle Bureau
P.O. Box 1178
Sumas, WA 98295-1178
(604) 820-7700
(www.cedarbureau.org)

Cellulose Insulation Manufacturers
   Association - CIMA
136 S. Keowee St.
Dayton, OH 45402
(888) 881-2462
(www.cellulose.org)

Ceramic Tile Distributors Association
800 Roosevelt Rd., Bldg. C Suite 312
Glen Ellyn, IL 60137
(800) 938-2832
(www.ctdahome.org)

Ceramic Tile Institute of America
12061 Jefferson Blvd.
Culver City, CA 90230-6219
(310) 574-7800
(www.ctioa.org)

Composite Panel Association (CPA)
19465 Deerfield Avenue, Suite 306
Leesburg, VA 20176
(703) 724.1128
(www.pbmdf.com)

Concrete Reinforcing Steel Institute
   (CRSI)
933 North Plum Grove Road
Schaumburg, IL 60173-4758
(847) 517-1200
(www.crsi.org)

Custom Electronic Design &
   Installation Association
7150 Winton Drive
Suite 300
Indianapolis, IN 46268
(800) 669-5329
(www.cedia.net)

Forest Products Laboratory (FPL)
United States Department of
   Agriculture
One Gifford Pinchot Drive
Madison, WI 53726
(608) 231-9200
(www.fpl.fs.fed.us)

Forest Products Society
2801 Marshall Ct.
Madison, WI 53705
(608) 231-1361
(www.forestprod.org)

DASMA Door and Access Systems
   Manufacturers Association
1300 Sumner Avenue
Cleveland, OH 44115-2851
(216) 241-7333
(www.doorandaccesssystems.com)

Glass Association of North America
2945 SW Wanamaker Drive Ste A
Topeka, KS 66614-5321
(785) 271-0208
(www.glasswebsite.com)

Gypsum Association
6525 Belcrest Road
Suite 480
Hyattsville, MD 20782
(301) 277-8686
(www.gypsum.org)

Hardwood Plywood & Veneer
   Association
1825 Michael Faraday Dr.
Reston, VA 20190
(703) 435-2900
(www.hpva.org)

Hardwood Plywood Manufacturer's
   Association (HPMA)
1825 Michael Faraday Drive
Reston, VA 22090-2789
(703) 435-2900

Home Ventilating Institute
1000 N Rand Rd, Ste 214
Wauconda, IL 60084
(847) 526-2010
(www.hvi.org)

Insulation Contractors Association
PO Box 26237
Alexandria, VA 22313
(703) 739-0356
(www.insulate.org)

International Code Council
500 New Jersey Avenue NW
6th Floor
Washington, DC 20001-2070
888-422-7233
(www.iccsafe.org)

International Interior Design
  Association (IIDA)
222 Merchandise Mart, Suite 567
Chicago, IL 60654
(888) 799 4432
(www.iida.org)

International Masonry Institute
The James Brice House
42 East Street
Annapolis, MD 21401
(410) 280-1305
(www.imiweb.org)

Manufactured Housing Institute (MHI)
101 Wilson Blvd., Suite 610
Arlington, VA 22201-3040
(703) 558-0400
(www.manufacturedhousing.org)

Marble Institute of America
28901 Clemens Rd, Ste 100
Cleveland, OH 44145
440-250-9222
(www.marble-institute.com)

Mason Contractors Association of
  America (MCAA)
33 South Roselle Road
Schaumburg, IL 60193
(800)-536-2225
(www.masoncontractors.org)

Metal Building Manufacturers
  Association (MBMA)
1300 Sumner Ave
Cleveland, OH 44115-2851
(216) 241-7333
(www.mbma.com)

National Association of Brick
  Distributors
11490 Commerce Park Dr # 300
Reston, VA 20191
(703) 620-0010
(www.manta.com)

National Association of Home
  Builders (NAHB)
1201 15th Street, NW
Washington, DC 20005
202-266-8200, 800-368-5242
(www.nahb.org)

National Association of the
  Remodeling Industry (NARI)
780 Lee Street, Suite 200
Des Plaines, Illinois 60016
(800) -611-6274
(www.nari.org)

National Institute of Standards (NIST)
100 Bureau Drive, Stop 1070
Gaithersburg, MD 20899-1070
(301) 975-6478
(www.nist.gov)

National Clay Pipe Institute (NCPI)
PO BOX 759
Lake Geneva, WI 53147
(262) 248-9094
(www.ncpi.org)

National Concrete Masonry
  Association (NCMA)
13750 Sunrise Valley Drive
Herndon, VA 20171-4662
(703) 713-1900
(www.ncma.org)

National Fire Protection Association
  (NFPA)
1 Batterymarch Park
Quincy, Massachusetts 02169-7471
(617) 770-3000
(www.nfpa.org)

National Kitchen & Bath Association
687 Willow Grove Street
Hackettstown, NJ 07840
(908) 852-0033
(www.nkba.org)

National Lime Association (NLA)
200 North Glebe Road, Suite 800
Arlington, Virginia 22203
(703) 243-5463
(www.lime.org)

National Pest Management
  Association (NPMA)
10460 North Street
Fairfax, VA 22030
(703) 352-6762
(www.pestworld.org)

National Propane Gas Association
1150 17th Street, NW, Suite 310
Washington, D.C. 20036-4623
(202) 466.7200
(www.npga.org)

National Tile Contractors Association
626 Lakeland East Drive
Jackson, MS 39232
(601) 939-2071
(www.tile-assn.com)

North American Association of Floor
  Covering Distributors (NAFCD)
401 N. Michigan Ave, Suite 2400
Chicago, IL 60611
(312) 321-6836
(www.nafcd.org)

National Fenestration Rating Council
6305 Ivy Lane, Suite 140
Greenbelt, MD 20770
(301) 589-1776
(www.nfrc.org)

Northeastern Lumber Manufacturers
  Association
PO Box 87A
Cumberland Center, ME 04021-0687
(207) 829-6901
(www.nelma.org)

Perlite Institute (PI)
4305 North Sixth Street Suite A
Harrisburg, PA 17110
(717) 238-9723
(www.perlite.org)

Plumbing Manufacturers Institute
1921 Rohlwing Rd., Unit G
Rolling Meadows, IL 60008
(847) 481-5500
(www.pmihome.org)

Portland Cement Association (PCA)
5420 Old Orchard Road
Skokie, IL 60077
(847) 966 6200
(www.cement.org)

Post-Tensioning Institute (PTI)
38800 Country Club Drive
Farmington Hills, MI 48331
(248) 848-3180
(www.post-tensioning.org)

Sheet Metal & Air Conditioning
  Contractor's National Association
  (SMACCNA)
4201 Lafayette Center Drive Chantilly
Virginia 20151-1209
(703) 803-2980
(www.smacna.org)

Society of Plastics Industry, Inc. (SPI)
1667 K St., NW, Suite 1000
Washington, DC 20006
(202) 974-5200
(www.plasticsindustry.org)

Solar Energy Industries Association
122 C St., N.W., Fourth Fl.
Washington, DC 20001
(202) 383-2600
(www.seia.org)

Southeastern Lumber Manufacturers
  Association
200 Greencastle Road
Tyrone, GA 30290
(770) 631-6701
(www.slma.org)

Southern Forest Products Association
  (SFPA)
2900 Indiana Ave.
Kenner, LA 70065
504/443-4464
(www.sfpa.org)

Southern Gas Association (SGA)
3030 LBJ Freeway
Suite 1300
Dallas, TX 75234
(972) 620-8505
(www.southerngas.org)

Southern Pine Council
2900 Indiana Avenue
Kenner, LA 70065-4605
(504) 443-4464
(www.southernpine.com)

Southern Pine Inspection Bureau (SPIB)
P. O. Box 10915
Pensacola, Fl. 32524-0915
(850) 434-2611
(www.spib.org)

Southwest Research Institute (SRI)
8500 Culebra Road
San Antonio, TX 78228
(210) 684-5111
(www.swri.org)

Steel Door Institute
30200 Detroit Road
Westlake, Ohio 44145
(440) 899-0010
www.steeldoor.org

Steel Joist Institute (SJI)
1173B London Links Drive
Forest, VA 24551
(434) 525.7377
(www.steeljoist.org)

Steel Window Institute (SWI)
1300 Sumner Avenue
Cleveland, OH 44115-2851
(216) 241 - 7333
(www.steelwindows.com)

Tile Council of America
100 Clemson Research Blvd.
Anderson, SC 29625
(864) 646-8453
(www.tileusa.com)

Truss Plate Institute (TPI)
218 North Lee Street, Suite 312
Alexandria, VA 22314
(703) 683-1010
(www.tpinst.org)

Underwriters' Laboratories, Inc. (ULI)
2600 N.W. Lake Rd.
Camas, WA 98607-8542
(877) 854.3577
(www.ul.com)

Western Wood Products Association
522 SW Fifth Ave., Suite 500
Portland, OR 97204-2122
(503) 224-3930
(www.wwpa.org)

Window & Door Manufacturers
  Association
2025 M Street, NW, Suite 800
Washington, DC 20036
(202) 367-1280
(www.wdma.com)

Wood Moulding & Millwork
  Producers Association
507 First St.
Woodland, CA 95695
(916) 661-9591
(www.wmmpa.com)

Woodwork Institute
P.O. Box 980247
West Sacramento, CA 95798
(916) 372-9943
(www.wicnet.org)

World Floor Covering Association
2211 E. Howell Ave.
Anaheim, CA 92806
(800) 624-6880
(www.wfca.org)

Fannie Mae Regional Offices
Home Office
FNMA
3900 Winconsin Avenue NW
Washington, DC 20016-2899
(202) 752-7000
(www.fanniemae.com)

Midwestern Regional Office
Fannie Mae
One South Wacker Drive, Suite 1400
Chicago, Ill. 60606-4667
Phone: (312) 368-6200

Northeastern Regional Office
Fannie Mae
1835 Market Street, Suite 2300
Philadelphia, Pa. 19103
Phone: (215) 575-1400

Southeastern Regional Office
Fannie Mae
950 East Paces Ferry Road, Suite 1800
Atlanta, Ga. 30326-1161
Phone: (404) 398-6000

Southwestern Regional Office
International Plaza II
14221 Dallas Parkway
Suite 1000
Dallas, TX 75254-2916
972-773-HOME (4663)

Western Regional Office
Fannie Mae
135 North Los Robles Ave., Suite 300
Pasadena, Calif. 91101-1707
Phone: (626) 396-5100

Freddie Mac Regional Offices
Corporate Headquarters
Federal Home Loan Mortgage Corp.
8200 Jones Branch Drive
McLean, VA 22102-3110
(703) 903-2000
(www.freddiemac.com)

Atlanta, Georgia
2300 Windy Ridge Prkwy SE
Ste 200 North Tower
Atlanta, GA 30339-5665
(770) 857-8800

New York City, New York
122 East 42nd Street
4th Floor
New York, NY 10168
(212) 418-8900

Dallas, Texas
5000 Plano Parkway
Carrollton, TX 75010 - 4902
(972) 395-4000

Los Angeles, California
444 South Flower Street, 44th Floor
Los Angeles, CA 90071
(213) 337-4200

Chicago, Illinois
333 West Wacker Drive, Suite 2500
Chicago, IL 60606-1287
(312) 407-7400

MATERIAL SUPPLIERS

Electrical Supplies
American Lighting
7660 East Jewell Avenue, Suite A
Denver, CO 80231
(303) 695-3019
(www.americanlighting.com)

Cooper Wiring Devices
203 Cooper Circle
Peachtree City, GA 30269
(770) 631-2100
(www.cooperwiringdevices.com)

Fasco America
105 Industrial Park Drive
PO Box 2389
Muscle Shoals, AL 35662
(256) 381-6364
(www.fascoamerica.com)

Hunter Fan Co.
2500 Frisco Ave.
Memphis, TN 38114
(901)-248-2235
(www.hunterfan.com)

Leviton
201 North Service Rd.
Melville, NY 11747
(800) 824.3005
(www.leviton.com)

Lightolier
631 Airport Rd.
Fall River, MA 02720
(214) 647-7880
(www.lightolier.com)

LiteTouch
3400 S. West Temple
Salt Lake City, UT 84119
(801) 486-8500
(www.litetouch.com)

Osram Sylvania Inc.
100 Endicott St.
Danvers, MA 01923
(508) 777-1900
(www.sylvania.com)

Sylvan Designs
8955 Quartz Ave.
Northridge, CA 91324
(818) 998-6868
(www.sylvandesigns.com)

WAC Lighting Co.
615 South Street
Garden City, NY 11530
(516) 515-5000
(www.waclighting.com)

Equipment
Ace Hardware Corp.
2200 Kensington Ct.
Oak Brook, IL 60521
(708) 990-6751
(www.acehardware.com)

Arrow Fastener Co.
271 Mayhill St.
Saddle Brook, NJ 07662
(201) 843-6900
(www.arrowfastener.com)

Campbell Hausfeld
100 Production Dr.
Harrison, OH 45030
(513) 367-4811
(www.cpocampbellhausfeld.com)

Caterpillar
100 N.W. Adams St.
Peoria, IL 61629
(309) 675-1000
(www.cat.com)

Cooper Tools
5925 McLaughlin Rd.
Mississauga, ONTARIO LSR 1B8
(905)-501-4785
(www.coopertools.com)

DeWalt Industrial Tool Co.
701 E. Joppa Road, TW425
Baltimore, MD 21286
(800) 4-DEWALT
(www.dewalt.com)

Deere & Co. (John Deere Industrial
   Equipment Co.)
8000 Jersey Ridge Rd.
Davenport, IA 52807
(309) 765-3195
(www.deere.com)

Fein Power Tools
3019 W. Carson St.
Pittsburgh, PA 15204
(412) 331-2325 (800) 441-9878
(www.fein.de)

Hitachi Koki USA
3950 Steve Reynolds Blvd.
Brisbane, CA 94005-1835
(650) 589-8300
(www.hitachipowertools.com)

Jepson Power Tools
20333 S. Western Ave.
Torrance, CA 90501
(310) 320-3890
(www.jepsonpowertools.com)

Klein Tools
450 Bond Street
PO Box 1418
Lincolnshire, IL 60069-1418
(800) 553-4676
(www.kleintools.com)

Komatsu Forklift (USA)
14481 Lochridge Boulevard
Covington, GA 30014
(800) 821-9365
www.komatsuforkliftusa.com

Kraft Tool Co.
8325 Hedge Lane Terrace
Shawnee, KS, 66227
(800) 422-2448
(www.krafttool.com)

Makita USA
14930 Northam St.
La Mirada, CA 90638
(714) 522-8088
(www.makita.com)

Milwaukee Electric Tool Corp.
13135 W. Lisbon Rd.
Brookfield, WI 53005
(414) 781-3811
(www.milwaukeetool.com)

Mitsubishi Caterpillar Forklift
2011 W. Sam Houston, Pkwy. N.
Houston, TX 77043-2421
(713) 365-1000
(www.mcfa.com)

Omaha Industrial Tools
14685 Grover St.
Omaha, NE 68144-5435
(402) 334-8185 (800) 228-2765
(www.omahatool.com)

Panasonic Power Tool Div.
1 Panasonic Way
Secaucus, NJ 07094
(201) 392-6442
(www.panasonic.com)

Porter-Cable Corp.
4825 Hwy. 45 N.
Jackson, TN 38302-2468
(901) 668-8600 (800) 4-US-TOOL
(www.porter-cable.com)

Quik Drive USA
7528 Hickory Hills Ct.
Whites Creek, TN 37189
(888) 487-7845
(www.quikdrive.com)

Quikspray
PO Box 327
Port Clinton, OH 43452
(419) 732-2601
(www.quikspray.com)

Ryobi America Corp.
5201 Pearman Dairy Rd.
Anderson, SC 29625
(803) 226-6511 (800) 525-2579
(www.ryobi.com)

Senco Products
8485 Broadwell Rd.
Cincinnati, OH 45244
(800) 543-4596
(www.senco.com)

Stanley Tools
600 Myrtle St.
New Britain, CT 06053
(203) 225-5111 (800) 648-7654
(www.stanleytools.com)

Stanley-Bostitch
Briggs Dr.
East Greenwich, RI 02818
(401) 884-2500 (800) 556-6696
(www.bostitch.com)

Wallboard Tool Co.
1697 Seabright Ave.
Long Beach, CA 90813-1146
(310) 437-0701
(www.wallboardtools.com)

Exterior Siding Materials

ABC Seamless Siding
3001 Fiechtner Dr.
Fargo,) 58103
(800) 732-6577
(www.abcseamless.com)

Alcoa Building Products
PO Box 716
Sidney, OH 45365
(513) 492-1111
(www.alcoa.com)

Alsco
3101 Poplarwood Ct., Suite 200
Raleigh, NC 27604
(919) 876-9333
(www.alsco.com)

Arriscraft Corp.
875 Speedsville Rd., PO Box 3190
Cambridge, ON N3H 4S8 Canada
(800) 265-8123
(www.arriscraft.com)

CertainTeed Siding
PO Box 860
Valley Forge, PA 19482
(800) 233-8990
(www.certainteed.com/products/siding)

Dryvit Systems
1 Energy Way
West Warwick, RI 02893
(401) 822-4100
(www.dryvit.ca)

Georgia-Pacific Corp.
PO Box 1763
Norcross, GA 30091
(800) BUILD-GP
(www.gp.com)

Kasko Industries
18 Turtle Creek Dr.
Tequesta, FL 33469
(407) 575-1193

Niagara Fiberboard
10 Stevens St.
Lockport, NY 14095
(716) 434-8881
(www.niagarafiberboard.com)

Royal Building Products
30A Vinyl Ct.
Woodbridge, ON L4L 4A3 Canada
(905) 850-9700
(www.royalbuildingproducts.com)

The Pacific Lumber Co.
PO Box 565
Scotia, CA 95565
(707) 764-8888
(www.palco.com)

Weyerhaeuser Composite Products Div.
CCB-100
Tacoma, WA 98477
(800) 458-7180
(www.weyerhaeuser.com)

Flooring

Buchtal Corp. USA
1325 Northmeadow Pkwy., Suite 114
Roswell, GA 30076
(404) 442-5500

Dalton Georgia Wholesale Carpet
3041 N. Dug Gap Rd.
Dalton, GA 30720
(706) 277-9559 (800) 476-9559

Direct Carpet Mills of America
PO Box 2759
Rome, GA 30164-2759
(706) 235-6009 (800) 424-1223

Direct Dalton Carpet Mills
PO Box 2645
Dalton, GA 30722
(706) 277-1200 (800) 892-6789

Mannington Mills
PO Box 30
Salem, NJ 08079-0030
(800) 952-1857
(www.mannington.com)

Quarry Tile Co.
6328 Utah Ave.
Spokane, WA 99212
(509) 536.2812 (800) 423-2608
(www.quarrytile.com)

The Georgia Marble Co.
Blue Ridge Ave., PO Box 9
Nelson, GA 30151
(404) 735-2591 (800) 334-0122

Heating, Ventilation, & Air Conditioning

American-Standard Air Conditioning
PO Box 9010
Tyler, TX 75711
(www.americanstandardair.com)

Bryant
PO Box 70
Indianapolis, IN 46206
(317) 243-0851 (800) 428-4326
(www.bryant.com)

Cool Attic – Ventamatic, Ltd.
100 Washington St.
Mineral Wells, TX 76067
(817) 325-7887 (800) 433-1626
(www.bvc.com)

Duralast Products Corp.
PO Box 15869
New Orleans, LA 70175
(504) 895-2068
(www.duralast.com)

Goodman Manufacturing Co.
1501 Seamist
Houston, TX 77008
(713) 861-2500
(www.goodmanmfg.com)

Heatway
3131 W. Chestnut Expwy
Springfield, MO 65802
(417) 864-6108 (800) 255-1996

Lennox Industries
PO Box 799900
Dallas, TX 75379-9900
(214) 497-5000
(www.lennox.com)

Payne
PO Box 70
Indianapolis, IN 46206
(317) 243-0851 888-417-2963)
(www.payne.com)

Rheem Mfg. Air Conditioning Div.
5600 Old Greenwood Rd.
Fort Smith, AR 72917-7010
(501) 646-4311
(www.rheem.com)

Ruud Air Conditioning Division
PO Box 17010
Fort Smith, AR 72917-7010
(501) 646-4311
(www.ruudac.com)

The Trane Co. Unitary Products Group
6200 Troup Hwy.
Tyler, TX 75707
(www.trane.com)

Insulation

CertainTeed Corp. Insulation Group
PO Box 860
Valley Forge, PA 19482-0860
(610) 341-7739
(www.certainteed.com)

Du Pont Tyvek
PO Box 80-705, Centre Rd.
Wilmington, DE 19880-0705
(800) 44-TYVEK
(www2.dupont.com/Tyvek)

Owens Corning
Fiberglas Tower
Toledo, OH 43659
(800) GET-PINK
(www.owenscorning.com)

SOLEC (Solar Energy Corp.)
129 Walters Ave.
Ewing Township, NJ 08638-1829
(609) 883-7700
(www.solec.org)

The Dow Chemical Co. Construction
  Materials
2020 Dow Center
Midland, MI 48674
(517) 636-2303
(www.dow.com)

Interior Finish

AJ Stairs
1095 Towbin Ave.
Lakewood, NJ 08701
(908) 905-8500 (800) 425-7824
(www.ajstairs.com)

Armstrong World Industries
PO Box 3001
Lancaster, PA 17604
(717) 397-0611 (800) 233-3823
(www.armstrong.com)

Builders Edge
PO Box 7739
Pittsburgh, PA 15215
(412) 782-4880 (800) 969-7245
(www.buildersedge.com)

Chesapeake Hardwood Products
201 W. Dexter St.
Chesapeake, VA 23324-3023
(804) 543-1601 (800) 446-8162
(www.chpi.com)

Classic Mouldings
226 Toryork Dr.
Weston, ON M9L 1Y1 Canada
(416) 745-5560
(www.classicmouldings.com)

Custom Doors & Drawers
232 St. Amaud St.
Amherstburg, ON N9V 2Z7 Canada
(519) 736-2195
(www.customdoorsanddrawers.com)

Driwood Moulding Co.
PO Box 1729
Florence, SC 29503-1729
(803) 669-2478
(www.driwood.com)

Duo-Fast Corp.
3702 River Rd.
Franklin Park, IL 60131
(800) 752-5207
(www.itwindfast.com)

Flexi-Wall Systems (A Div. of Wall
   & Floor Treatments)
PO Box 89
Liberty, SC 29657
(864) 843-3104 (800) 843-5394
(www.flexiwall.com)

Intl. Cellulose Corp.
PO Box 450006
Houston, TX 77245
(713) 433-6701
(www.ada-noisesolutions.com)

Kentucky Millwork
PO Box 33276
Louisville, KY 40232
(502) 451-1528, (800) 235-5235
(www.kentuckymill.com)

Kwikset Corp. (A Sub. of Black
   & Decker Co.)
1 Park Plaza, Suite 1000
Irvine, CA 92714
(714) 474-8800 (800) 327-LOCK
(www.kwikset.com)

Masco Corporation
21001 Van Born Rd.
Taylor, MI 48180
(313) 274-7400
(www.masco.com)

Milton W. Bosley Co.
151 Eighth Ave. N.W.
Glen Burnie, MD 21061
(410) 761-7727 (800) 638-5010
(www.bosleymouldings.com)

Prest-on Co.
316 N. Lookout Point
Hot Springs, AR 71913
(501) 525-4683 (800) 323-1813
(www.prest-on.com)

Schlage Lock Co.
2401 Bayshore Blvd.
San Francisco, CA 94134
(415) 467-1100
(www.schlage.com)

Spiral Stairs of America
1700 Spiral Ct., Franklin Ave.
Erie, PA 16510
(814) 898-3700 (800) 422-3700
(www.spiralstairsofamerica.com)

Stanley Hardware
480 Myrtle St.
New Britain, CT 06053
(860) 225-5111
(www.stanleyhardware.com)

United States Gypsum Co.
PO Box 806278
Chicago, IL 60680-4124
(312) 606-4000 (800) USG-4YOU
(www.usg.com)

Weiser Lock (A Div. of Masco Corp.)
6660 S. Broadmoor Rd.
Tucson, AZ 85746
(602) 741-6200 (800) 677-LOCK
(www.weiserlock.com)

Weslock National
13344 S. Main St.
Los Angeles, CA 90061
(310) 327-2770
(www.weslock.com)

Yale Locks & Hardware
PO Box 25288
Charlotte, NC 28229-8010
(704) 283-2101 (800) 438-1951
(www.yalecommercial.com)

Kitchen & Bath

Abet Laminati
60 W. Sheffield Ave.
Englewood, NJ 07631
(201) 541-0700 (800) 228-ABET
(www.abetlaminati.com)

Amana Refrigeration
2800 220th Trail, PO Box 8901
Amana, IA 52204
(319) 622-5511 (800) 843-0304
(www.amana.com)

American Marazzi Tile
359 Clay Rd.
Sunnyvale, TX 75182
(214) 226-0110
(www.marazzitile.com)

American Olean Tile
1000 Cannon Ave.
Lansdale, PA 19446
(215) 393-2237
(americanolean.com)

American Standard
1 Centennial Plaza
Piscataway, NJ 08855
(908) 980-2400 (800) 524-9797
(www.americanstandard-us.com)

American Whirlpool
3050 N. 29th Ct.
Hollywood, FL 33020
(305) 921-4400 (800) 327-1394
(www.americanwhirlpool.com)

Aristokraft
PO Box 420, 1 Aristokraft Sq.
Jasper, IN 47547-0420
(812) 482-2527
(www.aristokraft.com)

Avonite
1945 S. Hwy. 304
Belen, NM 87002
(505) 864-3800 (800) 428-6648
(www.avonitesurfaces.com)

Block Tops
4770 E. Wesley Dr.
Anaheim, CA 92807
(714) 978-5080
(www.blocktops.com)

Broan Mfg. Co.
P.O. Box 140
Hartford, WI 53027
(800) 558-1711
(www.broan.com)

Craft-Maid Custom Kitchens
501 S. Ninth St., Bldg. C
Reading, PA 19602
(610) 376-8686
(www.craft-maid.com)

Crossville Ceramics
PO Box 1168, Industrial Park
Crossville, TN 38557
(615) 484-2110
(www.crossvilleinc.com)

Delta Faucet Co.
55 E. 111th St.
Indianapolis, IN 46280-1000
(317) 848-1812 (800) 345-DELTA
(www.deltafaucet.com)

Designs In Tile
PO Box 358
Mt. Shasta, CA 96067
(916) 926-2629
(www.designsintile.com)

Du Pont Co. Corian Products
Chestnut Run Plaza, PO Box 80702
Wilmington, DE 19880
(800) 4CORIAN
(www.simplicity.dupont.com)

Elkay Manufacturing Company
2222 Camden Ct.
Oak Brook, IL 60521
(708) 574-8484
(www.elkay.com)

Formica Corp.
10155 Reading Rd.
Cincinnati, OH 45241
(513) 786-3400 (800) FORMICA
(www.formica.com)

Frigidaire Co. Frigidaire, Tappan
 & White-Westinghouse
6000 Perimeter Dr.
Dublin, OH 43017
(614) 792-4100 (800) 685-6005
(www.frigidaire.com)

GE Appliances
Appliance Park
Louisville, KY 40225-0001
(502) 452-4311 (800) 626-2000
(www.geappliances.com)

GROHE (A Sub. of Friedrich Grohe,
 Germany)
241 Covington Dr.
Bloomingdale, IL 60108
(708) 582-7711
(www.grohe.com)

Hearth Kitchens
711 10th Ave.
Hanover, ON N4N 2P7 Canada
(519) 364-1170 (800) 267-4524
(www.hearthkitchen.com)

Home Crest Corp.
1002 Eisenhower Dr. N.
Goshen, IN 46526
(219) 533-9571
(www.homecrestcab.com)

In-Sink-Erator Div. Emerson
 Electric Co.
4700 21st St.
Racine, WI 53406-5093
(414) 554-5432 (800) 558-5712
(www.insinkerator.com)

Jacuzzi Whirlpool Bath
PO Drawer J, 2121 N. California
 Blvd., Suite 475
Walnut Creek, CA 94596
(510) 938-7070 (800) 678-6889
(www.jacuzzi.com)

Jenn-Air
3035 N. Shadeland Ave.
Indianapolis, IN 46226
(317) 545-2271 (800) JENN-AIR
(www.jennair.com)

KitchenAid (A Brand of Whirlpool
 Corp.)
2000 M-63 N, Mail Drop 4302
Benton Harbor, MI 49022
(616) 923-5000 (800) 253-3977
(www.kitchenaid.com)

Kohler Co.
444 Highland Dr.
Kohler, WI 53044
(414) 457-4441
(www.kohler.com)

Kolson
653 Middle Neck Rd.
Great Neck, NY 11023
(516) 487-1224 (800) 783-1335
(www.kolson.com)

Kountry Kraft
PO Box 570
Newmanstown, PA 17073
(610) 589-4575 (800) 628-9061
(www.kountrykraft.com)

KraftMaid Cabinetry
PO Box 1055, 16052 Industrial Pkwy.
Middlefield, OH 44062
(216) 632-5333
(www.kraftmaid.com)

Magic Chef
3035 N. Shadeland Ave.
Indianapolis, IN 46226-0901
(800) 536-6247
(www.magicchef.com)

Maytag
1 Dependability Sq.
Newton, IA 50208
(515) 792-7000
(www.maytag.com)

Merillat Industries
5353 W. US 223
Adrian, MI 49221
(517) 263-0771
(www.merillat.com)

Monarch Tile
834 Rickwood Rd.
Florence, AL 35630
(205) 764-6181 (800) 289-8453
(www.monarchtile.net)

Peerless Faucet Co. (A Div. of Masco
  Corp. of Indiana)
55 E. 111th St.
Indianapolis, IN 46280
(317) 848-1812
(www.peerlessfaucet.com)

Price Pfister
13500 Paxton St.
Pacoima, CA 91331
(818) 896-1141 (800) PFAUCET
(www.pricepfister.com)

Quaker Maid
State Rt. 61, Box H
Leesport, PA 19533
(610) 926-3011
(www.quakermaid.com)

Quality Cabinets
515 Big Stone Gap
Duncanville, TX 75137
(800) 298-7020
(www.qualitycabinets.com)

Sub-Zero Freezer Co.
4717 Hammersley Rd.
Madison, WI 53711
(608) 271-2233 (800) 222-7820
(www.subzero.com)

Tappan (A Brand of Frigidaire Co.)
6000 Perimeter Dr.
Dublin, OH 43017
(614) 792-4100 (800) 685-6005
(www.frigidaire.com)

Vent-A-Hood
PO Box 830426
Richardson, TX 75083-0426
(214) 235-5201
(www.ventahood.com)

Whirlpool
2000 M-63
Benton Harbor, MI 49022
(616) 923-5000 (800) 253-1301
(www.whirlpool.com)

White-Westinghouse (A Brand of
  Frigidaire Co.)
6000 Perimeter Dr.
Dublin, OH 43017
(614) 792-4100 (800) 685-6005
(www.whitewestinghouse.com)

Wilsonart International
2400 Wilson Pl., PO Box 6110
Temple, TX 76503-6110
(817) 778-2711 (800) 433-3222
(www.wilsonart.com)

Landscaping Materials

Anchor Wall Systems
6101 Baker Rd., Suite 390
Minnetonka, MN 55345-5973
(612) 933-8855
(www.anchorwall.com)

Belden Brick Co.
700 W. Tuscarawas St.
Canton, OH 44702
(216) 456-0031
(www.beldenbrick.com)

Cangelosi Marble & Granite
14021 S. Gessner
Missouri City, TX 77489
(713) 499-7521
(www.cangelosi.com)

Cedarbrook Sauna
21326 Hwy. 9
Woodinville, WA 98072
(509) 782-2447
(www.cedarbrooksauna.com)

Moultrie Mfg.
PO Box 1179
Moultrie, GA 31776-1179
(800) 841-8674
(www.moultriemanufacturing.com)

Redland Brick (Cushwa Plant)
PO Box 160
Williamsport, MD 21795-0160
(301) 223-7700
(www.redlandbrick.com)

Risi Stone
Le Parc Office Tower, 8500 Leslie St.,
  Suite 390
Thornhill, ON L3T 7P1 Canada
(905) 882-5898
(www.risistone.com)

Rockwood Retaining Walls
7200 N. Hwy. 63
Rochester, MN 55906
(507) 288-8850
(www.retainingwall.com)

UltraGuard
3773 State Road
Cuyahoga Falls, OH 44223
(800) 457-4342
(www.ultraguardvinylfence.com)

Wausau Tile
PO Box 1520
Wausau, WI 54402-1520
(715) 359-3121
(www.wausautile.com)

Manufactured & Kit Homes

Appalachian Log Structures
PO Box 614
Ripley, WV 25271
(304) 372-6410
(www.applog.com)

Deck House
930 Main St.
Acton, MA 01720
(617) 259-9450
(www.deckhouse.com)

Deltec Homes
69 Bingham Road
Asheville, NC 28806
(800) 642-2508
(www.deltechomes.com)

Garland Homes
PO Box 12
Victor, MI 59875
(406) 642-3095

Insulspan
P P.O. Box 38
Blissfield, MI 49228
(517) 486-4844
(www.insulspan.com)

Lindal Cedar Homes/Sunrooms
4300 S. 104th Pl.
Seattle, WA 98178
(206) 725-0900
(www.lindal.com)

Simplex Industries
Keyser Valley Industrial Park, 1
  Simplex Dr.
Scranton, PA 18504
(717) 346-5113
(www.simplexhomes.com)

Southland Log Homes
PO Box 1668
Irmo, SC 29063-1668
(800) 641-4754
(southlandloghomes.com)

Sterling Building Systems
PO Box 8005
Wausau, WI 54402-8005
(800) 735.1812
(sterlingbldg.com)

Wausau Homes
PO Box 8005
Wausau, WI 54402-8005
(715) 359-7272
(www.wausauhomes.com)

Miscellaneous

BC Greenhouse Builders
115 Garfield St. #3434
Sumas WA 98295
(888) 391-4433
(www.bcgreenhouses.com)

Bang & Olufsen of America
1200 Business Center Dr.
Mount Prospect, IL 60056
(708) 299-9380 (800) 323-0378
(www.bang-olufsen.com)

Carolina Solar Structures
8 Loop Rd.
Arden, NC 28704
(828) 684-9900 (800) 241-9560
(www.carolinasolar.com)

Closet Maid
720 S.W. 17th St.
Ocala, FL 32674
(904) 351-6100 (800) 874-0008
(www.closetmaid.com)

First Alert Professional Security
  Systems
172 Michael Dr.
Syosset, NY 11791
(516) 921-6066
(www.firstalertpro.com)

Florian Solar Products
549 Aviation Blvd
Georgetown, SC 29440
(800)FLORIAN 800.356.7426

Four Seasons Sunrooms
5005 Veterans Memorial Hwy.
Holbrook, NY 11741
(800) FOUR SEASONS
(www.fourseasonssunrooms.com)

Hearth, Patio & Barbecue Association
1901 North Moore Street, Suite 600
Arlington, Va. 22209
(703) 522-0086
(www.hpba.org)

Heat-N-Glo
7571 215th Street West
Lakeville, MN 55044
(612) 890-8367 (800) 669-HEAT
(www.heatnglo.com)

Heatilator
1915 W. Saunders St.
Mt. Pleasant, IA 52641
(319) 385-9211 (800) 247-6798
(www.heatilator.com)

IntelliNet
Two Concourse Pkwy Suite 100
Atlanta, Ga 30328
(404) 442.8000
(www.intellinet.com)

Majestic Products Company
149 Cleveland Drive
Paris, Kentucky 40361
(219) 356-8000 (800) 525-1898
(www.majesticproducts.com)

Miracle Sealants & Abrasives
12806 Schabarum Ave., Bldg. A
Irwindale, CA 91706
(818) 814-8988 (800) 350-1901
(www.miraclesealants.com)

Rev-A-Shelf
2409 Plantside Dr.
Jeffersontown, KY 40299
(502) 499-5835 (800) 626-1126
(www.rev-a-shelf.com)

Sears Contract Sales Sears, Roebuck
  and Co.
3333 Beverly Rd., E3-363A
Hoffman Estates, IL 60179
(708) 286-2994 (800) 359-2000
(www.contractsales.sears.com)

Painting Materials

Benjamin Moore & Co.
101 Paragon Drive
Montvale NJ 07645
(201) 573-9600
(www.benjaminmoore.com)

Bostik Tile & Flooring Group
211 Boston St.
Middleton, MA 01949
(508) 777-0100 (800) 726-7845
(www.bostik-us.com)

DAP
2400 Boston Street, Suite 200
Baltimore, MD 21224
(410) 675-2100, (800) 543-3840
(www.dap.com)

Dur-A-Flex
PO Box 280166
East Hartford, CT 06128-0166
(860) 528-9838 (800) 253-3539
(www.dur-a-flex.com)

Dura Seal Div. of Minwax Co.
PO Box 438
Montvale, NJ 07645-1814
(201) 391-0253 (800) 526-0495
(www.duraseal.com)

Dutch Boy Professional Paints
101 Prospect Ave.
Cleveland, OH 44115
(216) 566-2929
(www.dutchboy.com)

Dyco Paints
5850 Ulmerton Rd.
Clearwater, FL 34620
(813) 536-6560 (800) 237-8232
(www.dycopaints.com)

GE Silicones
260 Hudson River Rd.
Waterford, NY 12188
(518) 233-3505 (800) 255-8886
(www.gesilicones.com)

Klean-Strip
PO Box 1879
Memphis, TN 38101
(901) 775-0100 (800) 238-2672
(www.kleanstrip.com)

Loctite Corporation
1001 Trout Brook Crossing
Rocky Hill, CT 06067-3910
(203) 571-5100
(www.loctite.com)

Minwax Co.
50 Chestnut Ridge Rd.
Montvale, NJ 07645
(201) 391-0253 (800) 526-0495
(www.minwax.com)

NPC Sealants
1208 S. Eighth Ave.
Maywood, IL 60153
(708) 681-1040 (800) 654-1042
(www.npcsealants.com)

Olympic Paints and Stains
1 PPG Plaza
Pittsburgh, PA 15272
(412) 434-3900 (800) 621-2024
(www.olympic.com)

Osmose Wood Preserving
1016 Everee Inn Rd.
Griffin, GA 30224
(404) 228-8434 (800) 522-9663
(www.osmosewood.com)

PPG Industries
1 PPG Pl.
Pittsburgh, PA 15272
(412) 434-3131 (800) 2-GETPPG
(www.ppg.com)

Plasti-Kote Co.
1000 Lake Rd.
Medina, OH 44256
(216) 725-4511 (800) 431-5928
(www.plastikote.com)

Pratt & Lambert
75 Tonawanda St.
Buffalo, NY 14207
800-289-7728 , (716) 873-6000
(www.prattandlambert.com)

Pro-Cote
DuPont Soy Polymers
4300 Duncan Avenue
St. Louis, MO 63110, USA
(314) 659.3818
(www.procote.com)

Red Devil
2400 Vauxhall Rd.
Union, NJ 07083-1933
(908) 688-6900 (800) 4-A-DEVIL
(www.reddevil.com)

Super-Tek Products
25-44 Borough Pl.
Woodside, NY 11377-7899
(718) 278-7900
(www.super-tek.com)

Surebond
2801 International Lane
Madison, WI 53704
(608) 237-7554
(www.surebond.com)

The Glidden Co.
925 Euclid Ave.
Cleveland, OH 44115
(216) 344-8491 (800) 984-5444
(www.glidden.com)

The Sherwin-Williams Co.
101 Prospect Ave. N.W.
Cleveland, OH 44115
(216) 566-2000 (800) 336-1110
(www.sherwin-williams.com)

Tufco
PO Box 23500, 3161 Ridge Rd.
Green Bay, WI 54305
(414) 336-0054 (800) 558-8145
(www.tufco.com)

Wagner Spray Tech Corp.
1770 Fernbrook Ln.
Plymouth, MN 55447-4663
(612) 553-7000 (800) 328-8251
  (www.wagnerspraytech.com)

Wood-Kote Products
8000 N.W. 14th Pl.
Portland, OR 97211
(503) 285-8371
(www.woodkote.com)

Roofing Materials
Air Vent
3000 W. Commerce
Dallas, TX 75212
(800) 247-8368
(www.airvent.com)

Alcoa Building Products
PO Box 716
Sidney, OH 45365
(513) 492-1111 (800) 962-6973
(www.alcoa.com)

Atlas Roofing Corp.
1775 The Exchange, Suite 160
Atlanta, GA 30339
(404) 933-4463
(www.atlasroofing.com)

Boral Concrete Products
4400 MacArthur Blvd., Suite 500
Newport Beach, CA 92660
(714) 955-4976
(ww.boral.com.au)

Cedar Plus
PO Box 515
Sumas, WA 98295
(800) 963-3388
(www.clarkegroup.com)

CertainTeed Corp. Roofing Products
  Group
PO Box 860
Valley Forge, PA 19482-0860
(800) 233-8990
(www.certainteed.com)

Cobra Ventilation Products GAF
  Materials Corp.
1361 Alps Rd.
Wayne, NJ 07470
(201) 628-3874
(ww.gaf.com)

GAF Materials Corp.
1361 Alps Rd.
Wayne, NJ 07470
(201) 628-3000
(www.gaf.com)

GS Roofing Co.
5525 MacArthur Blvd., Suite 900
Irving, TX 75038
(214) 580-5600
(www.gsroof.com)

IKO Mfg.
120 Hay Rd.
Wilmington, DE 19809
(302) 764-3100
(www.iko.com)

MFM Building Products Corp.
PO Box 340
Coshocton, OH 43812
(614) 622-2645
(www.mfmbp.com)

Revere Copper Products
PO Box 300
Rome, NY 13440
(315) 338-2022
(www.reverecopper.com)

SEMCO Southeastern Metals
11801 Industry Dr.
Jacksonville, FL 32218
(904) 757-4200
(www.semetals.com)

Tamko Roofing Products
220 W. Fourth St.
Joplin, MO 64801
(417) 624-6644 (800) 641-4691
(www.tamko.com)

Universal Marble & Granite
1919 Halethorpe Farms Rd.
Baltimore, MD 21227
(410) 247-2442
(www.umgrocks.com)

Software
Autodesk
111 McInnis Pkwy.
San Rafael, CA 94903
(415) 507-5000
(usa.autodesk.com)

BUILDSOFT
PO Box 13893
Research Triangle Park, NC 27560
(919) 941-6269
(www.buildsoft.com)

Cadkey
4 Griffin Rd. N.
Windsor, CT 06095-1511
(203) 298-8888
(www.cadkey.com)

Construction Data Control (CDCI)
4000 De Kalb Technology Pkwy,
  Suite 220
Atlanta, GA 30340
800-285-3929
(www.cdci.com)

Industry Specific Software
1200 Woodruff Rd., Suite B-20
Greenville, SC 29607
(864) 297-7086
(www.iss-software.com)

NEBS, Inc.
20 Industrial Park Dr.
Nashua, NH 03062
(603) 880-5100
(www.nebs.com)

Peachtree Software
1505 Pavilion Pl.
Norcross, GA 30093
(404) 564-5700
(www.peachtree.com)

RS Means
100 Construction Plaza, PO Box 800
Kingston, MA 02364
(617) 585-7880
(www.rsmeans.com)

Sage Master Builder
15195 NW Greenbrier Parkway
Beaverton, Oregon 97006
(503) 626-6775
(www.sagecre.com)

Win Estimator
19450 68th Ave S
Kent, WA 98032
(206) 395-3631
(www.winest.com)

Windows & Doors
Acadia Windows & Doors
9611 Pulaski Park Dr.
Baltimore, MD 21220-1435
(410) 780-9600 (800) 638-6084
(www.acadiawindows.com)

Andersen Windows
100 Fourth Ave. N.
Bayport, MN 55003
(612) 439-5150 (800) 426-4261
(www.andersenwindows.com)

Beveled Glass Works
23852 Pacific Coast Hwy, Ste 351
Malibu, CA
(800) 421-0518
(www.beveledglassworks.com)

CertainTeed Windows
PO Box 860
Valley Forge, PA 19482
(800) 233-8990
(www.certainteed.com)

Clopay Garage Doors
312 Walnut St., Suite 1600
Cincinnati, OH 45202
(513) 381-4800 (800) 225-6729
(www.clopaydoor.com)

Glass Block Designs
381 11th St.
San Francisco, CA 94103
(415) 626-5770
(www.glassblockdesigns.com)

Glass Blocks Unlimited
126 E. 16th St.
Costa Mesa, CA 92627-3707
(714) 548-8531
(www.glassblockdesigns.com)

Holmes Garage Door Co.
2601 W. Valley Hwy., PO Box 1976
Auburn, WA 98071-1976
(206) 931-8900
(www.holmesgaragedoor.com)

Karona
4100 Karona Court
Caledonia, MI 49316
(616) 554.3551
(www.karonadoor.com)

Lincoln Windows
1400 West Taylor Street
P.O. Box 375
Merrill, WI 54452
(715) 536-2461
(www.lincolnwindows.com)

Marvin Windows and Doors
PO Box 100
Warroad, MN 56763
(218) 386-1430 (800) 346-5128
(www.marvin.com)

Overhead Door Corp.
6750 LBJ Fwy., Suite 1200
Dallas, TX 75240
(214) 233-6611 (800) 929-DOOR
(www.overheaddoor.com)

Peachtree Doors
PO Box 5700
Norcross, GA 30091
(770) 497-2000 (800) 477-6544
(www.peachtreedoor.com)

Pella Corporation
102 Main St.
Pella, IA 50219
(515) 628-6992 (800) 84-PELLA
(web.pella.com)

Semco Windows & Doors Semling
  Menke Co.
PO Box 378
Merrill, WI 54452
(715) 536-9411 (800) 333-2206
(www.semcowindows.com)

Stanley Door Systems (A Sub. of
  Stanley Works)
1225 E. Maple Rd.
Troy, MI 48083-5600
(810) 528-1400 (800) 521-2752
(www.stanleyworks.com)

The Genie Co.
22790 Lake Park Blvd.
Alliance, OH 44601-3498
(216) 821-5360 (800) OK-GENIE
(www.geniecompany.com)

Therma-Tru Corp.
1750 Indian Wood Circle
Maumee, Ohio, 43537
(800) 843-7628
(www.thermatru.com)

Weather Shield Mfg.
PO Box 309, 1 Weather Shield Plaza
Medford, WI 54451
(715) 748-2100 (800) 477-6808
(www.weathershield.com)

Wood & Structural Materials

Boise Cascade
1111 W. Jefferson St.
Boise, ID 83702
(208) 384-6321 (800) 458-4631
(www.bc.com)

Georgia-Pacific Corp.
PO Box 1763
Norcross, GA 30091
(800) BUILD-GP
(www.gp.com)

Homasote Co.
PO Box 7240
West Trenton, NJ 08628-0240
(609) 883-3300 (800) 257-9491
(www.homasote.com)

Louisiana-Pacific Corp.
111 S.W. Fifth Ave.
Portland, OR 97204
(888) 820-0325
(www.lpcorp.com)

Norbord Industries
1 Toronto St., Suite 500
Toronto, ON M5C 2W4 Canada
(416) 365-0710
(www.norbord.com)

Weyerhaeuser Engineered Strand
  Products Division
2000 Frontis Plaza Blvd., Suite 101
Winston-Salem, NC 27103
(910) 760-7151 (800) 648-2566
(www.weyerhaeuser.com)

MASTERPLAN

STEP	DESCRIPTION	DATE STARTED	DATE COMPLETED	DATE INSPECTED	COMMENTS
PRE-CONSTRUCTION					
PC1	BEGIN construction project scrapbook				
PC2	DETERMINE size of home				
PC3	EVALUATE alternate house plans				
PC4	SELECT house plan				
PC5	MAKE design changes				
PC6	PERFORM material and labor takeoff				
PC7	CUT costs as required				
PC8	UPDATE material and labor takeoff				
PC9	ARRANGE temporary housing				
PC10	FORM separate corporation				
PC11	APPLY for business license				
PC12	ORDER business cards				
PC13	OBTAIN free and clear title to land				
PC14	APPLY for construction loan				
PC15	ESTABLISH project checking account				
PC16	CLOSE construction loan				
PC17	OPEN builder accounts				
PC18	OBTAIN purchase order forms				
PC19	OBTAIN building permit				
PC20	OBTAIN builder's risk policy				
PC21	OBTAIN workman's compensation policy				
PC22	ACQUIRE minimum tools				
PC23	VISIT construction sites				
PC24	DETERMINE need for on-site storage				
PC25	NOTIFY all subs and labor				
PC26	OBTAIN compliance bond				
PC27	DETERMINE minimum occupancy requirements				
PC28	CONDUCT spot survey of lot				
PC29	MARK all lot boundaries and corners				
PC30	POST construction signs				
PC31	POST building permit on site				
PC32	ARRANGE for portable toilet				
PC33	ARRANGE for site telephone				
EXCAVATION					
EX1	CONDUCT excavation bidding process				
EX2	LOCATE underground utilities				
EX3	DETERMINE exact dwelling location				
EX4	MARK area to be cleared				
EX5	MARK where curb is to be cut				
EX6	TAKE last picture of site area				
EX7	CLEAR site area				
EX8	EXCAVATE trash pit				
EX9	INSPECT cleared site				
EX10	STAKE out foundation area				

MASTERPLAN

STEP	DESCRIPTION	DATE STARTED	DATE COMPLETED	DATE INSPECTED	COMMENTS
EX11	EXCAVATE foundation/basement area				
EX12	INSPECT site and excavated area				

BATTER BOARDS

STEP	DESCRIPTION	DATE STARTED	DATE COMPLETED	DATE INSPECTED	COMMENTS
EX13	LAY OUT corner stakes for outer 4 corners of house				
EX14	DRIVE three batter board stacks at each corner				
EX15	NAIL 1×6 horizontal boards to stakes				
EX16	PULL batter board strings				
EX17	RE-CHECK distance of each string				
EX18	LAYOUT remaining batter boards and re-check				

DRAINAGE

STEP	DESCRIPTION	DATE STARTED	DATE COMPLETED	DATE INSPECTED	COMMENTS
EX19	APPLY crusher run to driveway area				
EX20	SET UP silt fence				
EX21	PAY excavator (initial)				

BACKFILL

STEP	DESCRIPTION	DATE STARTED	DATE COMPLETED	DATE INSPECTED	COMMENTS
EX22	PLACE supports on foundation wall				
EX23	INSPECT backfill area				
EX24	BACKFILL foundation				
EX25	CUT driveway area				
EX26	COVER septic tank and septic tank line				
EX27	COVER trash pit				
EX28	PERFORM final grade				
EX29	CONDUCT final inspection				
EX30	PAY excavator (final)				

PEST CONTROL

STEP	DESCRIPTION	DATE STARTED	DATE COMPLETED	DATE INSPECTED	COMMENTS
PS1	CONDUCT standard bidding process.				
PS2	NOTIFY pest control sub for pre-treat				
PS3	APPLY pesticide to foundation				
PS4	APPLY pesticide to perimeter				
PS5	OBTAIN signed pest control warranty				
PS6	PAY pest control sub				

CONCRETE

STEP	DESCRIPTION	DATE STARTED	DATE COMPLETED	DATE INSPECTED	COMMENTS
CN1	CONDUCT bidding process (concrete supplier)				
CN2	CONDUCT bidding process (formwork/finishing)				
CN3	CONDUCT bidding process (block masonry)				

FOOTINGS

STEP	DESCRIPTION	DATE STARTED	DATE COMPLETED	DATE INSPECTED	COMMENTS
CN4	INSPECT batter boards				
CN5	DIG footings				
CN6	SET footing forms				
CN7	INSPECT footing forms				
CN8	SCHEDULE and complete footing inspection				
CN9	CALL concrete supplier for concrete				

MASTERPLAN

STEP	DESCRIPTION	DATE STARTED	DATE COMPLETED	DATE INSPECTED	COMMENTS
CN10	POUR footings				
CN11	INSTALL key forms				
CN12	FINISH footings as needed				
CN13	REMOVE footing forms				
CN14	INSPECT footings				
CN15	PAY concrete supplier for footings				

POURED WALL

STEP	DESCRIPTION	DATE STARTED	DATE COMPLETED	DATE INSPECTED	COMMENTS
CN16	SET poured wall concrete forms				
CN17	SCHEDULE concrete supplier				
CN18	INSPECT poured wall concrete forms				
CN19	POUR foundation walls				
CN20	FINISH poured foundation walls				
CN21	REMOVE poured wall concrete forms				
CN22	INSPECT poured walls				
CN23	BREAK OFF tie ends from walls				
CN24	PAY concrete supplier for walls				

BLOCK WALL

STEP	DESCRIPTION	DATE STARTED	DATE COMPLETED	DATE INSPECTED	COMMENTS
CN25	ORDER blocks/schedule masons				
CN26	LAY concrete blocks				
CN27	FINISH concrete blocks				
CN28	INSPECT concrete block work				
CN29	PAY concrete block masons				
CN30	CLEAN up after block masons				

CONCRETE SLAB

STEP	DESCRIPTION	DATE STARTED	DATE COMPLETED	DATE INSPECTED	COMMENTS
CN31	SET slab forms				
CN32	PACK slab sub-soil				
CN33	INSTALL stub plumbing				
CN34	SCHEDULE HVAC sub for slab work				
CN35	INSPECT plumbing and gas lines				
CN36	POUR and spread crushed stone				
CN37	INSTALL vapor barrier				
CN38	LAY reinforcing wire as required				
CN39	INSTALL rigid insulation				
CN40	CALL concrete supplier for slab				
CN41	INSPECT slab forms				
CN42	POUR slab				
CN43	FINISH slab				
CN44	INSPECT slab				
CN45	PAY concrete supplier for slab				

DRIVEWAY

STEP	DESCRIPTION	DATE STARTED	DATE COMPLETED	DATE INSPECTED	COMMENTS
CN46	COMPACT drive and walkway soil				
CN47	SET drive and walkway forms				
CN48	CALL concrete supplier				

MASTERPLAN

STEP	DESCRIPTION	DATE STARTED	DATE COMPLETED	DATE INSPECTED	COMMENTS
CN49	INSPECT drive and walkway forms				
CN50	POUR drive and walkway				
CN51	FINISH drive and walkway areas				
CN52	INSPECT drive and walkway areas				
CN53	ROPE OFF drive and walkway areas				
CN54	PAY concrete supplier				
CN55	PAY concrete finishers				
CN56	PAY concrete finisher's retainage				
	WATERPROOFING				
WP1	CONDUCT standard bidding process				
WP2	SEAL footing and wall intersection				
WP3	APPLY first coat of portland cement				
WP4	APPLY second coat of portland cement				
WP5	APPLY waterproofing compound				
WP6	APPLY 6 mil poly vapor barrier				
WP7	INSPECT waterproofing (tar/poly)				
WP8	INSTALL gravel bed for drain tile				
WP9	INSTALL drain tile				
WP10	INSPECT drain tile				
WP11	APPLY top layer of gravel on drain tile				
WP12	PAY waterproofing sub				
WP13	PAY waterproofing sub retainage				
	FRAMING				
FR1	CONDUCT standard bidding process (material)				
FR2	CONDUCT standard bidding process (labor)				
FR3	DISCUSS aspects of framing with crew				
FR4	ORDER special materials				
FR5	ORDER first load of framing lumber				
FR6	INSTALL sill felt				
FR7	ATTACH sill plate with lag bolts				
FR8	INSTALL support columns				
FR9	SUPERVISE framing process				
FR10	FRAME first floor joists and subfloor				
FR11	FRAME basement stairs				
FR12	POSITION all first floor large items				
FR13	FRAME exterior walls (first floor)				
FR14	PLUMB and line first floor				
FR15	FRAME second floor joists and subfloor				
FR16	POSITION all large second floor items				
FR17	FRAME exterior walls (second floor)				
FR18	PLUMB and line second floor				
FR19	INSTALL second floor ceiling joists				
FR20	FRAME roof				
FR21	INSTALL roof decking				
FR22	PAY framing first installment				

MASTERPLAN

STEP	DESCRIPTION	DATE STARTED	DATE COMPLETED	DATE INSPECTED	COMMENTS
FR23	INSTALL lapped tar paper				
FR24	FRAME chimney chases				
FR25	INSTALL pre-fab fireplaces				
FR26	FRAME dormers and skylights				
FR27	FRAME tray ceilings, skylights and bays				
FR28	INSTALL sheathing on exterior walls				
FR29	INSPECT sheathing				
FR30	REMOVE temporary bracing				
FR31	INSTALL exterior windows and doors				
FR32	APPLY deadwood				
FR33	INSTALL roof ventilators				
FR34	FRAME decks				
FR35	INSPECT framing				
FR36	SCHEDULE loan draw inspection				
FR37	CORRECT any framing problems				
FR38	PAY framing labor (second payment)				
FR39	PAY framing retainage				
ROOFING					
RF1	SELECT shingle style, material and color				
RF2	PERFORM standard bidding process.				
RF3	ORDER shingles and roofing felt				
RF4	INSTALL metal drip edge				
RF5	INSTALL roofing felt				
RF6	INSTALL drip edge on rake				
RF7	INSTALL roofing shingles				
RF8	INSPECT roofing				
RF9	PAY roofing subcontractor				
RF10	PAY roofing subcontractor retainage				
GUTTERS					
GU1	SELECT gutter type and color				
GU2	PERFORM standard bidding process				
GU3	INSTALL underground drain pipe				
GU4	INSTALL gutters and downspouts				
GU5	INSTALL splashblocks				
GU6	INSPECT gutters				
GU7	INSTALL copper awnings				
GU8	PAY gutter subcontractor				
GU9	PAY gutter sub retainage				
PLUMBING					
PL1	DETERMINE type and quantity of fixtures				
PL2	CONDUCT standard bidding process				
PL3	WALK THROUGH site with plumber				
PL4	ORDER special plumbing fixtures				
PL5	APPLY for water connection and sewer tap				

MASTERPLAN

STEP	DESCRIPTION	DATE STARTED	DATE COMPLETED	DATE INSPECTED	COMMENTS
PL6	INSTALL stub plumbing for slabs				
PL7	MOVE large plumbing fixtures into place				
PL8	MARK location of all fixtures				
PL9	INSTALL rough-in plumbing				
PL10	INSTALL water meter and spigot				
PL11	INSTALL sewer tap				
PL12	SCHEDULE plumbing inspection				
PL13	INSTALL septic tank and line				
PL14	CONDUCT rough-in plumbing inspection				
PL15	CORRECT rough-in plumbing problems				
PL16	PAY plumbing sub for rough-in				
PLUMBING FINISH					
PL17	INSTALL finish plumbing				
PL18	TAP into water supply				
PL19	CONDUCT finish plumbing inspection				
PL20	CORRECT finish plumbing problems				
PL21	PAY plumbing sub for finish work				
PL22	PAY plumbing retainage				
HVAC					
HV1	CONDUCT energy audit				
HV2	SHOP for heating and cooling system				
HV3	CONDUCT standard bidding process				
HV4	FINALIZE HVAC design				
HV5	ROUGH-IN heating and air system				
HV6	INSPECT heating and air rough-in				
HV7	CORRECT rough-in problems				
HV8	PAY HVAC sub for rough-in				
HV9	INSTALL finish heating and air				
HV10	INSPECT final HVAC installation				
HV11	CORRECT all HVAC problems				
HV12	CALL gas company to install lines				
HV13	DRAW gas line on plat diagram				
HV14	PAY HVAC sub for finish work				
HV15	PAY HVAC retainage				
ELECTRICAL					
EL1	DETERMINE electrical requirements				
EL2	SELECT electrical fixtures and appliances				
EL3	DETERMINE phone installation requirements				
EL4	CONDUCT standard bidding process				
EL5	SCHEDULE phone wiring				
EL6	APPLY for temporary electrical hookup				
EL7	INSTALL temporary electric pole				
EL8	PERFORM rough-in electrical				
EL9	INSTALL phone wiring				

MASTERPLAN

STEP	DESCRIPTION	DATE STARTED	DATE COMPLETED	DATE INSPECTED	COMMENTS
EL10	SCHEDULE rough-in electrical inspection				
EL11	INSPECT rough-in electrical				
EL12	CORRECT rough-in electrical problems				
EL13	PAY electrical sub for rough-in work				
EL14	INSTALL garage door for opener installation				
EL15	INSTALL garage door opener				
EL16	PERFORM finish electrical work				
EL17	CALL phone co. to connect service				
EL18	INSPECT finish electrical				
EL19	CORRECT finish electrical problems				
EL20	CALL electrical utility to connect service				
EL21	PAY electrical sub for finish work				
EL22	PAY electrical sub retainage				
MASONRY					
MA1	DETERMINE brick pattern and color				
MA2	DETERMINE stucco pattern and color				
MA3	PERFORM standard bidding process (masonry)				
MA4	PERFORM standard bidding process (stucco)				
MA5	APPLY flashing				
MA6	ORDER brick, stone, & angle irons				
MA7	LAY bricks				
MA8	FINISH bricks and mortar				
MA9	INSPECT brickwork				
MA10	CORRECT brickwork deficiencies				
MA11	PAY masons				
MA12	CLEAN UP excess bricks				
MA13	INSTALL base for decorative stucco				
MA14	PREPARE stucco test area				
MA15	APPLY stucco lath				
MA16	APPLY first coat of stucco				
MA17	APPLY second coat of stucco				
MA18	INSPECT stucco work				
MA19	CORRECT stucco deficiencies				
MA20	PAY stucco sub				
MA21	PAY mason sub retainage				
MA22	PAY stucco sub retainage				
SIDING/CORNICE					
SC1	SELECT siding material and color				
SC2	CONDUCT standard bidding process				
SC3	ORDER window & doors				
SC4	INSTALL windows & doors				
SC5	INSTALL flashing				
SC6	INSTALL siding				
SC7	INSTALL siding trim around corners				
SC8	CAULK siding joints				

MASTERPLAN

STEP	DESCRIPTION	DATE STARTED	DATE COMPLETED	DATE INSPECTED	COMMENTS
SC9	INSPECT siding				
SC10	PAY sub for siding work				
SC11	INSTALL cornice				
SC12	INSPECT cornice work				
SC13	CORRECT any cornice problems				
SC14	PAY cornice sub				
SC15	CORRECT siding problems				
SC16	ARRANGE for painter to caulk & paint				
SC17	PAY siding retainage				
SC18	PAY cornice retainage				
INSULATION					
IN1	DETERMINE insulation requirements				
IN2	CONDUCT standard bidding process.				
IN3	INSTALL exterior wall insulation				
IN4	INSTALL soundproofing				
IN5	INSTALL floor insulation				
IN6	INSTALL attic insulation				
IN7	INSPECT insulation				
IN8	CORRECT insulation work as needed				
IN9	PAY insulation sub				
IN10	PAY insulation sub retainage				
DRYWALL					
DR1	PERFORM standard bidding process.				
DR2	ORDER and receive drywall				
DR3	HANG drywall on all walls				
DR4	FINISH drywall				
DR5	INSPECT drywall				
DR6	TOUCH UP and repair drywall				
DR7	PAY drywall sub				
DR8	PAY drywall sub retainage				
TRIM					
TR1	DETERMINE trim requirements				
TR2	SELECT moulding and millwork				
TR3	INSTALL windows after framing				
TR4	CONDUCT standard bidding process				
TR5	INSTALL interior doors				
TR6	INSTALL window casing and aprons				
TR7	INSTALL trim around cased openings				
TR8	INSTALL staircase openings				
TR9	INSTALL crown moulding				
TR10	INSTALL base and base cap moulding				
TR11	INSTALL chair rail moulding				
TR12	INSTALL picture moulding				
TR13	INSTALL, sand and stain paneling				

MASTERPLAN

STEP	DESCRIPTION	DATE STARTED	DATE COMPLETED	DATE INSPECTED	COMMENTS
TR14	CLEAN all sliding door tracks				
TR15	INSTALL thresholds and weather stripping				
TR16	INSTALL shoe moulding				
TR17	INSTALL door and window hardware				
TR18	INSPECT trimwork				
TR19	CORRECT trim and stain work				
TR20	PAY trim sub				
PAINTING AND WALLCOVERING					
PT1	SELECT paint schemes.				
PT2	PERFORM standard bidding process.				
PT3	PURCHASE all painting materials				
EXTERIOR					
PT4	PRIME and caulk exterior surfaces				
PT5	PAINT exterior siding and trim				
PT6	PAINT cornice work				
PT7	PAINT gutters				
INTERIOR					
PT8	PREPARE painting surfaces				
PT9	PRIME walls and trim				
PT10	PAINT or stipple ceilings				
PT11	PAINT walls				
PT12	PAINT or stain trim				
PT13	REMOVE paint from windows				
PT14	INSPECT paint job				
PT15	TOUCH UP paint job				
PT16	CLEAN UP				
PT17	PAY painter				
PT18	PAY retainage				
CABINETRY					
CB1	CONFIRM kitchen design				
CB2	SELECT cabinetry/countertop				
CB3	COMPLETE cabinet layout diagram				
CB4	PERFORM standard bidding process				
CB5	PURCHASE cabinetry and countertops				
CB6	STAIN or paint cabinetry				
CB7	INSTALL bathroom vanities				
CB8	INSTALL kitchen wall cabinets				
CB9	INSTALL kitchen base cabinets				
CB10	INSTALL kitchen and bath countertops				
CB11	INSTALL cabinet hardware				
CB12	INSTALL utility room cabinetry				
CB13	CAULK cabinet joints as needed				
CB14	INSPECT cabinetry and countertops				

MASTERPLAN

STEP	DESCRIPTION	DATE STARTED	DATE COMPLETED	DATE INSPECTED	COMMENTS
CB15	TOUCH UP and repair as needed				
CB16	PAY cabinet sub				

GLAZING

STEP	DESCRIPTION	DATE STARTED	DATE COMPLETED	DATE INSPECTED	COMMENTS
GL1	DETERMINE mirror and glass requirements				
GL2	SELECT mirrors and glass				
GL3	CONDUCT standard bidding process				
GL4	INSTALL fixed pane picture windows				
GL5	INSTALL shower doors				
GL6	INSTALL mirrors				
GL7	INSPECT all glazing installation				
GL8	CORRECT all glazing problems				
GL9	PAY glazier				

FLOORING AND TILE

STEP	DESCRIPTION	DATE STARTED	DATE COMPLETED	DATE INSPECTED	COMMENTS
FL1	SELECT carpet color, style and coverage				
FL2	SELECT hardwood floor type				
FL3	SELECT vinyl floor covering				
FL4	SELECT tile type and coverage				
FL5	CONDUCT standard bidding process				

TILE

STEP	DESCRIPTION	DATE STARTED	DATE COMPLETED	DATE INSPECTED	COMMENTS
FL6	ORDER tile and grout				
FL7	PREPARE area to be tiled				
FL8	INSTALL tile base in shower stalls				
FL9	APPLY tile adhesive				
FL10	INSTALL tile and marble thresholds				
FL11	APPLY grout over tile				
FL12	INSPECT tile				
FL13	CORRECT tile problems				
FL14	SEAL grout				
FL15	PAY tile sub				
FL16	PAY tile sub retainage				

HARDWOOD

STEP	DESCRIPTION	DATE STARTED	DATE COMPLETED	DATE INSPECTED	COMMENTS
FL17	ORDER hardwood flooring				
FL18	PREPARE sub floor for hardwood flooring				
FL19	INSTALL hardwood flooring				
FL20	SAND hardwood flooring				
FL21	INSPECT hardwood flooring				
FL22	CORRECT hardwood flooring problems				
FL23	STAIN hardwood flooring				
FL24	SEAL hardwood flooring				
FL25	INSPECT hardwood flooring (final)				
FL26	PAY hardwood flooring sub				
FL27	PAY hardwood flooring sub retainage				

MASTERPLAN

STEP	DESCRIPTION	DATE STARTED	DATE COMPLETED	DATE INSPECTED	COMMENTS
VINYL FLOOR COVERING					
FL28	PREPARE sub floor for vinyl floor covering				
FL29	INSTALL vinyl floor covering				
FL30	INSPECT vinyl floor covering				
FL31	CORRECT vinyl floor covering problems				
FL32	PAY vinyl floor covering sub				
FL33	PAY vinyl floor covering sub retainage				
CARPET					
FL34	PREPARE subfloor for carpeting				
FL35	INSTALL carpet stretcher strips				
FL36	INSTALL carpet padding				
FL37	INSTALL carpet				
FL38	INSPECT carpet				
FL39	CORRECT carpet problems				
FL40	PAY carpet sub				
FL41	PAY carpet retainage				
LANDSCAPING					
PRE CONSTRUCTION					
LD1	EVALUATE lot				
LD2	DETERMINE best house location				
LD3	DEVELOP site plan				
LD4	CONTACT landscape architect				
LD5	FINALIZE site plan				
LD6	SUBMIT site plan				
LD7	DELIVER fill dirt				
LD8	PAY landscape architect				
AFTER CONSTRUCTION					
LD9	CONDUCT soil test				
LD10	TILL topsoil				
LD11	APPLY soil treatments				
LD12	INSTALL underground sprinkling system				
LD13	PLANT flower bulbs				
LD14	APPLY seed or sod				
LD15	SOAK lawn with water				
LD16	INSTALL bushes and trees				
LD17	PREPARE landscaped islands				
LD18	INSTALL mailbox				
LD19	INSPECT landscaping job				
LD20	CORRECT any landscaping problems				
LD21	PAY landscaping sub				
LD22	PAY landscaping sub retainage				

MASTERPLAN

STEP	DESCRIPTION	DATE STARTED	DATE COMPLETED	DATE INSPECTED	COMMENTS
DECKING					
DE1	EXAMINE building site				
DE2	DRAW complete plan for deck				
DE3	PURCHASE materials for deck				
DE4	FINISH all grading and backfilling				
DE5	LAYOUT deck design				
DE6	DIG footings and pour concrete piers				
DE7	PULL leveling line from corner to corner				
DE8	INSTALL posts on pier anchors				
DE9	ATTACH deck beams to the posts				
DE11	MOUNT joist header to the house				
DE13	INSTALL joists between frame				
DE14	MOUNT cross bracing between posts				
DE15	APPLY deck boards to joists				
DE16	SAW the deck boards flush with header joist				
DE17	ATTACH stairway stringers to header joist				
DE18	INSTALL treads between the stringers				
DE19	INSTALL post rails around the perimeter				
DE20	TREAT deck with wood preservative				

Construction Contract

(Date) _____

This construction contract ("Contract") is made by and between
("Owner(s)") _____
(at address) _____
and ("Contractor") _____
(at address) _____
and is subject to the terms and conditions stated herein. The "Effective Date" of this contract shall be the date it is last signed by Owner(s) or Contractor.

Contractor agrees to construct and deliver to the Owner(s) a residential dwelling ("Residence") in accordance with the plans, specifications, and change orders attached ("Construction Documents"), on the tract or parcel of land ("the Property") described as follows:

Lot _____ of the _____ District
of _____ Country, _____(state), and also known as
_____

Contract Sum: In consideration of the mutual promises contained herein, the Owner(s) shall pay the Contractor in current funds for full performance of the Contract ("Contract Sum"), excluding the cost of the property and additional arranged work, in the amount of:
_____ DOLLARS ($ _____).

Payments: Before construction begins, Owner(s) will provide Contractor with necessary proof of sufficient funds to pay any and all construction costs specified in this contract and in the attached Construction Documents. Owner(s) agree to pay promptly all costs and fees in connection with the construction of the Residence as set forth in the payment schedule below:

Progress Payment Phases
1) Upon the completion of the foundation: $ _____
2) Upon the completion of framing, roof, windows, & doors: $ _____
3) Upon the completion of drywall installation and finishing: $ _____
4) Upon the completion of painting & trim: $ _____
5) Upon issuance of certificate of occupancy: $ _____

Contractor agrees to provide Owner(s) with an Application for Payment no later than ten (10) days in advance of each Progress Payment. Owner(s) agree to pay Contractor within ten (10) days of notification. Contractor agrees to pay all obligations for work done through the date covered by each Progress Payment within fifteen (15) days following receipt of each Progress Payment. Contractor may withhold Retainage from the payment due each subcontractor, not to exceed ten (10) percent.

Taxes and Fees: Contractor agrees to pay sales, use, and other taxes charged against labor or materials incorporated into the Residence. Contractor also agrees to pay for any governmental building permits, fees, licenses, or inspections necessary for the completion of the Construction Contract.

Final payment shall consist of the entire unpaid balance of the Contract Sum, minus Retainage, as set forth herein. Owner(s) agree to make the final payment (minus any Retainage) upon Substantial Completion of the terms of the Contract.

Retainage: Upon Substantial Completion of the Contract, Owner(s) may withhold from the Final Payment the amount of $ _____ (dollars), sufficient to cover the completion of Punch List items.

Punch List: Upon Substantial Completion of the Contract, Owner(s) will immediately inspect the Residence and supply the Contractor with a written "Punch List" of all deficiencies. Contractor will promptly correct all deficiencies.

After final approval, Owner(s) will promptly pay any monies retained. If the Contractor fails to complete the Punch List in a reasonable time, the Owner has the option to pay for the completion of the Punch List from the monies retained. Any balance remaining shall be promptly paid to the Contractor.

Construction Documents: The Owner(s) and/or Owner's architect or designer represents that the Construction Contract, plans, specifications, and change orders attached to this contract are adequate for project completion and contain no errors or omissions. Attached specifications have been developed and agreed upon by Owner(s) and Contractor, and they shall control in any area of conflict with the plans. The Construction Contract and attached Construction Documents constitute the sole and entire agreement between parties and no modification shall be binding unless attached and signed by all parties to the agreement. No representation, promise, or inducement to the agreement not included herein shall be binding upon any party.

Change Orders: The Owner(s), without invalidating the Contract, shall have the right to make changes to the specifications contained in the Construction Documents. The cost of any alterations, additions, omissions or deviations shall be added to the agreed Contract Sum. No such changes will be made unless stipulated in writing by the execution of a "Change Order" and signed by all parties. The Change Order will be an addendum to this Construction Contract, specifying any additional amount due along with provision for payment.

Allowances: The Contract Sum includes allowances from the Contractor for items listed below. The amounts are listed at contractor's cost. If the contractor's cost exceeds any allowance listed, the Contract Sum shall be increased by the amount above the stated allowance. If Owner(s) do not spend the allowance limit, the Contractor will lower the Contract Sum by the amount below the stated allowance. The allowances are as follows:
   1)   Lighting: $_____
   2)   Carpet: $_____
   3)   Tile and Flooring: $_____
   4)   Wallpaper and Paint: $_____

The Contractor shall notify the Owner(s) at least ten (10) days prior to the deadline for the selection of the above stated materials, colors and other selections. The Owner(s) shall indicate their selection or preference prior to the deadline date. The Owner(s) will make these selections in a timely manner and assist the Contractor in obtaining the necessary materials. Delays resulting from the Owner(s) failure to make these selections will be reasonable justification for a delay in the Completion Term.

Construction Term: Contractor shall begin construction of the Residence no later than the (date) _____ . The contractor shall have the Residence ready for occupancy no later than _____ calendar days ("Completion Deadline") from the time that construction is commenced. Owner(s) will not occupy the residence without the written consent of the Contractor, and until all payments due to the Contractor have been made.

Completion: The Construction Term shall be deemed fully completed when approved by the mortgagee for loan purposes and a certificate of occupancy is obtained. Contractor agrees to give full possession of the Residence to the Owner(s) immediately after the full payment of any remaining cash balance due. At that time, the contractor agrees to furnish the purchaser with all product warranties, a termite guarantee, and a notarized affidavit certifying that all materials, labor, and fees for the construction of the Residence have been paid in full and indemnifying the Owner and the Owner's title insurance company from any liens, costs, damages, or expenses.

Incentive Payment: If the Construction Term is completed prior to the Completion Deadline, Owner(s) shall pay to Contractor, in addition to the Contract Sum, an amount equal to $ _____ (dollars) per day for every day saved from the Completion Term.

Delays: Every effort will be exerted on the part of the contractor to complete the construction by the Completion Deadline. However, Contractor assumes no responsibility for delays occasioned by causes beyond his control. In the event of a delay of construction caused by weather, Owner(s) selections, etc., the period of delay is to be added to the Construction Term allotted to Contractor. If the work is not completed by the Completion Deadline, adjusted for any legitimate delays as set forth herein and including any delays from authorized Change Orders, then the Contract Sum stated above shall be reduced by $_____$ (dollars), per day as liquidated damages until the actual Completion Date.

Subcontractors: A subcontractor is any person, firm, or corporation hired by the Contractor to perform Construction Work on the Residence. As soon as practical after the execution of the Contract, the Contractor shall provide Owner(s) with a list in writing of the subcontractors to be used in the completion of the Construction Contract. Both parties shall have the right to object to the hiring of any particular subcontractor. Contracts between the Contractor and subcontractor shall require the subcontractor to be bound by the terms of the Construction Documents to the extent of the work performed by said subcontractor.

Termination of Contract: Owner(s) may terminate the services of the Contractor upon ten (10) days written notice. Any costs due to the Contractor must be paid by Owner(s) on or prior to the effective date of the termination. Owner(s) may issue a written order to stop construction if the Contractor fails to complete or correct work in accordance with the provisions of this Construction Contract. If the Contractor fails to perform the work, the Owner(s) may remedy the failure and deduct the reasonable cost from the remaining Contract Sum payable to the Contractor.

Contractor may terminate this Agreement upon Owner's failure to pay any Progress Payments within ten (10) days after the Progress Payment is due. Within ten (10) days of any termination of this Construction Contract by Owner(s) or Contractor, Owner(s) shall pay all construction costs that are due and payable as of termination, and Owner(s) shall pay within 10 days of receipt of an invoice any other construction costs that were incurred prior to termination but were not billed as of the termination.

Casualty Loss: Contractor shall purchase and maintain throughout the Construction Term, an "All Risk" insurance policy in the amount of the Residence's replacement cost. "Replacement Cost" shall mean the full cost to replace the Residence on the same site, using identical or equivalent materials and labor without deduction for depreciation. If, before completion, a substantial portion of the Residence is damaged by fire, lightning, or any causes covered by the Contractor's All Risk policy, then Contractor shall complete the residence provided that the Completion Deadline is appropriately extended.

At all times during the Construction Period, including completion of Punch List items, Contractor will maintain workers compensation insurance as required by law, as well as general liability insurance in the amount of $500,000 (dollars) covering personal injury and property damage. Contractor agrees to provide Owners(s) with certificates of insurance before the Construction Term commences. Owner(s) agree to maintain liability insurance for the property as deemed appropriate by the construction lender.

Warranty: The Contractor shall provide to owner, in writing, a twelve (12) month construction warranty. Contractor warrants that the house will be substantially free from defects in workmanship and materials and shall meet the terms of the Construction Documents. Contractor shall not be responsible for damages caused by actions not executed by the Contractor, including abuse, modifications by Owner(s), improper maintenance, or normal wear and tear. Owner(s) and Contractor acknowledge that radon and radon gases exist everywhere and that Contractor provides no warranty regarding radon presence.

Liens: The Contractor warrants good title to all materials, supplies, and equipment incorporated into the Residence. Contractor warrants that the Residence shall be delivered free from any liens, claims, and charges filed by any person, firm or corporation. The Contractor agrees to obtain and submit "lien waivers" to Owner(s) from all subcontractors and material suppliers, if permitted by state law.

Other Provisions

a) The Owner(s) shall provide and pay for all required surveys, easements, legal descriptions, and title insurance necessary for the construction to proceed.

b) The Contractor shall exercise proper precautions relating to the safe completion of construction and shall be responsible to Owner(s) for any and all acts or omissions of the Contractor's employees, material suppliers, subcontractors, or their agents and employees. Contractor shall comply with the provisions of the Occupational Safety and Health Act.

c) The Contractor shall deliver, store, and install all materials according to the manufacturer's instructions in a manner that will not void any implied or express warranties provided by said manufacturer.

d) Contractor agrees to maintain discipline and good order among laborers and other persons involved in the completion of the contract. Contractor agrees to provide secure storage for on-site materials and equipment.

e) Contractor agrees to keep the premises clean of trash, waste, and material that may pose a safety hazard. Upon Substantial Completion of the Residence, the Contractor shall remove all remaining trash, rubbish, tools, machinery and surplus material from the Property.

f) Both parties agree that the Contract Sum is predicated upon the Property being free from rock or shale that is expensive or impossible to excavate with a tracked front-end loader. This exception applies to any other conditions below ground or hidden from reasonable examination prior to excavation including, but not limited to, springs or hidden water streams, fill dirt, unknown soil conditions, or other special engineering requirements. Contractor will notify Owner(s) of any such conditions immediately so that a Change Order can be executed to compensate for additional costs incurred by the Contractor for said conditions. Owner(s) may use excessive excavation costs as reasonable justification for termination of the Construction Contract and will pay the Contractor for any expenses incurred in the discovery of these conditions.

Claims and Disputes: Both parties agree to mediate any disputes, unless the parties mutually agree otherwise, in accordance with the current Construction Industry Mediation Rules of the American Arbitration Association. In such case, a request for mediation shall be filed in writing with all parties and the American Arbitration Association. Claims that are not resolved by mediation shall be resolved by arbitration in accordance with the current Construction Industry Arbitration Rules of the American Arbitration Association. The demand for arbitration shall be filed in writing with all parties and the American Arbitration Association. The award rendered by the arbitrator(s) shall be deemed final in accordance with applicable law in any court having jurisdiction.

Severability: The invalidity of any provision of this Contract shall not affect the validity of any other provision, and all other provisions shall remain in full force and effect.

This contract shall inure to the benefit of, and shall be binding upon, the parties hereto, their heirs, successors, administrators, executors, and assigns. In the event one or more persons or entities are identified herein as Owners or Contractors, then they shall be jointly and severally liable hereunder. The signature of one Owner shall be binding on all Owners with respect to any changes in the contract documents. The signature of one Contractor or the Contractor's official representative(s) shall be binding on all Contractors.

The parties do set their hands and seals the day and year first above written.

(Owner)   (Date) _____   _____

(Owner)   (Date) _____   _____

(Contractor)   (Date) _____   _____

SUBCONTRACTOR'S AGREEMENT

Sub. Name _____   Date _____
Address _____   Job. No. _____
_____   Job Address _____
_____   _____

Phone _____
Have Worker's Comp.?   ☐ Yes   ☐ No

Worker's Comp. No. _____   Plan No. _____
Expiration Date _____   Bid Amount _____
Tax ID No. _____   Bid Good Until _____

Payment to me made as follows:
_____
_____
_____
_____

Deduct _____ % for worker's compensation
Deduct _____ % for retainage

All material is guaranteed to be as specified. Subcontractor agrees to complete work in a timely manner according to the Subcontractor Agreement General Conditions attached and made a part of this agreement, for the Bid Amount listed above. Any alteration or deviation from the General Conditions or Specifications below must be accompanied by a written Change Order submitted and approved by the General Contractor.

SPECIFICATIONS: (Attach extra sheet if necessary) _____
_____
_____
_____
_____
_____
_____
_____
_____
_____
_____

DATE	CHECK NO.	AMOUNT PAID	AMOUNT DUE

SUBCONTRACTOR AGREEMENT GENERAL CONDITIONS

Scope of Work: Subcontractor agrees to provide all labor, materials, tools, scaffolding, and other equipment necessary to fulfill the Scope of Work as described in the Subcontractor's submitted Specifications and the Construction Documents. It is understood that the Construction Documents shall consist of project plans, specifications, and change orders supplied by the General Contractor. In the event of a conflict, the General Contractor's specifications shall control. The Subcontractor will perform the work under the direction of the General Contractor in accordance with all documents provided.

Scheduling of Work: Time is of the essence. All work will be done in a prompt, thorough and efficient manner according to the Subcontractor's Specifications and the Construction Documents. Work must also meet the standards required by governmental building inspectors. Any substandard work will be promptly corrected by Subcontractor at its expense. Subcontractor agrees to coordinate with other subcontractors and suppliers who may be working on the project so as not to delay or impede their performance. Subcontractor will not seek damages for any delay with the sole remedy being a reasonable extension of time to complete the work. Subcontractor agrees to participate in meetings with the General Contractor and other contractors as deemed necessary by the General Contractor for the efficient Scheduling of Work.

Limitations: Both parties agree that the Subcontractor, who is the Service Provider, is an independent business entity and is not an employee of the General Contractor, who is the Service Recipient. General Contractor will not provide worker's compensation, liability insurance, health benefits, or other employee benefits to the Subcontractor or any persons in the Subcontractor's employ.

Change Orders: The General Contractor shall have the right to make changes to the Subcontractor Specifications and Construction Documents. The cost of any alterations, additions, omissions or deviations shall be added to the agreed Subcontractor Bid Amount. No such changes will be made unless stipulated in writing by the execution of a "Change Order" and signed by all parties. The Change Order will be an addendum to this Agreement, specifying any additional amount due along with provision for payment.

Punch List: Upon Substantial Completion of the Subcontractor's Scope of Work, the General Contractor will inspect the work and supply the Subcontractor with a written "Punch List" of all deficiencies. Subcontractor agrees to correct all deficiencies at a time deemed convenient by the General Contractor. General Contractor may withhold Retainage from the payment due Subcontractor, to cover any costs associated with completion of the Punch List, not to exceed ten (10) percent. After final approval, the General Contractor will promptly pay any monies retained. If the Subcontractor fails to complete the Punch List in a reasonable time, the General Contractor has the option to pay for the completion of the Punch List from the monies retained. Any balance remaining shall be promptly paid to the Subcontractor.

Term of Agreement: This Agreement will terminate automatically upon completion by Subcontractor of the Scope of Work and all Punch List items.

Warranty and Indemnification: Subcontractor warrants that work and materials supplied will comply with the Subcontractor Specifications, Construction Documents, and all applicable building codes. Subcontractor warrants that all materials will be installed according to the manufacturer's instructions in a manner that will not void any implied or express warranties provided by said manufacturer. Subcontractor agrees to indemnify and hold General Contractor harmless from all damages and expenses, including attorney fees and judgments, from any claims or damages resulting from the acts or omissions of Subcontractor and/or Subcontractor's employees, agents, or representatives. When requested by the General Contractor, Subcontractor will provide final lien waivers from all people or firms supplying Subcontractor with labor and/or materials, stating that all bills and obligations resulting from this Agreement have been paid in full, and that the Subcontractor has not filed a lien or has claim to any future lien against the General Contractor or the property where work was performed.

Cancellation and Default: General Contractor may cancel this Agreement at any time in the event of default. General Contractor will provide Subcontractor with a written notice of Cancellation. If Subcontractor fails to cure the default within forty-eight (48) hours of receiving the Notice of Cancellation, the General Contractor may engage other subcontractors to complete the Scope of Work. If the Agreement is terminated, Subcontractor will be paid for any work completed prior to termination. General Contractor may deduct any additional costs incurred as a consequence of the default from any monies due the Subcontractor. Subcontractor will be considered in Default if any of the following actions occur: (A) Subcontractor fails to perform the work in an expeditious and efficient manner; (B) Subcontractor fails to pay any of its bills when due, for labor or materials; (C) Subcontractor is subject to bankruptcy or receivership proceedings or commits any act of insolvency; (D) Subcontractor fails to comply with any of the terms of this Agreement. Nothing in this paragraph shall limit any other remedies or rights General Contractor has under any other provision of law or this Agreement.

Trash and Safety: Subcontractor agrees to keep the premises clean of trash, waste, and material that may pose a safety hazard. Waste materials must be removed or placed in a container or area provided by the General Contractor. Subcontractor will indemnify General Contractor from any OSHA claims or fines resulting from improper safety practices by Subcontractor or persons in Subcontractor's employ.

Conduct and Appearance: Subcontractor and its employees or representatives shall conduct themselves in a professional manner at all times while representing General Contractor. Subcontractor also agrees to make its employees available for drug tests and background checks on the request of the General Contractor.

Surety Bond: If requested by the General Contractor, Subcontractor agrees to furnish and maintain a Surety Bond for the Term of the Agreement in a form and with sureties acceptable to General Contractor, in an amount equal to the Bid Amount.

Insurance: Subcontractor agrees to maintain insurance coverage for the benefit of Subcontractor's employees and to provide a Certificate of Insurance to the General Contractor before Scope of Work commences. General Contractor reserves the right to deduct the cost of insurance if the Certificate of Insurance is not received prior to making any payments to Subcontractor. Subcontractor waives any rights to recovery from General Contractor for any injuries that Subcontractor may sustain while performing services under this agreement and that are a result of the negligence of Subcontractor or its employees. Subcontractor agrees to provide worker's compensation insurance, employer's liability insurance, and comprehensive general liability insurance in amounts satisfactory under state law. Except for worker's compensation insurance, the General Contractor should be named as an additional insured.

Severability and Assignment: The invalidity of any provision of this Contract shall not affect the validity of any other provision, and all other provisions shall remain in full force and effect. Neither party may assign or transfer this Agreement without the prior written consent of the non-assigning party, which approval shall not be unreasonably withheld.

Service Provider: Subcontractor Service   **Recipient:** General Contractor

By: _____   _____   _____   _____
  (Representative)   (Date)   (Representative)   (Date)

SUBCONTRACTOR'S AFFIDAVIT

STATE OF: ———————————————————

COUNTY OF: ———————————————————

Personally appeared before the undersigned attesting officer,

——

who being duly sworn, on oath says that he was the contractor in charge of improving the property
owned by ————————————————————————————
located at ————————————————————————————

Contractor says that all work, labor, services and materials used in such improvements were furnished and performed at the contractor's instance; that said contractor has been paid or partially paid the full contract price for such improvements; that all work done or material furnished in making the improvements have been paid for at the agreed price, or reasonable value, and there are no unpaid bills for labor and services performed or materials furnished; and that no person has any claim on lien by reason of said improvements.

————————————————————————————

(Seal)

Sworn to and subscribed before me,

this ————— day of ————————, 20 ————

————————————————————————————

Notary Public

Gantt chart — WEEK OF PROJECT (weeks 0–18)

MAJOR STEPS		0	1	2	3	4	5	6	7	8	9	10	11	12	13	14	15	16	17	18
PRECONSTRUCTION	PC	1-33																		
EXCAVATION	EX	1	2-17																	
PEST CONTROL	PS	1-2			3	18-20								21-30						
CONCRETE	CN	1-3		4-8		9-45									46-56					
WATERPROOFING	WP	1				5														
FRAMING	FR	1-4				2-13	6-33						34-39							
ROOFING	RF	1-2						3	4-9				4-6				10			
PLUMBING	PL	1-5		6				7	8-16									17-22		
HVAC	HV	1-4									5-8								9-15	
ELECTRICAL	EL	1-6		7							8-13							14-22		
MASONRY	MA	1-4									5-22									
SIDING & CORNICE	SC	1-2							3-4		5-9									
INSULATION	IN	1-2										10-18			6-10					
DRYWALL	DR	1										3-5		2-8						
TRIM	TR	1-2							3							4-12			13-20	
PAINTING	PT	1-3							4					5-6	7		8-18			
CABINETRY	CB	1-4							5								6-16			
FLOORING & TILE	FL	1-5														6-27			28-41	
GLAZING	GL	1-3							4											
GUTTERS	GU	1-2													3-9				5-9	
LANDSCAPING	LD	1-7															8-22	1-20		
DECKING	DE																			

SUPPLIER'S/SUBCONTRACTOR'S REFERENCE SHEET

NAME	PHONE	TYPE PRODUCT
ADDRESS	CALL WHEN	DATE
	PRODUCTS	
	WORKER'S COMP?	BONDED?

REFERENCE NAME	PHONE	COMMENTS

PRICES/COMMENTS

NAME	PHONE	TYPE PRODUCT
ADDRESS	CALL WHEN	DATE
	PRODUCTS	
	WORKER'S COMP?	BONDED?

REFERENCE NAME	PHONE	COMMENTS

PRICES/COMMENTS

PLAN ANALYSIS CHECKLIST

PROPERTY: _____

PLAN NUMBER: _____

DESCRIPTION	YES	NO
CIRCULATION: Is there access between the following without going through another room?		
Living room to bathroom		
Family room to bathroom		
Each bedroom to bathroom		
Kitchen to dining room		
Living room to dining room		
Kitchen to outside door		
Can get from auto to kitchen without getting wet if raining		
ROOM SIZE AND SHAPE		
Are all room adequate size and reasonable shape based on planned use?		
Can living room wall accommodate sofa and end tables (minimum 10 ft.)?		
Dan wall of den take sofa and end table (minimum 10 ft.)?		
Can wall of master bedroom(s) take double bed and end tables?		
STORAGE AND CLOSETS		
Guest closet near front entry		
Linen closet near baths		
Broom closet near kitchen		
Tool and lawnmower storage		
Space for washer and dryer		
EXTERIOR		
Is the house style "normal" architecture?		
Is the house style compatible wit surrounding architecture?		
STYLE ANALYSIS		
Does the house conform to the neighborhood?		
Is the house suited to the lot?		
Is the topography good?		
Will water drain away from the house?		
Will grade of driveway be reasonable?		
Is the lot wooded?		
Is the subject lot as good as other in the neighborhood?		
TOTAL		

LIGHTING AND APPLIANCE ORDER

JOB _____   STORE NAME _____

DATE _____   LIGHTING BUDGET _____

DESCRIPTION	STYLE NO.	QTY.	UNIT PRICE	TOTAL PRICE
Front Entrance				
Rear Entrance				
Dining Room				
Living Room				
Den				
Family Room				
Kitchen				
Kitchen Sink				
Breakfast Area				
Dinette				
Utility Room				
Basement				
Master Bath				
Hall Bath 1				
Guest Bath				
Hall 1				
Hall 2				
Stairway				
Master Bedroom				
Bedroom 2				
Bedroom 3				
Bedroom 4				
Closets				
OUTDOOR				
Front Porch				
Rear Porch				
Porch/Patio				
Carport/Garage				
Post/Lantern				
Floodlights				
Chimes				
Pushbuttons				
APPLIANCES				
Ovens				
Hood				

ITEM ESTIMATE WORKSHEET

VENDOR NAME	DESCRIPTION	MEASURING		CONVERSION FACTOR	ORDERING		PRICE EACH	COST	TAX	TOTAL COST	COST TYPE
		QTY.	UNIT		QTY.	UNIT					

EXPENSE CATEGORY	MATERIAL COST	LABOR COST	SUBCONTRACTOR COST	TOTAL COST

COST ESTIMATE SUMMARY

DESCRIPTION	MATERIAL	LABOR	SUBCONTRACTOR	TOTAL
EXCAVATION				
PEST CONTROL				
CONCRETE				
FRAMING				
ROOFING				
PLUMBING				
HVAC				
ELECTRIC				
MASONRY				
SIDING & CORNICE				
INSULATION				
DRYWALL				
TRIM				
PAINTING				
CABINETRY				
TILE				
CARPET				
OTHER FLOORING				
GLAZING				
GUTTERS				
LANDSCAPING				
SUBTOTAL				
APPLIANCES				
LUMBER				
OTHER MATERIALS				
TOOLS/RENTAL				
LOAN/LEGAL				
INSURANCE				
PERMITS/LICENSE				
SUBTOTAL				
			GRAND TOTAL	

COST ESTIMATE CHECKLIST

CODE NO.	DESCRIPTION	QTY.	MATERIAL		LABOR		SUBCONTRACTOR		TOTAL
			UNIT PRICE	TOTAL MATL.	UNIT PRICE	TOTAL LABOR	UNIT PRICE	TOTAL SUB	
		SUBTOTALS							
						GRAND TOTALS			

SUBCONTRACTOR BID CONTROL LOG

SUB/LABOR	NAME	BID	GOOD UNTIL	NAME	BID	GOOD UNTIL	NAME	BID	GOOD UNTIL	BEST BID	NAME
EXCAVATION											
PEST CONTROL											
CONCRETE											
FRAMING											
ROOFING											
PLUMBING											
HVAC											
ELECTRICAL											
MASONRY											
SIDING & CORNICE											
INSULATION											
DRYWALL											
TRIM											
PAINTING											
CABINETRY											
TILE											
GLASS											
LANDSCAPING											
GUTTERS											
ASPHALT											

PURCHASE ORDER CONTROL LOG

PO NUMBER	DATE ORDERED	DESCRIPTION	VENDOR NAME	DATE DELIVERED	DISCOUNT		PAYMENTS				TOTAL PAID
					RATE	RATE	CHK#	AMT.	CHK#	AMT.	

PURCHASE ORDER

TO: _____

_____

_____

_____

P/O NO.	PAGE NO.
Date of Order	Job No.
Salesman	Req. No.
Deliver to	

REQUESTED BY	DELIVERED BY	SHIP VIA	TERMS

B/O	QTY. ORDERED	QTY. RECEIVED	DESCRIPTION	UNIT PRICE	TOTAL

PLEASE SEND A COPY OF THIS PO WITH DELIVERED MATERIALS. CALL IF UNABLE TO MAKE DELIVERY DEADLINE.

PURCHASE ORDER cont.

P/O NO.	PAGE NO.

B/O	QTY. ORDERED	QTY. RECEIVED	DESCRIPTION	UNIT PRICE	TOTAL

PLEASE SEND A COPY OF THIS PO WITH DELIVERED MATERIALS. CALL IF UNABLE TO MAKE DELIVERY DEADLINE.

CHANGE ORDER

(blank box)

TO (CONTRACTOR): _____

_____

_____

_____

CHANGE ORDER NO.	
Date of Change Order	Job No.
Job Phone	Contract Date
Job Location	

YOU ARE DIRECTED TO MAKE THE FOLLOWING CHANGES IN THIS CONTRACT

Contract time will be ☐ increased ☐ decreased by:	Original Contract Price	
Days	Previously Authorized Change Orders	
New completion date	New Contract Price	
ACCEPTED — The above prices and specifications are satisfactory and are hereby accepted. All work to be done under same terms and conditions as original contract unless otherwise stipulated.	This Change Order	
	NEW AUTHORIZED CONTRACT PRICE	
AUTHORIZED BY:	SUBCONTRACTOR	
SIGNATURE DATE	SIGNATURE DATE	

CONSTRUCTION LOAN DRAW SCHEDULE

JOB DESCRIPTION _____ JOB NO. _____

BANK _____

ADDRESS _____

_____

_____

AMOUNT OF CONSTRUCTION LOANS	
CONSTRUCTION LOAN CLOSING COSTS	
TOTAL CONSTRUCTION LOAN COST	

DRAW NO.	DESCRIPTION	DATE	AMOUNT
1			
2			
3			
4			
5			
6			
7			
8			
		TOTAL OF DRAW LOANS	

BUILDING CHECKLIST

JOD _____

DATE STARTED _____ ESTIMATED COMPLETION DATE _____

(Bold faced numbered items indicate critical tasks that must be done before continuing.)

DATE DONE NOTES

A. PRE-CLOSING

☐ 1. Take-off

☐ 2. Spec sheet

☐ 3. Contract

☐ 4. Submit loan

☐ 5. Survey

☐ 6. Insurance

☐ 7. Closing

B. POST-CLOSING

☐ 1. Business license

☐ 2. Compliance bond

☐ 3. Sewer tap

☐ 4. Water tap

☐ 5. Perk test (if necessary)

☐ 6. Miscellaneous permits

☐ 7. Building permit

☐ 8. Temporary services:

☐   a. Electrical

☐   b. Water

☐   c. Phone

C. CONSTRUCTION

☐ 1. Rough layout

☐ 2. Rough grading

☐ 3. Water

☐   a. Meter

☐   b. Well

☐ 4. Set temporary pole

☐ 5. Crawl or basement

☐   a. batter boards

☐   b. Footings dug

☐   c. Footings poured

☐   d. Lay block/pour

☐ 6. Pre-treat

- [ ] 7. Slab _____ _____
- [ ]   a. Form boards _____ _____
- [ ]   b. **Plumbing** _____ _____
- [ ]   c. Miscellaneous pipes _____ _____
- [ ]   d. Gravel _____ _____
- [ ]   e. Poly _____ _____
- [ ]   f. **Re-wire** _____ _____
- [ ]   g. Pour concrete _____ _____
- [ ] 8. Spot survey _____ _____
- [ ] 9. Waterproof _____ _____
- [ ] 10. Fiberglass tubs _____ _____
- [ ] 11. Framing _____ _____
- [ ]   a. Basement walls _____ _____
- [ ]   b. First floor _____ _____
- [ ]   c. Walls — 1st floor _____ _____
- [ ]   d. Second Floor _____ _____
- [ ]   e. Walls — 2nd floor _____ _____
- [ ]   f. Sheathing _____ _____
- [ ]   g. Ceiling joints _____ _____
- [ ]   h. Rafter/trussed _____ _____
- [ ]   i. Decking _____ _____
- [ ]   j. Felt _____ _____
- [ ] 12. Roofing _____ _____
- [ ] 13. Set fireplace _____ _____
- [ ] 14. Set doors _____ _____
- [ ] 15. Set windows _____ _____
- [ ] 16. Siding and cornice _____ _____
- [ ] 17. **Rough plumbing** _____ _____
- [ ] 18. **Rough HVAC** _____ _____
- [ ] 19. Rough electrical _____ _____
- [ ] 20. Measure cabinets _____ _____
- [ ] 21. Rough phone/cable _____ _____
- [ ] 22. Rough miscellaneous _____ _____
- [ ] 23. Insulation _____ _____
- [ ] 24. Framing inspection _____ _____
- [ ] 25. Clean interior _____ _____
- [ ] 26. Drywall _____ _____
- [ ] 27. Clean interior _____ _____

- [ ] 28. Stucco _____ _____
- [ ] 29. Brick _____ _____
- [ ] 30. Clean exterior _____ _____
- [ ] 31. Septic tank _____ _____
- [ ] 32. Utility lines _____ _____
- [ ]   a. Gas line _____ _____
- [ ]   b. Sewer line _____ _____
- [ ]   c. Phone line _____ _____
- [ ]   d. Electric line _____ _____
- [ ] 33. Backfill _____ _____
- [ ] 34. Pour garage _____ _____
- [ ] 35. Drives and walks _____ _____
- [ ] 36. Landscape _____ _____
- [ ] 37. Gutters _____ _____
- [ ] 38. Garage door _____ _____
- [ ] 39. Paint/stain _____ _____
- [ ] 40. Cabinets _____ _____
- [ ] 41. Trim _____ _____
- [ ] 42. Tile _____ _____
- [ ] 43. Rock _____ _____
- [ ] 44. Hardwood floors _____ _____
- [ ] 45. Wallpaper _____ _____
- [ ] 46. **Plumbing (final)** _____ _____
- [ ] 47. **HVAC (final)** _____ _____
- [ ] 48. **Electrical (final)** _____ _____
- [ ] 49. Mirrors _____ _____
- [ ] 50. Clean interior _____ _____
- [ ] 51. Vinyl/carpet _____ _____
- [ ] 52. Final trim _____ _____
- [ ] 53. Bath accessories _____ _____
- [ ] 54. Paint touch up _____ _____
- [ ] 55. Final touch up _____ _____
- [ ] 56. Connect utilities _____ _____
- [ ] **DATE COMPLETED** _____ _____

Description of Materials

**U.S. Department of Housing and Urban Development
Department of Veterans Affairs
Farmers Home Administration**

OMB Control No. 2502-0313
(exp. 05/31/2011)

Public reporting burden for this collection of information is estimated to average 30 minutes per response, including the time for reviewing instructions, searching existing data sources, gathering and maintaining the data needed, and completing and reviewing the collection of information. This agency may not collect this information, and you are not required to complete this form, unless it displays a currently valid OMB control number.

The National Housing Act (12 USC 1703) authorizes insuring financial institutions against default losses on single family mortgages. HUD must evaluate the acceptability and value of properties to be insured. The information collected here will be used to determine if proposed construction meets regulatory requirements and if the property is suitable for mortgage insurance. Response to this information collection is mandatory. No assurance of confidentiality is provided.

☐ Proposed Construction   ☐ Under Construction   No. _____   (To be inserted by HUD, VA or FmHA)

Property address (Include City and State)

Name and address of Mortgagor or Sponsor

Name and address of Contractor or Builder

Instructions

1. For additional information on how this form is to be submitted, number of copies, etc., see the instructions applicable to the HUD Application for Mortgage Insurance, VA Request for Determination of Reasonable Value, or FmHA Property Information and Appraisal Report, as the case may be.
2. Describe all materials and equipment to be used, whether or not shown on the drawings, by marking an X in each appropriate check-box and entering the information called for each space. If space is inadequate, enter "See misc." and describe under item 27 or on an attached sheet. **The use of paint containing more than the percentage of lead by weight permitted by law is prohibited.**
3. Work not specifically described or shown will not be considered unless required, then the minimum acceptable will be assumed. Work exceeding minimum requirements cannot be considered unless specifically described.
4. Include no alternates, "or equal" phrases, or contradictory items. (Consideration of a request for acceptance of substitute materials or equipment is not thereby precluded.)
5. Include signatures required at the end of this form.
6. The construction shall be completed in compliance with the related drawings and specifications, as amended during processing. The specifications include this Description of Materials and the applicable Minimum Property Standards.

1. **Excavation**

   Bearing soil, type _____

2. **Foundations**

   Footings concrete mix _____ strength psi _____ Reinforcing _____

   Foundation wall material _____ Reinforcing _____

   Interior foundation wall material _____ Party foundation wall _____

   Columns material and sizes _____ Piers material and reinforcing _____

   Girders material and sizes _____ Sills material _____

   Basement entrance areaway _____ Window areaways _____

   Waterproofing _____ Footing drains _____

   Termite protection _____

   Basementless space ground cover _____ insulation _____ foundation vents _____

   Special foundations _____

   Additional information

3. **Chimneys**

   Material _____ Prefabricated (make and size) _____

   Flue lining material _____ Heater flue size _____ Fireplace flue size _____

   Vents (material and size) gas or oil heater _____ water heater _____

   Additional information

4. **Fireplaces**

   Type ☐ solid fuel   ☐ gas-burning   ☐ circulator (make and size) _____ Ash dump and clean-out _____

   Fireplace facing _____ lining _____ hearth _____ mantel _____

   Additional information

ref. HUD Handbook 4145.1 & 4950.1 form HUD-92005 (10/84)
VA Form 26-1852 and form FmHA 424-2

To download a usable HUD Description of Materials form, go to: www.selfmadehomes.com

5. Exterior Walls

Wood frame wood grade, and species _____ ☐ Corner bracing Building paper or felt _____

    Sheathing _____ thickness _____ width _____ ☐ solid ☐ spaced _____ o.c. ☐ diagonal _____

    Siding _____ grade _____ type _____ size _____ exposure _____ fastening _____

    Shingles _____ grade _____ type _____ size _____ exposure _____ fastening _____

    Stucco _____ thickness _____ Lath _____ weight _____ lb.

    Masonry veneer _____ Sills _____ Lintels _____ Base flashing _____

Masonry ☐ solid ☐ faced ☐ stuccoed total wall thickness _____ facing thickness _____ facing material _____

Backup material _____ thickness _____ bonding _____

Door sills _____ Window sills _____ Lintels _____ Base flashing _____

Interior surfaces dampproofing, _____ coats of _____ furring _____

Additional information

Exterior painting material _____ number of coats _____

Gable wall construction ☐ same as main walls ☐ other construction_____

6. Floor Framing

Joists wood, grade, and species _____ other _____ bridging _____ anchors _____

Concrete slab ☐ basement floor ☐ first floor ☐ ground supported ☐ self-supporting mix _____ thickness _____

    reinforcing _____ insulation _____ membrane _____

Fill under slab material _____ thickness _____

Additional information

7. Subflooring (Describe underflooring for special floors under item 21)

Material grade and species _____ size _____ type _____

Laid ☐ first floor ☐ second floor ☐ attic _____ sq. ft. ☐ diagonal ☐ right angles

Additional information

8. Finish Flooring (Wood only. Describe other finish flooring under item 21)

Location	Rooms	Grade	Species	Thickness	Width	Bldg. Paper	Finish
First floor							
Second floor							
Attic floor	sq. ft.						

Additional information

9. Partition Framing

Studs wood, grade, and species _____ size and spacing _____ Other _____

Additional information

10. Ceiling Framing

Joists wood, grade, and species _____ Other _____ Bridging _____

Additional information

11. Roof Framing

Rafters wood, grade, and species _____ Roof trusses (see detail) grade and species _____

Additional information

12. Roofing

Sheathing wood, grade, and species _____ ☐ solid ☐ spaced _____ o.c.

Roofing _____ grade _____ size _____ type _____

Underlay _____ weight or thickness _____ size _____ fastening _____

Built-up roofing _____ number of plies _____ surfacing material _____

Flashing material _____ gage or weight _____ ☐ gravel stops ☐ snow guards

Additional information

ref. HUD Handbook 4145.1 & 4950.1 form HUD-92005 (10/84)
VA Form 26-1852 and form FmHA 424-2

13. Gutters and Downspouts

Gutters material _____ gage or weight _____ size _____ shape _____

Downspouts material _____ gage or weight _____ size _____ shape _____ number _____

Downspouts connected to ☐ Storm sewer ☐ sanitary sewer ☐ dry-well ☐ Splash blocks material and size _____

Additional information

14. Lath and Plaster

Lath ☐ walls ☐ ceilings material _____ weight or thickness _____ Plaster coats _____ finish _____

Dry-wall ☐ walls ☐ ceilings material _____ thickness _____ finish _____

Joint treatment

15. Decorating (Paint, wallpaper, etc.)

Rooms	Wall Finish Material and Application	Ceiling Finish Material and Application
Kitchen		
Bath		
Other		

Additional information

16. Interior Doors and Trim

Doors type _____ material _____ thickness _____

Door trim type _____ material _____ Base type _____ material _____ size _____

Finish doors _____ trim _____

Other trim (item, type and location) _____

Additional information

17. Windows

Windows type _____ make _____ material _____ sash thickness _____

Glass grade _____ ☐ sash weights ☐ balances, type _____ head flashing _____

Trim type _____ material _____ Paint _____ number coats _____

Weatherstripping type _____ material _____ Storm sash, number _____

Screens ☐ full ☐ half type _____ number _____ screen cloth material _____

Basement windows type _____ material _____ screens, number _____ Storm sash, number _____

Special windows _____

Additional information

18. Entrances and Exterior Detail

Main entrance door material _____ width _____ thickness _____ Frame material _____ thickness _____

Other entrance doors material _____ width _____ thickness _____ Frame material _____ thickness _____

Head flashing _____ Weatherstripping type _____ saddles _____

Screen doors thickness _____ number _____ screen cloth material _____ Storm doors thickness _____ number _____

Combination storm and screen doors thickness _____ number _____ screen cloth material _____

Shutters ☐ hinged ☐ fixed Railings _____ Attic louvers _____

Exterior millwork grade and species _____ Paint _____ number coats _____

Additional information

19. Cabinets and Interior Detail

Kitchen cabinets, wall units material _____ lineal feet of shelves _____ shelf width _____

Base units material _____ counter top _____ edging _____

Back and end splash _____ Finish of cabinets _____ number coats _____

Medicine cabinets make _____ model _____

Other cabinets and built-in furniture _____

Additional information

ref. HUD Handbook 4145.1 & 4950.1 form HUD-92005 (10/84)
VA Form 26-1852 and form FmHA 424-2

20. Stairs

Stair	Treads		Risers		Strings		Handrail		Balusters	
	Material	Thickness	Material	Thickness	Material	Size	Material	Size	Material	Size
Basement										
Main										
Attic										

Disappearing make and model number _____

Additional information

21. Special Floors and Wainscot (Describe Carpet as listed in Certified Products Directory)

Floors

Location	Material, Color, Border, Sizes, Gage, Etc.	Threshold Material	Wall Base Material	Underfloor Material
Kitchen				
Bath				

Wainscot

Location	Material, Color, Border, Cap. Sizes, Gage, Etc.	Height	Height Over Tub	Height in Showers (From Floor)
Bath				

Additional information

22. Plumbing

Fixture	Number	Location	Make	MFR's Fixture Identification No.	Size	Color
Sink						
Lavatory						
Water closet						
Bathtub						
Shower over tub						
Stall shower						
Laundry trays						

Bathroom accessories ☐ Recessed material _____ number _____ ☐ Attached material _____ number _____

Additional information

☐ Curtain rod ☐ Door ☐ Shower pan material _____ * (Show and describe individual system in complete detail in separate drawings and specifications according to requirements.)

Water supply ☐ public ☐ community system ☐ individual (private) system*

Sewage disposal ☐ public ☐ community system ☐ individual (private) system*

House drain (inside) ☐ cast iron ☐ tile ☐ other _____ House sewer (outside) ☐ cast iron ☐ tile ☐ other _____

Water piping ☐ galvanized steel ☐ copper tubing ☐ other _____ _____ Sill cocks, number _____

Domestic water heater type _____ make and model _____ heating capacity _____ gph. 100° rise.

Storage tank material _____ capacity _____ gallons

Gas service ☐ utility company ☐ liq. pet. gas ☐ other _____ ☐ Gas piping ☐ cooking ☐ house heating

Footing drains connected to ☐ storm sewer ☐ sanitary sewer ☐ dry well ☐ Sump pump make and model _____

capacity _____ discharges into _____

Additional information

ref. HUD Handbook 4145.1 & 4950.1 form HUD-92005 (10/84)
VA Form 26-1852 and form FmHA 424-2

23. Heating

☐ Hot water ☐ Steam ☐ Vapor ☐ One-pipe system ☐ Two-pipe system

☐ Radiators ☐ Convectors ☐ Baseboard radiation   Make and model _____

☐ Radiant panel ☐ floor ☐ wall ☐ ceiling   Panel coil material _____

☐ Circulator ☐ Return pump Make and model _____   capacity _____ gpm.

Boiler make and model _____   Output _____ Btuh. net rating _____ Btuh.

Additional information

Warm air ☐ Gravity ☐ Forced   Type of system _____

Duct material supply _____   return _____   Insulation _____   thickness _____ ☐ Outside air intake

Furnace: make and model _____   Input _____ Btuh. output _____ Btuh.

Additional information

☐ Space heater ☐ floor furnace ☐ wall heater Input _____ Btuh. output _____ Btuh. number units _____

Make, model _____

Additional information

Controls make and types _____

Additional information

Fuel: ☐ Coal ☐ oil ☐ gas ☐ liq. pet. gas ☐ electric ☐ other _____   storage capacity _____

Additional information

Firing equipment furnished separately ☐ Gas burner, conversion type ☐ Stoker hopper feed ☐ bin feed

Oil burner ☐ pressure atomizing ☐ vaporizing _____

Make and model _____

Control _____

Additional information

Electric heating system type _____ Input _____ watts @ _____ volts output _____ Btuh.

Additional information

Ventilating equipment ☐ attic fan, make and model _____   capacity _____ cfm.

☐ kitchen exhaust fan, make and model _____

Other heating, ventilating, or cooling equipment _____

Additional information

24. Electric Wiring

Service ☐ overhead ☐ underground   Panel ☐ fuse box ☐ circuit-breaker make _____ AMP's _____ No. circuits ___

Wiring ☐ conduit ☐ armored cable ☐ nonmetallic cable ☐ knob and tube ☐ other _____

Special outlets ☐ range ☐ water heater ☐ other _____

☐ Doorbell ☐ Chimes ☐ Push-button locations _____

Additional information

25. Lighting Fixtures

Total number of fixtures _____   Total allowance for fixtures, typical installation, $ _____

Nontypical installation _____

Additional information

ref. HUD Handbook 4145.1 & 4950.1 form HUD-92005 (10/84)
VA Form 26-1852 and form FmHA 424-2

26. Insulation

Location	Thickness	Material, Type, and Method of Installation	Vapor Barrier
Roof			
Ceiling			
Wall			
Floor			

27. Miscellaneous : (Describe any main dwelling materials, equipment, or construction items not shown elsewhere; or use to provide additional information where the space provided was inadequate. Always reference by item number to correspond to numbering used on this form.)

Hardware   (make, material, and finish.)

Special Equipment   (State material or make, model and quantity. Include only equipment and appliances which are acceptable by local law, custom and applicable FHA standards. Do not include items which, by established custom, are supplied by occupant and removed when he vacates premises or chattles prohibited by law from becoming realty.)

Porches

Terraces

Garages

Walks and Driveways

Driveway width _____ base material _____ thickness _____ surfacing material _____ thickness _____

Front walk width _____ material _____ thickness _____ Service walk width _____ material _____ thickness _____

Steps material _____ treads _____ risers _____ Cheek walls _____

Other Onsite Improvements

(Specify all exterior onsite improvements not described elsewhere, including items such as unusual grading, drainage structures, retaining walls, fence, railings, and accessory structures.)

Landscaping, Planting, and Finish Grading

Topsoil _____ thick ☐ front yard ☐ side yards ☐ rear yard to _____   feet behind main building

Lawns (seeded, sodded, or sprigged) ☐ front yard _____ ☐ side yards _____ ☐ rear yard _____

Planting ☐ as specified and shown on drawings ☐ as follows:

_____ Shade trees deciduous _____ caliper   _____ Evergreen trees _____ to _____ B & B

_____ Low flowering trees deciduous _____ to _____   _____ Evergreen shrubs _____ to _____ B & B

_____ High-growing shrubs deciduous _____ to _____   _____ Vines, 2-year _____

_____ Medium-growing shrubs deciduous _____ to _____   Other

_____ Low-growing shrubs deciduous _____ to _____

Identification— This exhibit shall be identified by the signature of the builder, or sponsor, and/or the proposed mortgagor if the latter is known at the time of application.

Date (mm/dd/yyyy)_____   Signature

Signature

ref. HUD Handbook 4145.1 & 4950.1 form HUD-92005 (10/84)
VA Form 26-1852 and form FmHA 424-2

GLOSSARY

3-way switch. Type of switch that allows control of a lighting fixture from two switch locations such as both ends of a hallway.

Acid. Cleaning agent used to clean brick.

Acrylic resin. A thermoplastic resin used in latex coatings (see latex paint).

Aggregate. Irregular-shaped gravel suspended in cement.

Air chamber. Pipe appendage with trapped air added to a line to serve as a shock absorber to retard or eliminate air hammer.

Air-dried lumber. Dried by exposure to air, usually in a yard, without artificial heat. For the United States, the minimum moisture content of thoroughly air-dried lumber is 12 to 15 percent and average is somewhat higher.

Alkyd resin. One of a large group of synthetic resins used in making latex paints.

Amperage. The amount of current flow in a wire. Similar to the amount of water flowing in a pipe.

Anchor. Irons of special form used to fasten together timbers or masonry.

Anchor bolt. Bolt which secures a wooden sill plate to concrete or masonry floor or foundation wall.

Apron. Trim used at base of windows. Also used as base to build out crown moulding.

Asphalt. Base ingredient of asphalt shingles and roofing paper (felt composition saturated in asphalt base). Used widely as a waterproofing agent in the manufacture of waterproof roof coverings of many types, exterior wall coverings, flooring tile, and the like. Most native asphalt is a residue from evaporated petroleum. It is insoluble in water but soluble in gasoline and melts when heated.

Attic ventilator. A screened opening provided to ventilate an attic space. They are located in the soffit area as inlet ventilators and in the gable end or along the ridge as outlet ventilators. Attic ventilation can also be provided by means of power-driven fans. See also Louver.

Attic ventilators. In houses, screened openings provided to ventilate an attic space. They are located in the soffit area as inlet ventilators and in the gable ends or along the ridge as outlet ventilators. They can also consist of power-driven fans used as an exhaust system. See also "Louver".

BTU. British thermal unit. A standard unit of hot or cold air output.

Backfill. Process of placing soil up against foundation after all necessary foundation treatments have been performed. In many instances, foundation wall must have temporary interior bracing or house framed in to support weight of dirt until it has had time to settle.

Backflow. The flow of water or other fluids or materials into the distributing source of potable water from any source other than its intended source.

Backhoe. A machine that digs narrow, deep trenches for foundations, drain pipe, cable, etc.

Backing. The bevel on the top edge of a hip rafter that allows the roofing board to fit the top of the rafter without leaving a triangular space between it and the lower side of the roof covering.

Backsplash. A small strip (normally 3" or 4") placed against the wall and resting on the back of the countertop. This is normally used to protect the wall from water and stains.

Balloon frame. The lightest and most economical form of construction, in which the studding and corner posts are set up in continuous lengths from first floor line or sill to the roof plate.

Baluster. A vertical member in a railing used on the edge of stairs, balconies, and porches. A small pillar or column used to support a rail.

Balustrade. A railing made up of balusters, top rail, and sometimes bottom rail.

Band. A low, flat decorative trim around windows and doors and horizontal relief; normally 2×4 or 2×6.

Band joist. See under Joist, band.

Base. The bottom of a column; the finish of a room at the junction of the walls and floor.

Base cabinet. A cabinet resting on the floor at waist level, supporting a counter top or an appliance.

Base coat. First coat put on drywall. Normally a very light color. Should even be put on prior to wallpaper so that the paper can be easily removed.

Base moulding. Moulding applied at the intersection of the wall and floor. Moulding used to trim the upper edge of baseboard.

Base or Baseboard. A board placed against the wall around a room next to the floor.

Base or baseboard. A board placed against the wall around a room next to the floor to finish properly the area between the floor and wall.

Base shoe. Moulding used next to the floor on interior baseboards. Sometimes called a carpet strip.

Batten. Narrow strips of wood used to cover joints or as decorative vertical members over plywood or wide boards.

Batter board. A pair of horizontal boards nailed to vertical posts set at the corners of an excavation area used to indicate the desired level of excavation. Also used for fastening taut strings to indicate outlines of the foundation walls.

Batter pile. Pile driven at an angle to brace a structure against lateral thrust.

Bay window. Any window that projects out from the walls of the structure.

Bazooka. Term for a automated drywall application tool that tapes, applies mud and feathers in one pass. Gets its name because it looks like a bazooka. Expensive to buy or rent _ only used on large scale operations.

Bead. Any corner or edge that must be finished off with stucco.

Beam. A long piece of lumber or metal used to support a load placed at right angles to the beam — usually floor joists. An inclusive term for joists, girders, rafters, and purling.

Bearing partition. A partition that supports any vertical load in addition to its own weight.

Bearing wall. A wall that supports any vertical load in addition to its own weight.

Bed. A moulding used to cover the joint between the plancier and frieze (horizontal decorative band around the wall of a room); else used as a base moulding upon heavy work, and some

Bed moulding. The trim piece (moulding) that covers the intersection of the vertical wall and any overhanging horizontal surface, such as the soffit. In a series of mouldings, the lowest one.

Bedding. A layer of mortar into which brick or stone is set.

Belt course. A horizontal board across or around a building, usually made of a flat member and a moulding.

Bend. Any change in direction of a line.

Bending strength. The resistance of a member when loaded like a beam.

Bent. A single vertical framework consisting of horizontal and vertical members supporting the deck of a bridge or pier.

Berm. Built-up mound of dirt for drainage and landscaping. A raised area of earth such as earth pushed against a wall.

Bevel board (pitch board). A board used in framing a roof or stairway to lay out bevels.

Bibb. Special cover used where lines are stubbed through an exterior wall.

Blind-nailing. Nailing in such a way that the nail heads are not visible on the face of the work — usually at the tongue of matched boards.

Board. Lumber less than 2 inches thick.

Board foot. A unit of measurement equal to a 1" thick piece of wood one foot square. Thickness × Length × Width equals board feet.

Boarding in. The process of nailing boards on the outside studding of a house.

Bolster. A short horizontal wood or steel beam on top of a column to support and decrease the span of beams or girders.

Boston Fidget. A method of applying shingles at the ridge or hips of a roof as a finish.

Boston ridge. A method of applying asphalt or wood shingles at the ridge or at the hips of a roof as a finish.

Brace. Diagonally framed member used to temporarily hold wall in place during framing. An inclined piece of framing lumber applied to wall or floor to stiffen the structure.

Bracket. A projecting support for a shelf or other structure.

Break joints. To arrange joints so that they do not come directly under or over the joints of adjoining pieces, as in shingling, siding, etc.

Breather paper. A paper that lets water vapor pass through, often used on the outer face of walls to stop wind and rain while not trapping water vapor.

Brick veneer. A facing of brick laid against and fastened to the sheathing of a frame wall or tile wall construction.

Bridging. Diagonal metal or wood cross braces installed between joists to prevent twisting and to spread the load to adjoining joists.

Btu. British Thermal Unit. A standard unit of hot or cold air output.

Builder's level. A surveying tool consisting of an optical siting scope and a measuring stick. It is used to check the level of batter boards and foundation.

Building drain. The common artery of the drainage system receiving the discharge from other drainage pipes inside the building conveying it to the sewer system outside the building.

Building paper. Cheap, thick paper, used to insulate a building before the siding or roofing is put on; sometimes placed between double floors.

Building sewer. That part of the drainage system extending from the building drain to a public or private sewer system.

Building supply. The pipe carrying potable water from the water meter or other water source to points of distribution throughout the building and lot.

Built-up member. A single structural component made from several pieces fastened together.

Built-up roof. A roofing composed of three to five layers of asphalt felt laminated with coat tar, pitch or asphalt. The top is finished with crushed slag or gravel. Generally used on flat or low-pitched roofs.

Built-up timber. A timber made of several pieces fastened together, and forming one of larger dimension.

Bulkhead. Vertical drop in footing when changing from one depth to another.

Bull. A covering for a wide soil stack so that rain will not enter.

Butt joint. Junction of the ends of two framing members such as on a sill. Normally a square cut joint.

CPVC. Chlorinated Poly Vinyl Chloride. A flexible form of water line recently introduced. No soldering involved. Plastic.

Can. Recessed lighting fixture.

Cant strip. A piece of lumber triangular in cross section, used at the junction of a flat deck and a wall to avoid a sharp bend and possible cracking of the covering which is applied over it.

Cantilever. A horizontal structural component that projects beyond its support, such as a second-story floor that projects out from the wall of the first floor.

Cap. Hardware used to terminate any plumbing line. The upper member of a column, pilaster, door cornice, or moulding.

Cap moulding. Trim applied to the top of base moulding.

Carriage. The supports for the steps and risers of a flight of stairs. See Stringer.

Casement. A window in which the sash opens upon hinges.

Casement frame and sash. A frame of wood or metal enclosing part or all of a sash, which can be opened by means of hinges affixed to the vertical edge.

Casement window. A window that swings out to the side on hinges.

Casing. Moulding of various widths used around door and window openings.

Casing nails. Used to apply finish trim and millwork. Nail head is small and is set below the surface of the wood to hide it.

Caulk. To fill or close a joint with a seal to make it watertight and airtight. The material used to seal a joint.

Cement mortar. A mixture of cement with sand and water used as a bonding agent between bricks or stones.

Center set. Standard holes for a standard faucet set.

Center-hung sash. A sash hung on its centers so that it swings on a horizontal axis.

Chair rail. Moulding applied to the walls, normally at hip level.

Chalking string. A string covered with chalk that, when stretched between two points and snapped, marks a straight line between the points.

Chamfer. The beveled edge of a board.

Check rails. Also called meeting rails. The upper rail of the lower sash and the lower rail of the upper sash of a double-hung window. Meeting rails are made sufficiently thicker than the rest of the sash frame to close the opening between the two sashes. Check rails are usually beveled to insure a tight fit between the two sashes.

Checking. Fissures that appear with age in many exterior paint coats. They are at first superficial, but in time they may penetrate entirely through the coating.

Checks. Splits or cracks in a board, ordinarily caused by seasoning.

Chock. Heavy timber fitted between fender piles along wheel guard of a pier or wharf.

Chord. The principal member of a truss on either the top or bottom.

Circuit breaker. A device to ensure that current overloads do not occur. A circuit breaker "breaks the circuit" when a dangerous overload or short circuit occurs.

Clamp. A mechanical device used to hold two or more pieces together.

Clapboards. A special form of outside covering of a house; siding.

Clean-out. A sealed opening in a pipe which can be screwed off to unclog the line if necessary.

Clear. Millwork with fine texture and no knots or other major imperfections.

Clear wood. Wood that has no knots.

Cleat. A length of wood fixed to a surface, as a ramp, to give a firm foothold or to maintain an object in place.

Collar beam. Nominal 1" or 2"-thick members connecting opposite roof rafters at or near the ridge board. They serve to stiffen the roof structure.

Column. In architecture, a perpendicular supporting member, circular or rectangular in section, usually consisting of a base, shaft and capital. In engineering, a vertical structural compression member which supports loads acting in the direction of its longitudinal axis.

Combination frame. A combination of the principal features of the full and balloon frames.

Compressor. Component of the central air-conditioning system which sits outside of dwelling.

Concrete. An artificial building material made by mixing cement and sand with gravel, broken stone, or other aggregate, and sufficient water to cause the cement to set and bind the entire mass.

Condensation. In a building, beads or drops of water (and frequently frost in extremely cold weather) that accumulate on the inside of the exterior covering of a building when warm, moisture-laden air from the interior reaches a point where the temperature no longer permits the air to sustain the moisture it holds. Use of louvers or attic ventilators will reduce moisture condensation in attics. A vapor barrier under the gypsum lath or dry wall on exposed walls will reduce condensation in them.

Conductors. Pipes for conducting water from a roof to the ground or to a receptacle or downspout.

Conduit. Metal pipe used to run wiring through when extra protection of wiring is needed or wiring is to be exposed.

Construction, frame. A type of construction in which the structural parts are wood or depend upon a wood frame for support. In codes, if masonry veneer is applied to the exterior walls, the classification of this type of construction is usually unchanged.

Control joint. A joint that penetrates only partially through a concrete slab or wall so that if cracking occurs it will be a straight line at that joint.

Coped joint. See "Scribing".

Corbel. Extending a course or courses of bricks beyond the face of a wall. No course should extend more than 2" beyond the course below it. Total corbeling projection should not exceed wall thickness.

Corian. An artificial material simulating marble manufactured by Corning.

Corner bead. A strip of formed sheet metal, sometimes combined with a strip of metal lath, placed on corners before plastering to reinforce them. Also, a strip of wood finish three-quarters-round or angular placed over a plastered corner for protection.

Corner board. A board used as trim for the external corner of a house or other frame structure, against which the ends of siding are butted.

Corner boards. Used as trim for the external corners of a house or other frame structure against which the ends of the siding are finished.

Corner brace. A diagonal brace placed at the corner of a frame structure to stiffen and strengthen the wall.

Corner braces. Diagonal bracing in the corners of a structure used to improve rigidity.

Cornice. When used on the outside of the house, the trimwork that finishes off the intersection of the roof and siding. The cornice may be flush with the siding or may overhang the siding by as much as 2'. It usually consists of a fascia board, a soffit for a closed cornice, and appropriate mouldings.

Cornice return. The underside of the cornice at the corner of the roof where the walls meet the gable end roofline. The cornice return serves as trim rather than as a structural element, providing a transition from the horizontal eave line to the sloped roofline of the gable.

Counterflashing. A flashing usually used on chimneys at the roofline to cover shingle flashing and to prevent moisture entry. They also allow expansion and contraction without danger of breaking the flashing.

Countersink. To set the head of a nail or screw at or below the surface.

Course. A horizontally laid set of bricks. A 32 course wall with ⅜" mortar joints stands 7' tall. A horizontal row of shingles.

Cove moulding. A moulding with a concave face used as trim or to finish interior corners.

Coverage. The maximum number of shingles that overlap in any one spot. This determines the degree of weather performance a roof has.

Crawl space. A shallow space below the living quarters of a house with no basement, normally enclosed by the foundation wall.

Creosote. A distillate of coal tar produced by high temperature carbonization of bituminous coal; it consists principally of liquid and solid aromatic hydrocarbons used as a wood preservative.

Cricket. A sloped area at the intersection of a vertical surface and the roof, such as a chimney. Used to channel off water that might otherwise get trapped behind the vertical structure. See saddle.

Crimp. A crease formed in sheet metal for fastening purposes or to make the material less flexible.

Cripple. A short stud used as bracing under windows and other structural framing.

Cross brace. Bracing with two intersecting diagonals.

Crown moulding. The trim piece that tops off the trim on a vertical structure. Usually refers to the more ornamental pieces of cornice trim. Moulding applied at the intersection of the ceiling and walls. Normally associated with more formal areas. Can also refer to the ornamental trim applied between the fascia and the roof. If a moulding has a concave face, it is called a cove moulding.

Crusher run. Crushed stone, normally up to 2" or 3" in size. Sharp edges. Used for drive and foundation support as a base. Very stable surface, as opposed to gravel, which is not very stable.

Cube. A standard ordering unit for masonry block units (6 × 6 × 8).

Curing. Process of maintaining proper moisture level and temperature (about 73° Fahrenheit) until the design strength is achieved. Curing methods include adding moisture (sprinkling with water, applying wet coverings such as burlap or straw) and retaining moisture (covering with waterproof materials such as polyurethane).

d. See "Penny".

Dado. A rectangular groove across the width of a board or plank. In interior decoration, a special type of wall treatment.

Damper. A metal flap controlling the flow of conditioned air through ductwork.

Dead load. The weight, expressed in pounds per square foot, of elements that are part of the structure.

Dead wood. Wood used as backing for drywall.

Deadening. Construction intended to prevent the passage of sound.

Deadman timber. A large buried timber used as an anchor as for anchoring a retaining wall.

Decay. Disintegration of wood or other substance through the action of fungi, as opposed to insect damage.

Deck paint. An enamel with a high degree of resistance to mechanical wear, designed for use on such surfaces as porch floors.

Decking. Heavy plank floor of a pier or bridge.

Deformed shank nail. A nail with ridges on the shank to provide better withdrawal resistance. See under **Nail.**

Den paneling. Paneling composed normally of sections of stain grade materials, joined with a number of vertical and/or horizontal strips of moulding.

Density. The mass of a substance in unit volume. When expressed in the metric system, it is numerically equal to the specific gravity of the same substance.

Dentil moulding. A special type of crown moulding with an even pattern of "teeth".

Dewpoint. Temperature at which a vapor begins to deposit as a liquid. Applies especially to water in the atmosphere.

Diagonal. Inclined member of a truss or bracing system used for stiffening and wind bracing.

Dimension. See "Lumber dimension".

Direct nailing. To nail perpendicular to the initial surface or to the junction of the pieces joined. Also termed "face nailing".

Door jam, interior. The surrounding case into which and out of which a door closes and opens. It consists of two upright pieces, called side jambs, and a horizontal head jamb.

Door jamb, interior. The surrounding case into which and out of which a door closes and opens. It consists of two upright pieces, called side jambs, and a horizontal head jamb.

Dormer. An opening in a sloping roof, the framing of which projects out to form a vertical wall suitable for windows or other opening. See eye dormer and shed dormer.

Dovetail. Joint made by cutting pins the shape of dove tails which fit between dovetails upon another piece.

Downspout. A pipe, usually of metal, for carrying rainwater from roof gutters.

Drag time. Time required to haul heavy excavation equipment to and from the site.

Drawboard. A mortise-and-tenon joint with holes so bored that when a pin is driven through, the joint becomes tighter.

Dressed and Matched. See Tongue and groove.

Dressed and matched (tongued and grooved). Boards or planks machined in such a manner that there is a groove on one edge and a corresponding tongue on the other.

Dressed size. Dimensions of lumber after planing smooth.

Dried in. Term describing the framed structure after the roof deck and protective tar paper have been installed.

Drip. (1) A structural member of a cornice or other horizontal exterior-finish course that has a projection beyond the parts for water runoff.

(2) A groove in the underside of a sill or drip cap to cause water to run off on the outer edge.

Drip cap. A moulding placed on the exterior top side of a door or window frame to cause water to drip beyond the outside of the frame.

Drip edge. Metal flashing normally 3" wide which goes on the eave and rakes to provide a precise point for water to drip from so that the cornice does not rot.

Drywall. Interior covering material which is applied in large sheets or panels. The term has become basically synonymous with gypsum wallboard.

Ducts. In a house, usually round or rectangular metal pipes for distributing warm air from the heating plant to rooms, or air from a conditioning device or as cold air returns.

EER (Energy Efficiency Rating). A national rating required to be displayed on appliances measuring their efficient use of electrical power.

Eave. Lowest point on a roof that overhangs edge of house.

Eaves. The portion of the roof that extends beyond the outside walls of the house. The main function of an overhanging eave is to provide visual separation of the roof and wall and to shelter the siding and windows from rain.

Edgenailing. Nailing into the edge of a board. See Nail.

Elastomeric. Having elastic, rubber-like properties.

Elbow. Trough corner extending outward from roof. Ductwork joint used to turn supply or return at any angle. A section of line which is used to change directions. Normally at right angles.

Enamel. Oil base paint used in high soil areas such as trim and doors.

End nailing. Nailing into the end of a board, which results in very poor withdrawal resistance. See Nail.

Exhaust. Air saturated with carbon dioxide. The byproduct of natural gas combustion in a forced air gas system. This exhaust gas should be vented directly out the top of the roof. It is dangerous to breathe this gas.

Expansion joint. A bituminous fiber strip used to separate blocks or units of concrete to prevent cracking caused by expansion as a result of temperature changes. Also used on concrete slabs.

Exposure. The vertical length of exposed shingle (portion not lapped by the shingle above).

Eye dormer. A dormer that has a gable roof.

Face nailing. Nailing applied perpendicular to the members. Also known as direct nailing.

Face nailing. Nailing perpendicular to the initial surface being penetrated. Also termed direct nailing.

Fascia. A flat board, band, or face, used by itself or, more often, in combination with mouldings, that covers the end of the rafter or the board that connects the top of the siding to the bottom of the soffit. The board of the cornice to which the gutter is fastened.

Fascia backer. The main structural support member to which the fascia is nailed.

Feathering. Successive coats of drywall compound applied to joints. Each successive pass should widen the compound joint.

Felt. Typical shingle underlayment. Also known as roofing felt or tar paper. See Asphalt.

Ferrule. Aluminum sleeve used in attaching trough to gutter spike.

Fill dirt. Loose dirt. Normally dirt brought in from another location to fill a void. Sturdier than topsoil used under slabs, drives, and sidewalks.

Filler. Putty or other pasty material used to fill nail holes prior to painting or staining. A heavily pigmented preparation used for filling and leveling off the pores in open-pored woods.

Filler (wood). A heavily pigmented preparation used for filling and leveling off the pores in open-pored woods.

Finger joint. Trim composed of many small scrap pieces by a joint resembling two sets of interlocking fingers. This trim is often used to reduce costs where such trim will be painted.

Finish grade. Final process of leveling and smoothing topsoil into final position prior to landscaping.

Fire stop. A solid, tight closure of a concealed space, placed to prevent the spread of fire and smoke through such a space. In a frame wall, this will usually consist of 2×4 cross blocking between studs.

Fished. An end butt splice strengthened by pieces nailed on the sides.

Fishplate. A wood or plywood piece used to fasten the ends of two members together at a butt joint with nails or bolts. Sometimes used at the junction of opposite rafters near the ridge line.

Fixture. Any end point in a plumbing system used as a source of potable water. Fixtures normally include sinks, tubs, showers, spigots, sprinkler systems, washer connections and other related items.

Flagstone (flagging or flags). Flat stones, from 1" to 4" thick, used for rustic walks, steps, and floors

Flashing. Galvanized sheet metal used as a lining around joints between shingles and chimneys, exhaust and ventilation vents and other protrusions in the roof deck. Flashing helps prevent water from seeping under the shingles.

Flat paint. An interior paint that contains a high proportion of pigment and dries to a flat or luster-less finish.

Float. To spread drywall compound smooth.

Floating. A process used after screeding to provide a smoother surface. The process normally involves embedding larger aggregate below the surface by vibrating, removing imperfections, high and low spots and compacting the surface concrete.

Flue. The space or passage in a chimney through which smoke, gas or fumes ascend. Each passage is called a flue, which together with any others

and the surrounding masonry make up the chimney.

Flue. The opening in a chimney through which smoke passes.

Flue lining. Fire clay or terra-cotta pipe, round or square, usually made in all ordinary flue sizes and in 2' lengths, used for the inner lining of chimneys with the brick or masonry work around the outside. Flue lining in chimney runs from about a foot below the flue connection to the top of the chimney.

Flush. Adjacent surfaces even, or in same plane (with reference to two structural pieces).

Fly rafters. End rafters of the gable overhang supported by roof sheathing and lookouts.

Footing. Lowest perimeter portion of a structure resting on firm soil or rock that supports the weight of the structure. With a pressure-treated wood foundation, a gravel footing may be used in place of concrete.

Footing ditch. Trough area dug to accommodate concrete or footing forms.

Footing form. A wooden or steel structure, placed around the footing that will hold the concrete to the desired shape and size.

Formwork. A temporary mold for giving a desired shape to poured concrete.

Foundation. The supporting portion of a structure below the first-floor construction, or below grade, including the footings.

Frame. The surrounding or enclosing woodwork of windows, doors, etc., and the timber skeleton of a building.

Framing. The rough timber structure of a building, including interior and exterior walls, floor, roof, and ceilings.

Framing, balloon. A system of framing a building in which all vertical structural elements of the bearing walls and partitions consist of single pieces extending from the top of the foundation sill plate to the roofplate to which all floor joists are fastened.

Framing, ladder. Framing for the roof overhang at a gable. Cross pieces

are used similar to a ladder to support the overhang.

Framing, platform. A system of framing a building in which floor joists of each story rest on top plates of the story below or on the foundation sill for the first story, and the bearing walls and partitions rest on the subfloor of each story.

Freon. A special liquid used by an air-conditioning compressor to move heat in or out of the dwelling. This fluid circulates in a closed system. This fluid is also used in refrigerators and freezers.

Frieze. A vertical piece of wood used with or without moulding to top off the intersection of the siding and the cornice. Frieze boards may be anywhere from 4" to 12" wide. A horizontal member connecting the top of the siding with the soffit of the cornice.

Frost line. The depth of frost penetration in soil. This depth varies in different parts of the country.

Frostline. The depth to which frost penetrates the soil. Footings should always be poured below this line to prevent cracking.

Full frame. The old fashioned mortised-and-tenoned frame, in which every joint was mortised and tenoned. Rarely used at the present time.

Fungi, wood. Microscopic plants that live in damp wood and cause mold, stain and decay.

Fungicide. A chemical that is poisonous to fungi.

Furring. Long strips of wood attached to walls or ceilings to allow attachment of drywall or ceiling tiles. Furring out refers to adding furring strips to a wall to bring it out further into a room. Furring down refers to using furring strips to lower a ceiling.

GFI (Ground Fault Circuit Interrupter). An extra sensitive circuit breaker usually installed in outlets in bathrooms and exterior locations to provide additional protection against shock. Required now by most building codes.

Gable. The vertical part of the exterior

wall that extends from the eaves upward to the peak or ridge of the roof. The gable may be covered with the same siding material as the exterior wall or may be trimmed with gable trim material.

Gable end. An end wall having a gable.

Gage. A tool used by carpenters to strike a line parallel to the edge of a board.

Gambrel. A roof that slopes steeply at the edge of the building, but changes to a shallower slope across the center of the building. This allows the attic to be used as a second story.

Gem box. A metal box installed in electrical rough-in that holds outlets, receptacles and other electrical units.

Girder. A large or principal beam of wood or steel used to support concentrated loads at isolated points along its length.

Girt (ribband). The horizontal member which supports the floor joists or is flush with the top of the joists. A horizontal member used as a stiffener between studs.

Gloss enamel. A finishing material made of varnish and sufficient pigments to provide opacity and color, but little or no pigment of low opacity. Such an enamel forms a hard coating with maximum smoothness of surface and a high degree of gloss.

Gloss paint, Gloss enamel. A paint or enamel that contains a relatively low proportion of pigment and dries to a sheen or luster.

Glue. A joint held together with glue.

Glueline, exterior. Waterproof glue at the interface of two veneers of plywood.

Gooseneck. Section of staircase trim with a curve in it.

Grade. The ground level around a building. The natural grade is the original level. Finished grade is the level after the building is complete and final grading is done.

Grading. Process of shaping the surface of a lot to give it the desired contours. See Finish Grade.

Grain. The direction of the fibers in the wood. Edge grain wood has been sawed parallel to the growth rings. Flat grain lumber is sawed perpendicular to the growth rings. Studs are flat grain lumber. The direction, size, arrangement, appearance, or quality of the fibers in wood.

Grain, edge (vertical). Edge-grain lumber has been sawed parallel to the pitch of the log and approximately at right angles to the growth rings; i.e., the rings form an angle of 45° or more with the surface of the piece.

Grain, edge or vertical. Edge-grain lumber has been sawed parallel to the pith of the log and approximately at right angles to the growth rings; i.e., the rings form an angle of 45° or more with the wide surface of the piece.

Grain, flat. Flat-grain lumber has been sawed parallel to the pitch of the log and approximately tangent to the growth rings; i.e., the rings form an angle of less than 45° with the surface of the piece.

Grain, quartersawn. Another term for edge grain.

Groove. A long hollow channel cut by a tool, into which a piece fits or in which it works. Two special types of grooves are the dado, a rectangular groove cut across the full width of a piece, and the housing, a groove cut at any angle with the grain and part way across a piece. Dadoes are used in sliding doors, window frames, etc. and housings are used for framing stair risers and threads in a string.

Ground. A strip of wood assisting the plasterer in making a straight wall and in giving a place to which the finish of the room may be nailed.

Grounds. Guides used around openings and at the floorline to strike off plaster. They can consist of narrow strips of wood or of wide subjambs at interior doorways. They provide a level plaster line for installation of casing and other trim.

Grout. Mortar made of such consistency (by adding water) that it will just flow into the joints and cavities of the masonry work and fill them solid.

Grouted. Filled with a mortar thin enough to fill the spaces in the concrete or ground around the object being set.

Gusset. A flat wood, plywood, or similar type member used to provide a connection at intersections of wood members. Most commonly used in joints of wood trusses.

Gutter or eave trough. A shallow channel or conduit of metal of wood set below and along the eaves of a house to catch and carry off rainwater from the roof.

Gypsum board. Same as DRYWALL.

Gypsum plaster. Gypsum formulated to be used with the addition of sand and water for base-coat plaster.

H-clip. A metal clip into which edges of adjacent plywood sheets are inserted to hold edges in alignment.

Halved. A joint made by cutting half the wood away from each piece so as to bring the sides flush.

Hang. To hang drywall is to nail it in place.

Hanger. Vertical-tension member supporting a load.

Head lap. Length (inches) of the amount of overlap between shingles (measured vertically).

Header. One or two pieces of lumber installed over doors and windows to support the load above the opening. A beam placed perpendicular to joists, to which joists are nailed in framing for chimneys, stairways, or other openings. A wood lintel.

Headroom. The clear space between floor line and ceiling, as in a stairway.

Hearth. The inner or outer floor of a fireplace, usually made of brick, tile, or stone.

Heartwood. Older wood from the central portion of the tree. As this wood dies, it undergoes chemical changes that often impart a resistance to decay and a darkening in color.

Heel of a rafter. The end or foot that rests on the wall plate.

Heel wedges. Triangular shaped pieces of wood that can be driven into gaps between rough framing and finished items, such as window frames, to provide a solid backing for these items

Hip. The external angle formed by the meeting of two sloping sides of a roof.

Hip roof. A roof that rises by inclined planes from all four sides of a building.

Hip roof. A roof which slopes up toward the center from all sides, necessitating a hip rafter at each corner.

Hopper window. A window that is hinged at the bottom to swing inward.

Hot. A wire carrying current.

Housed. A joint in which a piece is grooved to receive the piece which is to form the other part of the joint.

Humidifier. A device designed to increase the humidity within a room or a house by means of the discharge of water vapor It may consist of individual room-size units or larger units attached to the heating plant to condition the entire house.

Hydration. The chemical process wherein portland cement becomes a bonding agent as water is slowly removed from the mixture. The rate of hydration determines the strength of the bond and hence the strength of the concrete. Hydration stops when all the water has been removed. Once hydration stops, it cannot be restarted.

I-beam. A steel beam with a cross section resembling the letter I. I-beams are used for long spans as basement beams or over wide wall openings, such as a double garage door, when wall and roof loads are imposed on the opening.

Inside miter. Trough corner extending toward roof.

Insulation board, rigid. A structural building board made of foam or coarse wood or cane fiber impregnated with asphalt or given other treatment to provide water resistance. It can be obtained in various size sheets, in various thicknesses, and in various densities.

Insulation, thermal. Any material high in resistance to heat transmission that, when placed in the walls, ceiling or floors of a structure, will reduce the rate of heat flow.

Interior finish. Material used to cover the interior framed areas or materials of walls and ceilings.

Involute. Curved portion of trim used to terminate a piece of staircase railing. Normally used on traditional homes.

Isolation joint. A joint in which two incompatible materials are isolated from each other to prevent chemical action between the two.

Jack post. A hollow metal post with a jack screw in one end so that it can be adjusted to the desired height.

Jack rafter. A rafter that spans the distance from the wallplate to a hip, or from a valley to a ridge.

Jack stud. A short stud that does not extend from floor to ceiling, for example, a stud that extends from the floor to a window.

Jamb. Exterior frame of a door. The side and head lining of a doorway, window, or other opening.

Joint. Mortar in between bricks or blocks. The space between the adjacent surfaces of two members or components that are held together by nails, glue, cement, mortar, or other means. See Control joint, Coped joint, Expansion joint, and Isolation joint.

Joint butt. Squared ends or ends and edges adjoining each other:

Joint cement. A powder that is usually mixed with water and used for joint treatment in gypsum-wallboard finish. Often called "spackle".

Joist. A long piece of lumber used to support the load of a floor or ceiling and supported in turn by larger beams, girders, or bearing walls. Joists are always positioned on edge. See Band joist, Header, Tail beam, and Trimmer.

Kerf. The area of a board removed by the saw when cutting. Vertical notch or cut made in a batter board where a string is fastened tightly.

Key. Fancy decorative lintel above window made of brick, normally placed on various angles for a flared effect. Also known as a keystone.

Keyways. A tongue-and-groove type connection where perpendicular planes of concrete meet to prevent relative movement between the two components.

Kiln-dried. Dried in a kiln with the use of artificial heat.

Kiln-dried lumber. Lumber that has been dried by means of controlled heat and humidity, in ovens or kilns, to specified ranges of moisture content. See also Air-dried lumber and Lumber, moisture content.

Knee brace. A corner brace, fastened at an angle from wall stud to rafter, stiffening a wood or steel frame to prevent angular movement.

Knee wall. A short wall extending from the floor to the roof in the second story of a 1 History house.

Knot. In lumber, the portion of a branch or limb of a tree that appears on the edge or face of the piece.

Lag screws. Large screws with heads designed to be turned with a wrench.

Laminate. Any thin material, such as plastic or fine wood, glued to the exterior of the cabinet.

Laminated beam. A very strong beam created from several smaller pieces of wood that have been glued together under heat and pressure.

Landing. A platform between flights of stairs or at the termination of a flight of stairs.

Lap. A joint of two pieces lapping over each other.

Latex paint. A coating in which the vehicle is a water emulsion of rubber or synthetic resin. Water soluble paint. Normally recommended because of ease of use for interior work.

Lath. A grid of some sort (normally metal or fiberglass) applied to exterior sheathing as a base for stucco. Expanded metal or wood strips are commonly used.

Lattice. Crossed wood, iron plate, or bars.

Lay up. To place materials together in the relative positions they will have in the finished building.

Ledger strip. A strip of lumber nailed along the bottom of the side of a girder on which joists rest.

Ledgerboard. The support for the second-floor joists of a balloon-frame house.

Let-in brace. Nominal 1"-thick boards applied into notched studs diagonally.

Level. A term describing the position of a line or plane when parallel to the surface of still water; an instrument or tool used in testing for horizontal and vertical surfaces, and in determining differences of elevation.

Light. Space in a window sash for a single pane of glass. Also, a pane of glass.

Line. Any section of plumbing whether it is copper, PVC, CPVC or cast iron. A string pulled tight. Normally used to check or establish straightness.

Lineal foot. A measure of lumber based on the actual length of the piece.

Lintel. A structural member placed above doors and window openings to support the weight of the bricks above the opening, usually precast concrete. Also called a header.

Liquefied gas. A carrier of wood preservatives, this is a hydrocarbon that is a gas at atmospheric pressure but one that can be liquefied at moderate pressures (similar to propane).

Live load. The load, expressed in pounds per square foot, of people, furniture, snow, etc., that are in addition to the weight of the structure itself.

Load bearing wall. Any wall that supports the weight of other structural members.

Lookout. The horizontal board (usually a 2×4 or 1×4) that connects the ends of the rafters to the siding. This board becomes the base for nailing on the soffit covering. The end of a rafter, or the construction which projects beyond the sides of a house to support the eaves; also the projecting timbers at the gables which support the verge boards.

Louver. An opening with a series of horizontal slats so arranged as to permit ventilation, but to exclude rain, sunlight or vision. See also "Attic ventilators".

Lumber timbers. Yard lumber 5 or more inches in least dimension. Includes beams, stringer, posts, caps, sills, girders and purlins.

Lumber, board. Yard lumber less than 2" thick and 2" or more wide.

Lumber, boards. Lumber less than 2" thick and 2" wide.

Lumber, dimension. Yard lumber from 2" to, but not including, 5" thick and 2" wide. Includes joists, rafter, studs, plank and small timbers.

Lumber, dressed size. The dimension of lumber after shrinking from green dimension and after machining to size or pattern.

Lumber, matched. Lumber that is dressed and shaped on one edge in a grooved pattern and on the other in a tongued pattern. See Tongue and groove.

Lumber, moisture content. The weight of water contained in wood, expressed as a percentage of the total weight of the wood. See also Air-dried lumber and Kiln-dried lumber.

Lumber, pressure-treated. Lumber that has had a preservative chemical forced into the wood under pressure to resist decay and insect attack.

Lumber, shiplap. Lumber that has been milled along the edge to make a close rabbet or lap joint.

Lumber, yard. Lumber of those grades, sizes and patterns which are generally intended for ordinary construction; such as, framework and rough coverage of houses.

Mansard. A type of roof that slopes very steeply around the perimeter of the building to full wall height, providing space for a complete story. The center portion of the roof is either flat or very low sloped.

Mantel. The shelf above a fireplace. Also used in referring to the decorative trim around a fireplace opening.

Masonry. Stone, brick, concrete, hollow-tile, concrete block, gypsum block or other similar building units or materials or a combination of the same, bonded together with mortar to form a wall, pier, buttress or similar mass.

Mastic. Any pasty material used as a cement in such applications as setting tile or as a protective coating for thermal insulation or waterproofing. Usually comes in caulking tubes or 5 gallon cans.

Matching, or tonguing and grooving. The method used in cutting the edges of a board to make a tongue on one edge and a groove on the other.

Meeting rail. The bottom rail of the upper sash of a double-hung window. Sometimes called the check rail.

Member. A single piece in a structure, complete in itself.

Metal lath. Sheets of metal that are slit and drawn out to form openings. Used as a plaster base for walls and ceilings and as reinforcing over other forms of plaster base.

Mildewcide. A chemical that is poisonous specifically to mildew fungi. A specific type of fungicide.

Millwork. Generally all building materials made of finished wood and manufactured in millwork plants and planing mills are included under the term "millwork". It includes such items as inside and outside doors, window and door frames, blinds, porchwork, mantels, panelwork, stairways, mouldings and interior trim. It normally does not include flooring, ceiling or siding.

Miter. The joint formed by two abutting pieces meeting at an angle.

Miter box. Special guide and saw used to cut trim lumber at precise angles.

Miter joint. A diagonal joint formed at the intersection of two pieces of moulding. For example, the miter joint at the side and head casing at a door opening is made at a 45° angle.

Moisture content of wood. Weight of the water contained in the wood, usually expressed as a percentage of the weight of oven dry wood. See Lumber, moisture content.

Moulding. Decorative strips of wood or other material applied to wall joints and surfaces as a decorative accent. Moulding does not have any structural value.

Moulding Base. The moulding on the top of a baseboard.

Mortar. See Cement mortar.

Mortise. A slot cut into a board, plank, or amber, usually edgewise, to receive a tenon of another board, plank, or timber to form a joint.

Mud. Slang for spackle, or drywall compound. Used to seal joints and hide nail head dimples.

Mullion. A vertical bar or divider in the frame between windows, doors or other openings. The construction between the openings of a window frame to accommodate two or more windows.

Muntin. A short bar, horizontal or vertical, separating panes of glass in a window sash. The vertical member between two panels of the same piece of panel work.

Natural finish. A transparent finish which does not seriously alter the original color or grain of the natural wood. Natural finishes are usually provided by sealers, oils, varnishes, water-repellent preservatives and other similar materials.

Neoprene. A synthetic rubber characterized by superior resistance to oils, gasoline and sunlight.

Newel. A post to which the end of a stair railing or balustrade is fastened. Also, any post to which a railing or balustrade is fastened.

Nominal. Of wood dimension, the approximate size of a sawn wood section before it is planed.

Nominal size. Original size of lumber when cut.

Non-leachable. Not dissolved and removed by the action of rain or other water.

Nonbearing wall. A wall supporting no load other than its own weight.

Nosing. The projecting edge of a moulding or drip. Usually applied to the projecting moulding on the edge of a stair tread.

Notch. A crosswise rabbet at the end of a board.

O.C., on center. The measurement of spacing for studs, rafters, joists and the like in a building from the center of one member to the center of the next. Normally refers to wall or joist framing of 12", 16" or 18" O.C.

On center (O.C.). The measurement of spacing for elements such as studs, rafters, and joists, from the center of one member to the center of the next.

Oriented strand board (OSB). A type of structural flakeboard composed of layers, with each layer consisting of compressed strand-like wood particles in one direction, and with layers oriented at right angles to each other. The layers are bonded together with a phenolic resin.

Outrigger. An extension of a rafter beyond the wall line. Usually a smaller member nailed to a larger rafter to form a cornice or roof overhang.

PVC (Poly Vinyl Chloride). A form of plastic line used primarily for sewer and cold water supply.

Paint. A combination of pigments with suitable thinners or oils to provide decorative and protective coatings.

Paint grade. Millwork of quality intended for a painted finish. Not as fine as stain grade.

Panel. In house construction, a thin flat piece of wood, plywood or similar material, framed by stiles and rails as in a door, or fitted into grooves of thicker material with molded edges for decorative wall treatment. A sheet of plywood, fiberboard, structural flakeboard, or similar material.

Paper, building. A general term, without reference to properties or uses, for papers, felts and similar sheet materials used in buildings.

Paper, sheathing. A building material, generally paper or felt, used in wall and roof construction as a protection against the passage of air and sometimes moisture.

Parging. Thin coatings (¼") of mortar applied to the exterior face of concrete block where block wall and footing meet; serves as a waterproofing mechanism.

Parquet. A floor with inlaid design. For wood flooring it is often laid in blocks with boards at angles to each other to form patterns.

Particleboard. Panels composed of small wood particles usually arranged in layers without a particular orientation and bonded together with a phenolic resin. Some particleboards are structurally rated. See also Structural flakeboard.

Parting stop or strip. A small wood piece used in the side and head jambs of double-hung windows to separate upper and lower sashes.

Partition. An interior wall in a framed structure dividing two spaces.

Penny. As applied to nails, it originally indicated the price per hundred. The term now serves as a measure of nail length and is abbreviated by the letter "d".

Penta grease. A penta-petroleum emulsion system suspended in water by the use of emulsifiers and dispersing agents.

Pentachlorophenol (penta). A chlorinated phenol, usually in petroleum oil, used as a wood preservative.

Perm. A measure of water vapor movement through a material (grains per square foot per hour per inch of mercury difference in vapor pressure).

Picture. A moulding shaped to form a support for picture hooks, often placed at some distance from the ceiling upon the wall to form the lower edge of the frieze.

Pier. A column of masonry, used to support other structural members, such as concrete supports for a floor beam.

Piers. Masonry supports, set independently of the main foundation.

Pigment. A powdered solid in suitable degree of subdivision for use in paint or enamel.

Pilaster. A projection from a wall forming a column to support the end of a beam framing into the wall.

Piles. Long posts driven into the soil in swampy locations or whenever it is difficult to secure a firm foundation, upon which the footing course of masonry or other timbers are laid.

Piling. Large timbers or poles driven into the ground or the bed of a stream to make a firm foundation.

Pitch. The measure of the steepness of the slope of a roof, expressed as the ratio of the rise of the slope over a corresponding horizontal distance. Roof slope is expressed in the inches of rise per foot of run, such as 4/12.

Pitch board. Board sawed to the exact shape formed by the stair tread, riser, and slope of the stairs and used to lay out the carriage and stringers.

Pith. The small, soft core at the original center of a tree around which wood formation takes place.

Plan. A horizontal geometrical section of a building, showing the walls, doors, windows, stairs, chimneys, columns, etc.

Plank. A wide piece of sawed timber, usually ½" to 4" thick and 6" or more wide.

Plaster. A mixture of lime, hair, and sand, or of lime, cement, and sand, used to cover outside and inside wall surfaces.

Plaster grounds. Strips of wood used as guides or strike-off edges around window and door openings and at base of walls.

Plastic. Term interchangeable with "wet" as in "plastic cement".

Plate. A horizontal member used to anchor studs to the floor or ceiling. Sill plate: a horizontal member anchored to a masonry wall. Sole plate: bottom horizontal member of a frame wall. Top plate: top horizontal member of a frame wall supporting ceiling joists, rafters, or other members.

Plate cut. The cut in a rafter which rests upon the plate; sometimes called the seat cut.

Plenum. Chamber immediately outside of the HVAC unit where conditioned air feeds into all of the supplies. A space in which air is contained under slightly greater than atmospheric pressure. In a house, it is used to distribute heated or cooled air.

Plough. To cut a lengthwise groove in a board or plank.

Plow. To cut a groove running in the same direction as the grain of the wood.

Plugged Exterior. A grade of plywood used for subfloor underlayment. The knot holes in the face plys are plugged and the surface is touch-sanded.

Plumb. The condition when something is exactly vertical to the ground, such as the wall of a house.

Plumb bob. A weight attached to a string used to indicate a plumb (vertical) condition.

Plumb cut. Any cut made in a vertical plane; the vertical cut at the top end of a rafter.

Ply. A term to denote the number of thicknesses or layers of roofing felt, veneer in plywood, or layers in built-up materials, in any finished piece of such material.

Plywood. A piece of wood made of three or more layers of veneer joined with glue, and usually laid with the grain of adjoining plies at right angles. Almost always an odd number of plies are used to provide balanced construction.

Poly. Polyethylene. A heavy gauge plastic used for vapor barriers and material protection. The accepted thickness is 6 mil.

Porch. An ornamental entrance way.

Post. A timber set on end to support a wall, girder, or other member of the structure.

Post and beam roof. A roof consisting of thick planks spanning between beams that are supported on posts. This construction has no attic or air space between the ceiling and roof.

Potable water. Water satisfactory for human consumption and domestic use, meeting the local health authority requirements.

Preservative. Any substance that, for a reasonable length of time, will prevent the action of wood-destroying fungi, borers of various kinds and similar destructive agents when the wood has been properly coated or impregnated with it.

Primer. The first coat of paint in a paint job that consists two or more

coats; also the paint used for such a first coat.

Primer or prime coat. The first coat in a paint job that consists of two or more coats. The primer may have special properties that provide an improved base for the finish coat.

Pulley stile. The member of a window frame which contains the pulleys and between which the edges of the sash slide.

Purlin. A horizontal board the supports a roof rafter or stud to prevent bowing of the member by weight.

Putty. A type of cement usually made of whiting and boiled linseed oil, beaten or kneaded to the consistency of dough, and used in sealing glass in sash, filling small holes and crevices in wood and for similar purposes.

PVC (Poly Vinyl Chloride). A form of plastic line used primarily for sewer and cold water supply.

Quarter round. A small strip of moulding whose cross-section is similar to a quarter of a circle. Used with or without base moulding. May be applied elsewhere.

Quartersawn. Another term for edge grain, which see.

Quoin. Fancy edging on outside corners made of brick veneer or stucco.

Rabbet. A rectangular longitudinal groove cut in the corner edge of a board or plank.

Racking resistance. A resistance to forces in the plane of a structure that tend to force it out of shape.

Radiant heating. A method of heating, usually consisting of a forced hot water system with pipes placed in the floor, wall or ceiling; or with electrically heated panels.

Rafter. One of a series of structural members of a roof designed to support roof loads. The rafters of a flat roof are sometimes called roof joists. See also Fly rafter and Jack rafter.

Rafter, hip. A rafter that forms the intersection of an external roof angle.

Rafter, valley. A rafter that forms the intersection of an internal roof angle. The valley rafter is normally made of double 2"-thick members.

Rafters, common. Those which run square with the plate and extend to the ridge. Cripple-Those which cut between valley and hip rafters. Hip-Those extending from the outside angle of the plates toward the apex of the roof. Jacks-Those square with the plate and intersecting the hip rafter.

Rail. Cross members of panel doors or of a sash. Also the upper and lower members of a balustrade or staircase extending from one vertical support, such as a post, to another.

Rake. The angled edge of a roof located at the end of a roof that extends past gable. Trim members that run parallel to the roof slope and form the finish between the wall and a gable roof extension.

Re-bar. Metal rods used to improve the strength of concrete structures.

Reflective insulation. Sheet material with one or both surfaces of comparatively low heat emissivity, such as aluminum foil. When used in building construction the surfaces face air spaces, reducing the radiation across the air space.

Register. Metal facing plate on wall where supply air is released into room and where air enters returns. Registers can be used to direct the flow of air.

Reinforcing. Steel rods or metal fabric placed in concrete slabs, beams or columns to increase their strength.

Relative humidity. The amount of water vapor in the atmosphere, expressed as a percentage of the maximum quantity that could be present at a given temperature. The actual amount of water vapor that can be held in space increases with the temperature.

Resorcinol. An adhesive that is high in both wet and dry strength and resistant to high temperatures. It is used for gluing lumber or assembly joints that must withstand severe conditions.

Return. Ductwork leading back to the HVAC unit to be reconditioned. The

continuation of a moulding or finish of any kind in a different direction.

Reverse board and batten. Siding in which narrow battens are nailed vertically to wall framing and wider boards are nailed over these so that the edges of boards lap battens. A slight space is left between adjacent boards. This pattern is simulated with plywood by cutting wide vertical grooves in the face ply at uniform spacing.

Ribband. (See Ledgerboard.)

Ribbon (Girt). Normally a 1×4 board let into the studs horizontally to support ceiling or second-floor joists.

Ridge. Intersection of any two roofing planes where water drains away from the intersection. Special shingles are applied to ridges.

Ridge board. The board placed on edge at the ridge of the roof into which the upper ends of the rafters are fastened.

Ridge cut. (See Plumb cut.)

Ridge vent. Opening at the point where roof decking normally intersects along the highest point on a roof where air is allowed to flow from the attic. A small cap covers this opening to prevent rain from entering. When these are long, they are normally known as continuous ridge vents.

Ring shank nail. A nail with ridges forming rings around the shank to provide better withdrawal resistance.

Ripping. Cutting lumber parallel to the grain.

Rise. In stairs, the vertical height of a step or flight of stairs. The vertical distance through which anything rises, as the rise of a roof or stair.

Riser. The vertical board between two treads of a flight of stairs.

Roll roofing. Roofing material, composed of fiber and saturated with asphalt, that is supplied in 36"-wide rolls with 108 square feet of material. Weights are generally 45 to 90 pounds per roll.

Rolled roofing. Roofing material composed of fiber and saturated with asphalt, that is supplied in

36"-wide rolls with 108 square feet of material. Weights are generally 45 to 90 pounds per roll.

Roof sheathing. The boards or sheet material fastened to the roof rafters on which the shingle or other roof covering is laid.

Roof, built-up. See Built-up roof.

Roof, sheathing. The boards or sheet material fastened to the rafters, on which shingles or other roof covering is laid.

Roof, valley. See Valley.

Roofing. The material put on a roof to make it wind and waterproof.

Rottenstone. A slightly abrasive stone used to rub a transparent interior finish to achieve a smooth surface.

Rough grade. First grading effort used to level terrain to approximate shape for drainage and landscaping.

Rout. The removal of material, by cutting, milling or gouging, to form a groove.

Row lock. Intersecting bricks which overlap on outside corners.

Rubble. Roughly broken quarry stone.

Rubble masonry. Uncut stone, used for rough work, foundations, backing, and the like.

Run. In stairs, the net front-to-back width of a step or the horizontal distance covered by a flight of stairs. The length of the horizontal projection of a piece such as a rafter when in position.

STC (Sound Transmission Class). A numerical measure of the ability of a material or assembly to resist the passage of sound. Materials with higher STC numbers have greater resistance to sound transmission.

Saddle. A sloped area at the intersection of a vertical surface and the roof, such as a chimney. Used to channel off water that might otherwise get trapped behind the vertical structure. See cricket.

Saddle board. The finish of the ridge of a pitch-roof house. Sometimes called comb board.

Sagging. Slow dripping of excessively heavy coats of paint.

Sapwood. The outer zone of wood,

next to the bark. In the living tree it contains some living cells (the heartwood contains none), as well as dead and dying cells. In most species it is lighter colored than the heartwood. In all species, it is lacking in decay resistance.

Sash. A single light frame containing one or more lights of glass.

Saturated felt. A felt which is impregnated with tar or asphalt.

Saw kerf. See Kerf.

Scab. A short length of board nailed over the joint of two boards butted end to end to transfer tensile stresses between the two boards.

Scaffold or staging. A temporary structure or platform enabling workmen to reach high places.

Scale. A short measurement used as a proportionate part of a larger dimension. The scale of a drawing is expressed as ¼"= 1'.

Scaling. Loss of smooth surface of concrete as a result of flaking or scaling.

Scantling. Lumber with a cross-section ranging from 2" by 4" to 4" by 4".

Scarfed. A timber spliced by cutting various shapes of shoulders, or jogs, which fit each other.

Scarfing. A joint between two pieces of wood which allows them to be spliced lengthwise.

Scotia. A hollow moulding used as a part of a cornice, and often under the nosing of a stair tread.

Scratch cost. The first coat of plaster, which is scratched to form a bond for the second coat.

Screed. A small strip of wood, usually the thickness of the plaster coat, used as a guide for plastering.

Screeding. The process of running a straightedge over the top of forms to produce a smooth surface on wet (plastic) cement. The screeding proceeds in one direction, normally in a sawing motion.

Screen. Metallic or vinyl grid used to keep troughs free of leaves and other debris. Some screens are hinged.

Scribing. Fitting woodwork to an irregular surface. With mouldings, scribing means cutting the end of

one piece to fit the molded face of the other at an interior angle, in place of a miter joint.

Scuttle Hole. An opening in the ceiling to provide access to the attic. It is covered by a closure panel when not in use.

Sealant. See Caulk.

Sealer. A finishing material, either clear or pigmented, that is usually applied directly over uncoated wood for the purpose of sealing the surface.

Seam, standing. A joint between two adjacent sheets of metal roofing in which the edges are bent up to prevent leakage and the joint between the raised edges is covered.

Seasoning. Removing moisture from green wood to improve its serviceability.

Seat cut or plate cut. The cut at the bottom end of a rafter to allow it to fit upon the plate.

Section. A drawing showing the kind, arrangement, and proportions of the various parts of a structure. It is assumed that the structure is cut by a plane, and the section is the view gained by looking in one direction.

Self-rimming. Term used to describe a type of sink that has a heavy rim around the edge that automatically seals the sink against the Formica top.

Semi-gloss. A paint or enamel with insufficient nonvolatile vehicle such that the dried coating has a luster but does not look glossy.

Semigloss paint or enamel. A paint or enamel made with a slight insufficiency of nonvolatile vehicle so that its coating, when dry, has some luster but is not very glossy.

Septic tank. A receptacle used for storage of water, retained solids and digesting organic matter through bacteria and discharging liquids into the soil (subsurface, disposal fields or seepage pits), as permitted by local health authorities.

Service panel. Junction where main electrical service to the home is split among the many circuits internal to the home. Circuit breakers should

exist on each internal circuit.

Settling. Movement of unstable dirt over time. Fill dirt normally settles downward as it is compacted by its own weight or a structure above it.

Shake. A thick hand split shingle, resawn to form two shakes, usually edge-grained.

Shakes. Imperfections in timber caused during the growth of the timber by high winds or imperfect conditions of growth.

Sheathing. The structural covering, usually wood boards or plywood, used over studs or rafters of a structure. Structural building board is normally used only as wall sheathing.

Sheathing paper. See "Paper, sheathing". A building material, generally paper or felt, used in wall and roof construction as a protection against the passage of air and water.

Shed dormer. A dormer that has a roof sloping only one direction at a much shallower slope than the main roof of the house.

Sheet metal work. All components of a house employing sheet metal, such as flashing, gutters and downspouts.

Sheetrock. A term commonly applied to gypsum board.

Shellac. A transparent coating made by dissolving in alcohol "lac", a resinous secretion of the lac bug (a scale insect that thrives in tropical countries, especially India).

Shim. A thin wedge of wood for driving into crevices to bring parts into alignment.

Shingle. Uniform unit used in the coverage of a roofing deck. Shingles are manufactured from many materials including asphalt, fiberglass, wood shakes, tile and slate. Sizes normally vary from 12" × 36" to 12¼" × 36¼" with many sizes in between.

Shingle butt. The lower, exposed side of a shingle.

Shingles. Roof covering of asphalt, fiberglass, asbestos, wood, tile, slate, or other material or combinations of materials such as asphalt and felt, cut to stock lengths, widths, and thicknesses.

Shingles, siding. Various kinds of shingles, such as wood shingles or shakes and non-wood shingles, that are used over sheathing for exterior sidewall covering of a structure.

Shiplap. See "Lumber, shiplap".

Shutter. A lightweight louvered, flush wood or non-wood frame in the form of a door, located at each side of a window. Some are made to close over the window for protection; others are fastened to the wall for decorative purposes.

Side lap. Length (inches) of the amount of overlap between two horizontally adjoining shingles.

Siding. The finish covering of the outside wall of a frame building, whether made of horizontal weatherboards, vertical boards with battens, shingles or other material. The outside finish between the casings.

Siding, Dolly Varden. Beveled wood siding which is rabbeted on the bottom edge.

Siding, bevel. In lap siding, wedge-shaped boards used as horizontal siding in a lapped pattern. Bevel siding varies in butt thickness from ½" to ¼" and is available in widths up to 12". Normally used over some type of sheathing.

Siding, bevel (lap siding). Wedge-shaped boards used as horizontal siding in a lapped pattern,. This siding varies in butt thickness from ½- to ¾" and in widths up to 12". It is normally used over some type of sheathing.

Siding, drop. Siding that is usually ¼" thick and 6 or 8" wide, with tongue-and-groove or shiplap edges. Often used as siding without sheathing in secondary buildings.

Silicone. One of a large group of polymerized organic siloxanes that are available as resins, coatings, sealants, etc., with excellent waterproofing characteristics.

Sill. (1) The lowest member of the frame of a structure, resting on the foundation and supporting the floor joists or the uprights of the wall. (2) The member forming the lower side of an opening such as a door sill or window sill.

Sill caulk. Mastic placed between top of foundation wall and sill studs to make an airtight seal.

Silt fence. A barrier constructed of burlap, plastic or bales of hay used to prevent the washing away of mud and silt from a cleared lot onto street or adjacent lots.

Sizing. Working material to the desired size; a coating of glue, shellac, or other substance applied to a surface to prepare it for painting or other method of finish.

Slab. A concrete floor poured on the ground.

Sleeper. A wood member embedded in or resting directly on concrete, as in a floor, that serves to support and to fasten subfloor or flooring.

Slip tongue. A spline used to connect two adjacent boards that have grooves facing each other.

Smokepipe thimble. See Thimble.

Soffit. The underside of the cornice or any part of the rood that overhangs the siding.

Soil cover (ground cover). A light covering of plastic film, roll roofing or similar material used over the soil in crawl spaces of buildings to minimize moisture permeation of the area.

Soil cover or ground cover. A light covering of plastic film, roll roofing, or similar material, used over the soil in crawl spaces of buildings to minimize movement of moisture from the soil into the crawl space.

Soil stack. A vent opening out to the roof, which allows the plumbing system to equalize with external air pressure and allows the sewer system to "breathe".

Sole or sole plate. See "Plate".

Solid bridging. A solid member placed between adjacent floor joists near the center of the span to prevent joists from twisting.

Solids. Solid bricks used for fireplace hearths, stoops, patios or driveways.

Spackle. Soft putty-like compound used for drywall patching and touch-up that does not shrink as much as regular joint compound. Allows painting immediately after application. See Joint cement.

Spalling. Chips or splinters breaking loose from the surface of concrete because of moisture moving through from the reverse side.

Span. The distance between two supporting members of a joist or beam. The longest unsupported distance along a joist.

Specifications. The written or printed directions regarding the details of a building or other construction.

Spike. A long nail used to attach the trough to the roof. Used in conjunction with ferrule. See Ferrule.)

Splash Block. A small masonry block laid with the top close to the ground surface, to receive roof drainage from downspouts and carry it away from the building.

Splash block. A small masonry block laid with the top close to the ground surface to receive roof drainage from downspouts and to carry it away from the building

Splice. Joining of two similar members in a straight line.

Spline. A long, narrow, thin strip of wood or metal often inserted into the edges of adjacent boards to form a tight joint.

Spread set. Faucet set that requires three holes to be cut wider than normal.

Square. A unit of measure usually applied to roofing material, denoting a sufficient quantity to cover 100 square feet of surface. A tool used by mechanics to obtain accuracy.

Stack. Also known as vent stack. A ventilation pipe coming out of the roofing deck.

Stain grade. Millwork of finest quality intended for a stain finish. Capable of receiving and absorbing stain easily.

Stain, shingle. A form of oil paint, very thin in consistency, intended for coloring wood with rough surfaces, such as shingles, without forming a coating of significant thickness or gloss.

Stair carriage. Supporting member for stair treads. Usually a 2" plank notched to receive the treads; sometimes called a "rough horse".

Stair landing. See "Landing".

Stair rise. See "Rise".

Stairs, box. Those built between walls, and usually with no support except the wall.

Staking. To lay out the position of a home, the batter boards, excavation lines and depth(s).

Starter strip. A continuous strip of asphalt roofing used as the first course, applied to hang over the eave.

Stiffness. Resistance to deformation by loads that cause bending stresses.

Stile. An upright framing member in a panel door.

Stipple. Rough and textured coatings applied to ceilings. This process makes it easier to finish ceilings since they do not have to be taped and sanded as many times as the walls.

Stool. A flat moulding fitted over the window sill between jambs and contacting the bottom rail of the lower sash.

Stoop arms. Section of foundation wall extending out perpendicular to exterior wall used to support a masonry or stone stoop.

Stoop iron. Corrugated iron sheeting used as a base for the tip of a brick or stone stoop. This is to eliminate the need for completely filling the stoop area with fill dirt (which may settle) or concrete.

Stop valve. A shut-off valve allowing water to be cutoff at a particular point in the system.

Stop, gravel. A raised ridge of metal at the edge of a tar and gravel roof that keeps the gravel from falling off the roof.

Stop, trim. The trim member on the jambs of an opening that a door or window closes against.

Storm sash or storm window. An extra window usually placed on the outside of an existing one as additional protection against cold weather.

Story. That part of a building between any floor and the floor or roof next above.

Strike plate. A metal plate mortised into or fastened to the face of a door-frame side jamb to receive the latch or bolt when the door is closed.

String, stringer. A timber or other support for cross members in floors or ceilings. In stairs, the support on which the stair treads rest; also stringboard.

Stringer. A timber or other support for cross members in floors or ceilings. In stairs, the stringer (or stair carriage) supports the stair treads.

Strip flooring. Wood flooring consisting of narrow, matched strips.

Structural flakeboard. A panel material made of specially produced flakes that are compressed and bonded together with phenolic resin. Popular types include waferboard and OSB (oriented strand board). Structural flakeboards are used for many of the same applications as plywood.

Stucco. A fine plaster used for interior decoration and fine work; also for rough outside wall coverings.

Stud. Standard 2×4 lumber normally cut to 8' or 10' nominal lengths used for framing walls. (Plural = studs or studding.)

Stub-in. A term applied to the process of installing rough plumbing "stubs" to simplify the installation of plumbing at a later date.

Studding. The framework of a partition or the wall of a house; usually referred to as 2×4s.

Studwall. A wall consisting of spaced vertical structural members with thin facing material applied to each side.

Subfloor. Boards or plywood laid on joists over which a finish floor is to be laid.

Superstructure. The structural part of the deck above the posts or supports

Supply. Ductwork leading from the HVAC unit to the registers.

Suspended ceiling. A ceiling system supported by hanging it from the overhead structural framing.

T&G. Type of wood joint machined with a tongue or protrusion on one side and a groove on the other. This allows the two pieces to be joined snugly together. The term T&G is commonly used to refer to tongued and grooved flooring or roofing.

Tail beam. A relatively short beam or joist supported by a wall at one end and by a header at the other.

Tape. Paper used to cover the joints between sheets of gypsum. Tape joints are then sealed with mud (spackle).

Tenon. A projection at the end of a board, plank, or timber for insertion into a mortise.

Termite shield. A shield, normally of galvanized sheet metal, placed between footing and foundation wall to prevent the passage of termites.

Termites. Insects that superficially resemble ants in size, general appearance and habit of living in colonies; hence, they are frequently called "white ants". Subterranean termites establish themselves in buildings not by being carried in with lumber, but by entering from ground nests after the building has been constructed. If unmolested, they eat out the woodwork, leaving a shell of sound wood to conceal their activities; damage may proceed so far as to cause collapse of parts of a structure before discovery. There are about 56 species of termites known in the United States; but the two major ones, classified by the manner in which they attack wood, are ground-inhabiting or subterranean termites (the most common) and drywood termites, which are found almost exclusively along the extreme southern border and the Gulf of Mexico in the United States.

Thimble. The section of a vitreous clay flue that passes through a wall.

Thinwall. Thin flexible conduit used between outlet boxes.

Threshold. A strip of wood or metal with beveled edges used over the finish floor and the sill or exterior doors.

Tie beam (collar beam). A beam so situated that it ties the principal rafters of a roof together.

Tieback member. A timber, oriented perpendicular to a retaining wall, that ties the wall to a deadman buried behind the wall.

Tin shingle. A small piece of tin used in flashing and repairing a shingle roof.

To the weather. A term applied to the projecting of shingles or siding beyond the course above.

Toe nailing. Nailing at an angle (normally 45°) to bind two or more members. Normally used in nailing studs to plates.

Toenailing. Driving a nail at a slant with the initial surface to permit it to penetrate into a second member.

Ton. An industry standard measure to express a quantity of cold air produced by an air conditioning system.

Tongue and groove. Boards that join on edge with a groove on one unit and a corresponding tongue on the other to interlock. Certain plywoods and hardwood flooring are tongue and grooved. "Dressed and matched" is an alternative term with the same meaning.

Top plate. Piece of lumber supporting ends of rafters.

Topsoil. Two or three inch layer of rich, loose soil. This must be removed from areas to be cleared or excavated and replaced in other areas later. Not for load bearing areas.

Touch sanding. Very light sanding of prime paint coat.

Transit. Similar to a builder's level except that the instrument can be adjusted vertically. Used for testing walls for plumb and laying out batter board, and establishing degree of a slope.

Trap. A device providing a liquid seal which prevents the backflow of air without materially affecting the flow of sewage or waste water. "S" shaped drain traps are required in most building codes.

Tray ceiling. Raised area in a ceiling. Looks like a small vaulted ceiling.

Tray moulding. Special type of crown moulding where a large portion of the moulding is applied to the ceiling as opposed to the wall.

Tread. The horizontal board in a stairway on which the foot is placed.

Trig. A string support for guide lines to prevent sagging and wind disturbances on long expanses of wall.

Trim. The finish materials in a building, such as mouldings, applied around openings (window trim, door trim) or at the floor and ceiling of rooms (baseboard, cornice and other mouldings).

Trimmer. A beam or joist to which a header is nailed in framing for a chimney, stairway or other opening.

Trimming. Putting the inside and outside finish and hardware upon a building.

Troweling. A process used after floating to provide an even smoother surface.

Truss. A frame or jointed structure designed to act as a beam of long span, while each member is usually subjected to longitudinal stress only, either tension or compression.

Truss plate. A heavy-gauge, pronged metal plate that is pressed into the sides of a wood truss at the point where two more members are to be joined together.

Turpentine. A volatile oil used as a thinner in paints and as a solvent in varnishes. Chemically, it is a mixture of terpenes.

Undercoat. A coating applied prior to the finishing or top coats of a paint job. It may be the first of two or the second of three coats. In some uses of the word it may become synonymous with priming coat.

Underlayment. Any paper or felt composition used to separate the roofing deck from the shingles. Underlayments such as asphalt felt (tar paper) are common. A material placed under flexible flooring materials such as carpet, vinyl tile, or linoleum to provide a smooth base over which to lay such materials.

Underlayment Exterior. See Plugged Exterior.

Valley. The internal angle formed by the junction of two sloping sides of a roof where water drains at the intersection.

Vapor barrier. Material used to retard the movement of water vapor into walls and prevent condensation in them. Usually considered as having a perm value of less than 1.0. Applied separately over the warm side of exposed walls or as a part of batt or blanket insulation.

Vapor retarder. Material used to retard the movement of water vapor into walls. Vapor retarders are applied over the warm side of exposed walls or as a part of batt or blanket insulation. They usually have perm value of less than 1.0.

Varnish. A thickened preparation of drying oil or drying oil and resin suitable for spreading on surfaces to form continuous, transparent coatings, or for mixing with pigments to make enamels.

Vehicle. The liquid portion of a finishing material; it consist of the binder (nonvolatile) and volatile thinners.

Veneer. Thin sheets of wood made by rotary cutting or slicing of a log. Veneer is glued in plys or on top other wood to improve appearance or strength.

Vent. A pipe or duct, or a screened or louvered opening, which provides an inlet or outlet for the flow of air. Common types of roof vents include ridge vents, soffit vents, and gable end vents.

Vent system. A pipe or network of pipes providing a flow of air to or from a drainage system to protect trap seals from siphonage or back pressure.

Vestibule. An entrance to a house; usually enclosed.

Volatile thinner. A liquid that evaporates readily, used to thin or reduce the consistency of finishes without altering the relative volumes of pigments and nonvolatile vehicles.

Voltage. The "force" of electrical potential. Similar to the pressure of water in a pipe.

Waferboard. A type of structural flakeboard made of compressed, wafer-like wood particles or flakes bonded together with a phenolic resin. The flakes may vary in size and thickness may be either randomly or directionally oriented.

Wainscoting. Paneling and trim applied from the floor to a height of about 3'. Used in dining areas to protect against marks from dining chairs. Popular decorative touch in modern homes.

Wale. A horizontal beam.

Wallplate. The cover over an electrical outlet or switch on the wall.

Wane. Bark, or lack of wood from any cause, on the edge or corner, of a piece of wood. Hence, waney.

Wash. The slant upon a sill, capping, etc., to allow the water to run off easily.

Water line. Decorative relief line around foundation approximately 3' from the ground.

Water table. The finish at the bottom of a house which carries water away from the foundation.

Water-repellent preservative. A liquid designed to penetrate into wood and impart water repellency and a moderate preservative protection. It is used for millwork, such as sash and frames, and is usually applied by dipping.

Wattage. The product of the amperage times the voltage. A good indicator of the amount of electrical power needed to run the particular appliance. The higher the wattage, the more electricity used per hour.

Weatherstrip. Narrow or jamb-width sections of thin metal or other material to prevent infiltration of air and moisture around windows and doors. Compression weather stripping prevents air infiltration, provides tension and acts as counterbalance.

Weatherstripping. Strips of thin metal or other material, that prevent infiltration of air and moisture around windows and doors. Compression weatherstripping on single and double-hung windows performs the additional function of holding such windows in place in any position.

Web. The thin center portion of a beam that connects the wider top and bottom flanges.

Weep hole. Small gap in brick wall, normally on garage, that allows water to drain.

Whaler. A large structural member placed horizontally against foundation forms to which braces are temporarily attached to prevent forms from moving horizontally under the pressure of concrete.

Wharf. A structure that provides berthing space for vessels, to facilitate loading and discharge of cargo.

Wind. ("i" pronounced as in "kind") A term used to describe the surface of a board when twisted (winding) or when resting upon two diagonally opposite corners, if laid upon a perfectly flat surface.

Wire mesh. A heavy gauge steel mesh sold in rolls for providing reinforcing in concrete slabs.

Withe. A vertical layer of bricks, one brick thick.

Wood mold. A brick making process in which bricks are actually molded instead of extruded. Fancy shapes are the result of this process.

Wooden brick. Piece of seasoned wood, made the size of a brick, and laid where it is necessary to provide a nailing space in masonry walls.